Migration
Changing Concepts, Critical Approaches

Edited by
Doris Bachmann-Medick and Jens Kugele

DE GRUYTER

ISBN 978-3-11-068280-9
e-ISBN (PDF) 978-3-11-060048-3
e-ISBN (EPUB) 978-3-11-059903-9
ISSN 2190-3433

Library of Congress Control Number: 2018946964.

Bibliographic information published by the Deutsche Nationalbibliothek
The Deutsche Nationalbibliothek lists this publication in the Deutsche Nationalbibliografie; detailed bibliographic data are available on the Internet at http://dnb.dnb.de.

This volume is text- and page-identical with the hardback published in 2018.
Typesetting: Simon Ottersbach, Gießen
Printing and binding: CPI books GmbH, Leck

www.degruyter.com

Migration

Concepts for the Study of Culture

Edited by
Doris Bachmann-Medick, Horst Carl,
Wolfgang Hallet and Ansgar Nünning

Volume 7

Acknowledgements

Concepts in the study of culture provide a common ground for transdisciplinary and transnational research. But the work of concepts extends well beyond scholarly discussions. In the case of migration, it becomes especially evident that a concept can hold social and political power outside the academic realm – as a “cultural force” of its own and a “site of debate” (Mieke Bal). Applying a concept-based research approach to the study of culture in general and to the study of migration in particular, this volume explores changing concepts and critical approaches while attending to the important role of conceptual assumptions in academic research, political discourse, and people’s personal experiences.

Some of the contributions originate from an international lecture series on migration, which we held at the International Graduate Centre for the Study of Culture (GCSC) at Justus Liebig University Giessen during the summer semester of 2015. Little did we know at the time that the topic would gain such relevance and political force in the months to follow. It made the discussion of nuanced and critical concepts in dealing with our topic all the more necessary.

In addition to the presenters of the 2015 lecture series (Bachmann-Medick, Eigler, Gutiérrez Rodríguez, Konuk), we invited experts from the fields of cultural anthropology/sociology (Bischoff, Friese, Greschke, Hess, Karagiannis, Randeria, Schiffauer), Literary Studies (Payne), migration and education (Mecheril), and cultural history (Kansteiner) to contribute to this volume. We would like to thank all of these scholars for sharing their research as well as for their cooperative spirit.

We would not have met the deadline for submission without the valuable help of our colleague and PhD candidate, Simon Ottersbach, and our student assistant, Franziska Eick (both at the GCSC). We very much appreciate their support in formatting and typesetting the manuscript. Anne Wheeler and Marie Schlingmann were a tremendous help as English language proofreaders of the manuscript. Finally, our thanks go to De Gruyter, in particular Manuela Gerlof and Stella Diedrich, for seeing the project swiftly through the publication process; to the series editors for their support of the project; and last but not least to the GCSC not only for generously supporting the publication of this volume but also for providing the intellectually stimulating environment that has been most fruitful for this endeavor.

Giessen, June 2018
Doris Bachmann-Medick and Jens Kugele

https://doi.org/9783110600483-001

Contents

Doris Bachmann-Medick and Jens Kugele

Introduction

Migration – Frames, Regimes, Concepts

How do people think about migration? What patterns of perception – collective memories, images, and historical experiences – does such thinking activate? And how might scholars in the study of culture analyze the shifting social and political aspects of migration that are of such heightened concern today? Recent debates have demonstrated the deleterious effects of stereotypes on how migration is imagined, framed, and analyzed. Such preconceptions are often unconscious, unspoken, and unexplored, but they nonetheless exert a powerful effect in social life, in academic and political statements, in policies and decisions, and in the lives of migrants themselves. On the one hand, these framings determine the changing conditions, complex realities, and transnational entanglements of migration in the 21st century. On the other hand, these changing conditions are themselves a challenge in the discursive and scholarly fields, demanding a reconsideration of the dominant migration-related concepts and their powerful work in academic analyses as well as in public and political discourses.

The explicitly conceptual perspective of this volume is not entirely new: set in motion already in the field of migration studies (among others, see Schittenhelm 2007; Bartram, Poros, and Monforte 2014), it still demands further elaboration. In a constantly changing world of global migration we must develop new frameworks that allow reflection more specifically upon conditions, laws, borders, and cultural and political encounters, and on exclusions, inclusions, and unequal power relationships in transnational migration processes. We need to theorize these issues and explore them in their constructedness and changeability. And on a smaller scale, we also need to develop modified concepts suitable to and adequate for these changing situations of migration/immigration in contemporary societies: today, "earlier conceptions of immigrant and migrant no longer suffice [...] Now, a new kind of migrating population is emerging, composed of those whose networks, activities and patterns of life encompass both their host and home societies" (Glick Schiller, Basch, and Blanc-Szanton 1992, 1).

In the study of migration as "cross-border-studies" (Amelina, Nergiz, Faist, and Glick Schiller 2012, 1–9) it is no longer sufficient to work with framings and concepts based on naturalizing and essentializing assumptions (such as 'ethnicity') or on the presumption of a linear migration process. Conceptual frameworks that correspond with the approach criticized as 'methodological nationalism' no longer seem appropriate either. Today, we face emerging global and

https://doi.org/9783110600483-002

interconnected conditions of migration entangled with lives in transnational communities that are infused by insecurities and persecution. We are confronted with numerous manifestations of refuge: ranging from violent conflicts and wars, displacements to multiple places, and economic asymmetries, all of which are influenced and increasingly dependent upon new telecommunication technologies and transnational and social media. These new entanglements have given rise to conceptual frameworks that demand a methodological transnationalism with multiple references to multilayered network structures. They have elicited "a global power perspective," as Nina Glick Schiller suggests, that explores "structures and processes of unequal capital flow which influence the experience of people who reside in particular localities" (Glick Schiller 2007, 63).

Transnational approaches beyond methodological nationalism are certainly better equipped to capture broader links and connectivities of migration processes in their cross-cultural global relations – not least with regard to family bonds, digital communication, and network formation, as Heike Greschke discusses in her contribution to this volume. Addressing such new contexts and their changing practices in the contemporary field of migration in any case seems crucial. A simple step from 'methodological nationalism' to 'methodological transnationalism' is not sufficient, however; remaining under the umbrella term of 'methodological transnationalism' will not allow the full picture to emerge. If we seek to include perspectives on concrete migratory encounters, even more specific and more suitable frames demand our attention, including boundary-making, biographical analysis, group formation, mobility, medialization, network-building, impact of collective memory, Orientalization, and visualization.

Rather than concentrating on specific historical or contemporary case studies, this volume seeks to critically address migration's pressing social and political engagements by reflecting upon the overarching norms, policies, governance practices, and experiences of migration that provide basic dimensions for analyzing migration processes. By taking a distancing analytical standpoint – following the perspective of a second-order observation – this approach enables a stronger contextualized analysis of actual migration phenomena and individual experience. This proves to be methodologically productive when reconsidering recent migration issues: for instance, the enforced migration of academics from Turkey that Kader Konuk discusses in these pages, or the enormous impact of collective memory on public perception of the refugee situation in today's Germany, which Wulf Kansteiner examines in his contribution herein. We must seek such critical analytical standpoints that are not directly involved in actual political processes to avoid the danger of reifying scholarly concepts according to propagated concepts current in our societies – e.g., the notion of 'identity.'

1 'Frames' of Migration

The conceptual approach chosen in this volume, however, by no means shies away from migratory realities, but instead directly references these realities throughout. At the same time, it seeks to dig deeper into contemporary frames and framings of discourses on migration and their cultural perceptions by considering the analytical terms and conceptual assumptions of cultural engagements with migration. These can be described, in short, as "schemata of interpretation," as in Erving Goffman's *Frame Analysis* (1986, 21). 'Frames' constitute methodologically and epistemologically self-reflexive approaches to the complex field of migration, but they are also effective in shaping the field of socio-political experience and behavior that directly impacts the lives of migrants. As frames render events, issues, and situations meaningful, they become effective as tools for the organization of experience (Goffman 1986, 11); in the field of migration, this certainly holds true. The United Nations' *World Migration Report 2018* specifically addresses the power of 'framing migration' in public discourse through media coverage: "Migration coverage [...] presents a variety of different issues, narratives and viewpoints [...] Identifying how matters are framed is important because [...] media frames affect how people think about migration" (*World Migration Report 2018*, 194). Identifying how matters of migration are framed – and in this way, how such matters are shaped and changed through narratives, images, prejudices, preconceptions and false conceptions, collective memories, political predeterminations, and policy-making – should thus be a primary endeavor in the study of culture in general and in the study of migration in particular. For this reason, the present volume focuses on framings in an effort to open them up and deconstruct them into more operative concepts, considering their uses, functions, and changes. In concord with Mieke Bal's assertion, this approach strives to keep frames and their concepts dynamic, actively conceptualizing and re-conceptualizing them as an outcome of critical self-reflection (see Bal 2009, 17).

The notion of 'frames' expounded by Judith Butler in her book *Frames of War* also proved inspiring in our attempt to examine the bedrock of migration discourses by shedding light on framing concepts that determine and shape these discourses. This broader endeavor of a critical attention towards framing, however, also implies the danger of certain epistemological pitfalls. Frames, Butler contends, "are themselves operations of power" (Butler 2009, 1). The framing of migration, too, is necessarily entangled within its own operations of power. By examining concepts of person, citizenship, violence, race, and sexuality in this context, we can explore and analyze the powerful "normative schemes" (4) that shape our understanding of migration. The framing of migration is never innocent. In the context of precarious existential conditions of war – analogous to Butler's fundamental question "What is a life?" – we should search for a reflective and self-reflective frame that provides new answers to the old question "What is migration?" We might also look for frames as 'operative' dimensions in the research process, as they determine our perceptions of migration with respect to imprisonment, torture, and war, but also to the politics of immigration (24). Here, too, it makes a difference if "certain lives are perceived as lives, while others, though apparently living, fail to assume perceptual form as such" (24). Butler's position encourages us to consider frames of migration in an extended sense – seeing elements of war at work even in immigration processes through a "framing of immigration issues as a 'war at home'" (26). This tendency can be observed, for instance, in the mobilization of feminist sexual politics not only "to rationalize war against predominantly Muslim populations, but also to argue for limits to immigration to Europe from predominantly Muslim countries" (26).

2 'Regimes' of Migration

This admittedly broad analytical perspective reveals that attention to 'framings' cannot do the analytical work alone. It must connect to an entire cluster of concepts and regulations to capture the complex institutional arrangements, laws, conventions, and restrictions that determine and govern the broader field of migration. It is not by accident that we often speak of 'regimes' (border regime, regime of the gaze, mobility regime, etc.) when we want to address determining factors of migration processes. Not confined to specific models or types of migration, such 'regimes' are tools for conceiving, forming, and developing complex phenomena using an explicitly constructivist approach (see Casas-Cortes et al. 2014, 15). A multidimensional ensemble and complex interweaving of practices, patterns, and regulations of governmentality and control, of power and inequality, security, and transnational collaboration come into view, not least as regards human rights issues. A 'regime' approach could be useful for linking a micro- or actor-oriented perspective on migration to overarching regional or transnational relations, hierarchies, and other framings that conduct the local and individual behavior of migrants. As a matter of fact, migrants themselves are often caught in this web of interwoven migration factors (Tamas and Palme 2016, 3–19). Thus, the 'regime' lens encompasses analytical approaches as well as actual patterns of behavior and social norms, and can therefore be used for an actor-oriented analysis while remaining close to the field of political dynamics. An approach like this can be developed from a clustering or intersection of concepts of migration such as border control, discourses on human rights, and securitization. This goes along with a widened conception of migration research that reaches beyond the scholarly sphere, connecting to the realm of politics and focusing on the complexities of the effective norms, discourses, and regulations:

> [U]sing a 'regime' perspective might [...] add sociological and ethnographic stances to the analysis of political dynamics, hence bridging a long-standing divide in migration research between those focusing on the politics of migration and those dealing with the practices of mobility and settlement. (Horvath, Amelina, and Peters 2017)

This volume and its contributions aim to differentiate and specify the all-too-general notion of 'regime' that has characterized various recent approaches in migration studies by paying more attention to implicit or explicit concepts that constitute such regimes in the first place. Even more, the articles herein show that a regime approach is especially relevant when it not only synthesizes different migration factors but tries to identify them through actor-related concepts as a differentiated scenario of negotiations between various actors, as in contributions to this volume by Sabine Hess and Christine Bischoff. They highlight the importance of conceptual nuance in dealing with questions of migration, not least with critical reference to the dynamics and politics of contemporary discourses on migration and the latest developments in transnational migration.

3 'Concepts' of Migration

This volume's concern with nuance is introduced in a discussion of migration from the perspective of a concept-based study of culture, which claims that working with 'concepts' is a reciprocal endeavor. On one hand, the conceptual approach to migration influences the dynamic development of concepts themselves; on the other hand, a fundamental conceptual practice in the study of culture aims at fostering continued reflection upon the field's analytical tools as central elements of our knowledge construction. Still, our conceptual apparatus provides more than merely a collection of key concepts (as, for example, in Bartram, Poros, and Monforte 2014) or an address of some analytical concepts on the level of discursivity. Meditating upon concepts in order to transform or change them can be an effective way to involve the study of culture in the work of important conceptual developments that act as determining forces for shaping cultural perceptions in the social world. As Mieke Bal reflects about "working with concepts," concepts are never harmless, never purely descriptive, but normative and programmatic. They unfurl a power of their own by operating as tools that trigger debates about their meaning, impacting common language use as well as cultural perceptions (Bal 2007, 35, 38):

> Even those concepts that are tenuously established, suspended between questioning and certainty, hovering between ordinary word and theoretical tool, constitute the backbone of the interdisciplinary study of culture – primarily because of their potential intersubjectivity. Not because they mean the same thing for everyone, but because they do not. (Bal 2009, 17)

Concepts, in Mieke Bal's sense, thus "help us see" (Bal 2009, 22) and "focus interest" (Bal 2007, 40), making them effective as a critical cultural force even prior to any direct societal activism. Social change, however, does not result from mere analytical concepts, but rather from disputes about concepts, disagreements in interpretations and practices, or stronger political issues. As an example from the concept of climate migration indicates: "Climate migration is a weak analytical concept, but it has a particularly strong political currency" (Mayer 2016, 37).

The operative role of concepts, though, seems uncontested, as historical anthropologist Ann Laura Stoler highlights and historically reconsiders in her recent book, *Duress*. In a subchapter called "What is Concept Work?" Stoler points to the necessity of concept transformation in the humanities. Stoler argues against the assumed stability and security of conventional concepts in the study of imperial history (Stoler 2016, 17–23), and contends that keeping concepts "provisional, active, and subject to change" (19) would allow access to "occluded histories" (10). Critical concept work that explores "occluded histories" and unspoken assumptions could perhaps be fruitful for the field of migration studies. For this purpose, we accept the challenge to "better look to the unmarked space between their (the concepts' [the editors]) porous and policed peripheries, to that which hovers as not quite 'covered' by a concept [...] that which cannot be quite encompassed by its received attributes" (9).

What is required for exploring such "peripheries" is a creative and reflective work with concepts in migration studies that exposes familiar concepts, and revisits and reframes them to uncover the hidden dimensions, "backstage discourses," and "hidden transcripts" (Scott 1990, xii) of this complex field.[1] But this endeavor

1 We find a similar approach in the *New Keywords Collective*, a network of collective writing that seeks a rethinking of keywords: "This collaborative project of collective authorship emerges from an acute sense of the necessity of rethinking the conceptual and discursive categories that govern borders, migration, and asylum and simultaneously overshadow how scholarship and research on these topics commonly come to recapitulate both these dominant discourses and re-reify them." This project, too, starts from "an epistemological destabilization and theoretical questioning of the very meaning and function of certain key concepts and categories, such as

is not sufficient in itself. It is also necessary to expand the realm of concepts by connecting it to social practices, experiences, and mobilizing concepts operative in society and among migrants themselves. Indeed, a critical reconsideration of common concepts such as family, knowledge, mobility, and religion also could benefit from the perspective of migratory experiences, as various contributions to this volume demonstrate. In the case of family, for instance, one can observe a change from the Western concept of 'nuclear family' to the transnational concept of family as an institution of 'care circulation' kept alive by digital communication technologies in deterritorialized migration contexts (see Heike Greschke's contribution to this volume and Baldassar and Merla 2014). Such a conceptual approach to migration explores the dynamics and changing potentials of concepts through an expansion of traditional (Western) concepts and a continued reconceptualization for which Nina Glick Schiller and others have argued (Glick Schiller, Basch, and Blanc-Szanton 1992). Such reconceptualizations seem of particular importance as the heuristic instruments that we employ in our analyses fundamentally shape the outcome of our research. Our conceptions of religion might serve as a case in point here. When we use, for instance, Western-centric conceptions of religion, we not only categorically narrow the spectrum of possible insights but we also run the risk of implicitly universalizing these Western-centric conceptions. As Peggy Levitt suggests in her critique of contemporary migration studies, it would be far more fruitful to try to think outside the conventional boxes and to avoid universalizing conceptions of religion since "the traditions that immigrants bring with them do not fit into one, closed, Christian-like container" (Levitt 2012, 495).

The intention to develop or reimagine concepts, however, can be quite precarious, as exemplified by the conceptual shift from 'identity' to 'belonging' (see Pfaff-Czarnecka 2013). Even if we used a concept of multiple 'belongings' as a starting point seemingly closer to the actors' experiences, additional nuanced conceptual work would still be necessary to explore the conceptual peripheries and contact zones between 'belonging' and 'identification,' 'affiliation,' or 'origin.' As analysis of migration depends on such concepts to provide the framework of perception, the continuous reflection upon and rethinking of concepts constitutes a challenging yet fundamentally necessary task. It impacts not only the analysis of migration and migrants, but our cultural self-perception and self-interpretation in general.

'humanitarianism,' 'refugee,' 'migrant,' 'mobility,' and so forth" (Tazzioli and De Genova 2015). For a specific example of revisiting the conceptual framework of (return-)migration, see Cassarino (2004).

As forms and types of migration in a globalized world have become ever more differentiated throughout recent decades, conceptions of migration must also be readjusted to reach beyond the standard typologies of migration (refugees, labor migration, forced migration, etc.). Rather, switching back and forth between the scales of analysis, between an actor-oriented micro-perspective and a more structural macro-perspective seems indispensable. As both dimensions are central to exploring deeper structures within the migration process, a concept-based study should definitely be complemented by an actor-based approach. Aiming at such dynamic concept work, we also need to discuss and expand the scope of our very patterns of interpretation. As Saskia Sassen reminds us, it does not suffice to conceive of people on the move merely as "migrants in search of a better life who hope to send money and perhaps return to the family left behind," if we take seriously the fact that some of these migrants "are people in search of bare life, with no home to return to" (Sassen 2016, 205).

In this sense, in the 1980s and 1990s, Abdelmalek Sayad's work on Algerian immigration to France marked a radical departure from former approaches, categories, and labels in the study of migration. Sayad directed our attention to a non-ethnocentric dimension, that of the migratory experiences of individuals and groups – in their irritations, pain, and suffering – as related to a history of colonization, a topic which Encarnación Gutiérrez Rodríguez takes up in her contribution to this volume. This critical break with established approaches was accompanied by a departure from 'methodological nationalism' that relied upon 'state' and 'nation' as its operating categories, plus a further departure from 'methodological ethnicity' that concentrated on migrants as bounded collective groups, and has been followed by stereotyping ethnic generalizations. More recently, this conceptual frame has been supplanted by a rather individualistic and biographical approach that puts "individual persons in the forefront" (Pfaff-Czarnecka 2013, abstract) and investigates the "constellations of multiple belonging" (12) of individual migrants. What we can observe in the development of migratory concepts over the course of the last decade is exactly the shift from categorical and systemic top-down concepts to practical concepts of "social relating" (15), and to the dynamic and processual character of concepts (Brockmeyer and Harders 2016, 4).

4 Working with Concepts: Critical Contributions

The contributions to this collection engage with this dynamic character of concepts as they revisit established analytical categories in the study of migration such as border regimes, orders of belonging, coloniality, translation, trans/national digital culture, and memory. The perspectives on migration offered by the authors are related to the fields of history, visual studies, pedagogy, literary and cultural studies, and the social sciences. As they open up an interdisciplinary dialogue and discuss new conceptual notions of migration in light of changed realities, they also demonstrate the powerful work of frames and the relevance of conceptual nuance extending beyond academic discourse. In four main conceptual clusters, the contributions explore various frameworks of migration and reflect on the appropriate language for academic analysis.

Cluster I: The Power of Images and the Imaginary

Imagination, as the interplay of individual and collective ideas, mental images, visualizations, and fantasies, plays a powerful role in the experience and construction of realities. It thus also provides an important framework for conceptualizations of migration. The work of imagination informs subjective perception and collective memory, habits, social belonging, ritual performance, and political action. Imagination presents such a powerful framing of cultural history that French historian Jacques Le Goff famously called for an entire new field of research in historical studies to be dedicated to the complex cultural "imaginary" (Le Goff 1990, 12). In this vein, academic analyses would investigate the collective imagery that has been concocted in changing historical contexts, expressed in words, iconographic production, and material culture, and circulating across times, social strata, and societies. The contributions to this section engage in aspects of this complex cultural imaginary, drawing our attention to visuality, media representation, and imagined community in conceptualizations of migration.

As Christine Bischoff argues in her contribution here, in today's highly visual culture, our perceptions and interpretations of migration are often determined by images. Bischoff approaches processes of migration through a discussion of visuality that becomes effective in sociocultural representations of individuals, groups, generations, gender, and religions. Focusing on the 'regime of the gaze,' she directs attention to the effectiveness of practices of visibility and visualization. Bischoff's analysis is concerned with both the 'image of the Other' and the gaze of the invisible subjects constituted in the definition of the 'deviating Other.' Shedding light on 'visual regimes,' Bischoff demonstrates their potential for the analysis of 'images of migration' as well as the 'migration of images,' and points to their interrelations with image discourses understood as everyday media practice.

Focusing on refugee discourse, Heidrun Friese offers a slightly different take on the nexus of imagery and social imagination. Friese lists the numerous artistic representations of refugees at Lampedusa and draws our attention to the production and reproduction of emotion, designed, as she argues, to move the spectator, and to inscribe the migrant's destiny onto a powerful narration of misery, pity, indignation, and victimhood. Dominant media representations evoke catastrophe and contamination. Her discussion of representations and frames of mobility demonstrates the highly political dimension of this construction of the social imaginary: mobility ultimately challenges democracy as it incessantly points towards the paradox of democracy and to its inclusive-exclusive structure.

Approaching migration from an anthropological perspective, Werner Schiffauer introduces the concept of 'imaginary opportunity space' as a composite that connects (objectively given) opportunities with (subjective) imaginations. Suggesting that imaginations become (emotionally) powerful when they relate to opportunities, Schiffauer's systematic exploration of the concepts 'imaginary' and 'imaginary opportunity space' seeks to contribute to a better understanding of three key aspects of migration theory: the migratory decision, transnational communities, and irregular migration. Schiffauer suggests a focus on 'transnational imagined communities' and the concept of an 'imaginary opportunity space' to shed light on aspects such as emotions of guilt among others in the context of migration networks. As Schiffauer argues, this conceptual focus allows us to take a fresh look at the migration process and its emotional-economic implications.

Cluster II: (Border) Regimes

Constructions of borders and boundaries in their relation to movements have counted among the most central frames for analyzing migration. Recent critical scholarship has suggested exploring "border regimes" to foster an understanding of the complex interplay of various agents, discourses, and practices in this context of mental mapping, politics of migration, and national imagination (Hess 2012). One of the challenges for scholars of migration who study border regimes is finding a way to incorporate migration as a powerful force in the very process of theorizing the border itself and the activity of bordering, rather than regarding migration merely as being an object of the border (Hess 2012).

Analyzing conceptualizations that are customary in international border studies, Sabine Hess, in her contribution to this volume, criticizes the top-down structure underlying their conceptual and theoretical considerations. As Hess argues, conceptual tools in border studies often run the risk of producing a representational regime in which migrants appear as the dependent variable, as structurally powerless and as 'victims,' which can lead – even in critical research work – to a reification of the controlling power of the border regime. Hess asks how scholars of migration can re-conceptualize the relation between borders and movements, especially the movement that is politically labeled 'international migration.' Through her discussion, Hess offers a critical approach to the nexus among the concepts 'border,' 'movement,' and 'migration' that seeks to do justice to the agency of migration, while at the same time being capable of analytically examining the expansion of the border regime without minimizing the border regime's militarization and brutality. In this context, the concept of 'the autonomy of migration' as a prism for academic analysis resituates migration within the history of labor, capitalism, and modern forms of government.

Constructions of boundaries and (national) borders are inextricably connected with politics of documentation. Charlton Payne's contribution focuses on this connection, exploring the heuristic potential of the concept of 'documentality.' Payne investigates the links between the history of migration, displacement, and practices of documentation, and takes up the idea that the ability to leave and record traces supplies the performative basis of social institutions. Payne complements his discussion of the concept of 'documentality' by turning to literary and theoretical writings that link the identity document to reductive identity ascriptions as well as to a militarization of society. He calls attention to the mechanisms of exclusion, structures of complicity, and inherent instabilities that this system of reference engenders.

The section concludes with a migration pedagogy perspective on the realm of border regimes. Paul Mecheril highlights the importance of discursive regimes and draws our attention to the political, aesthetic, and educational labeling at work in these regimes. It is on this very discourse level that Mecheril sees the specific point of entry for a migration pedagogy perspective. Migration pedagogy as a critical practice recognizes the power of institutional and discursive orders of belonging and furthermore explores the question of how the capacity to act with dignity might be cultivated under the given conditions without uncritically affirming and accepting these conditions. Critical migration pedagogy thus focuses on the analysis of possibilities for and forms of the changing orders of belonging and hegemonic structures of domination, as well as resistance to and within these orders and structures.

Cluster III: Stories, Histories, and Politics

One of the central frames of migration has been the political realm of the nation-state with its founding myths, narratives, stories, and histories, and its individual and collective memories. Rogers Brubaker, in his book *Nationalism Reframed*, has prominently discussed the ways in which the term "nation" itself works as a theoretical and practical category but particularly as a cognitive frame. We should not ask: "what is a nation," but rather: "how is nationhood as a political and cultural form institutionalized within and among states? How does nation work as practical category, as classificatory scheme, as cognitive frame?" (Brubaker 1996, 16). All four contributions in this section engage in questions pertinent to this political framing. They critically reflect upon the category of the nation in the study of migration and address it as a realm of practice and politics rather than a unified object of research. Discussing aspects such as cultural memory, colonial history, and exilic displacement, they draw our attention to the role of narratives, myths, and processes of knowledge formation in the context of trans/national politics.

Suggesting a conceptualization of nations as "highly contingent cultural and psychological phenomena involving a wide range of heterogeneous and contradictory communications," Wulf Kansteiner focuses his case study on the German television series *Scene of the Crime*, Germany's Sunday evening primetime flagship program, and its engagements with migration. In light of its prominence in the German media landscape, Kansteiner interprets the series as a central site of German cultural memory and thus as part of the media framework for the society's conceptualizations of migration. Given that German television has played an important role in the remapping of collective national-ethnic identities after the fall of the Berlin Wall, in Kansteiner's view, TV networks and critics have not paid enough attention to the problematic narrative and aesthetic strategies with which crime shows influence perceptions of foreigners and processes of social integration. In his discussion of the conceptual frameworks at work in societies, Kansteiner advocates a mobility turn in memory studies and the politics of memory. Such a turn to the concept of mobility should, in Kansteiner's view, conceptualize moving people not as deficient *vis-à-vis* an allegedly static location from which they emerge (emigration) or to which they aspire (immigration). Instead, through the employment of mobility, movement itself would become the focus of analytical and narrative interest. This would provide a foundation for a wide spectrum of multi-directional memories instead of homogenizing memory regimes and their inherent constructions of alterity.

The larger conceptual framework of national memory politics is also central for Friederike Eigler's contribution. While Kansteiner's focus lies on the German media landscape and the framework of memory politics, Eigler's considerations engage predominantly with the concepts 'postmemory' and 'postgeneration' in the context of literary works. She discusses the heuristic value of these concepts for the study of flight and expulsion, and the role that literary texts play in their cultural remembrance. In light of the European history of contested public discourses on flight and expulsion it is not surprising that a considerable part of the new scholarship draws on memory studies as a methodological framework. Based on her discussion of this recent scholarship she asks how literary texts referencing national and transnational contexts can foster our understanding of particular aspects of 'memory' and 'postmemory' and thus contribute to a further refinement of methodological approaches and theoretical frameworks in the scholarship on migration.

Both Kansteiner and Eigler draw our attention to societies' interpretation and remembrance of the past as a powerful frame for their contemporary conceptualization of migration and their future politics of migration. Encarnación Gutiérrez Rodríguez, in her contribution to this section, shares this outlook and adds her perspective on the power of Europe's entangled colonial histories and the matrix of objectification and racialization informing local and national conceptualizations of migration and asylum. Proposing the concept 'coloniality of migration,' Gutiérrez Rodríguez develops an approach to migration within the theoretical framework of decolonial theory. In light of prevailing European myths about cultural homogeneity, racial and ethnic endogamy, immobility and territorial rootedness, the discussion of migration as an articulation of a modern world-system needs to be connected, as Gutiérrez Rodríguez argues, to the cultural, political, economic, and social legacies of European colonialism. Her discussion of the nexus between migration and asylum challenges the rupture among asylum, forced migration, and labor migration, and emphasizes instead the differential logic of racialization in the complex interplay amid the social, the economic, and the political.

The section concludes with Kader Konuk's contribution, which highlights the political framework of forced migration and exile studies in the context of migrant academics. Using recent political developments in Turkey forcing academics into exile as a case study, her conceptual discussion of exilic displacement offers an engaged analysis of academic knowledge formation in its interrelatedness to state power and to national politics. To this end, Konuk argues for a reconceptualization of academics and intellectuals who have been forced to leave their academic institutions and countries. More nuanced and more adequate concepts are needed that reach beyond the individual hardship of what has been called the "conscious pariah" or "exilic intellectual." In Konuk's view, these concepts run the risk of too readily reducing the notion of "exile" to merely a metaphor for "uprootedness." Consequently, this view of exile potentially distorts the historical record, diminishing the existential plight of those forced to migrate. Against this view of exile qua detachment, Konuk proposes the concept of multiple attachments. A critical investigation of these attachments teases out their implications both for the individual and for the respective societies at large, shedding light on the ambivalent liaison of nation-state and higher education.

Cluster IV: New Contexts – Changing Concepts

If concepts, in Mieke Bal's sense, "help us see" (Bal 2009, 22) and "focus interest" (Bal 2007, 40), we ought to be attentive to the changing character of concepts and their situatedness in specific contexts. This is especially true in light of the mobility of concepts that Mieke Bal famously pointed to:

> Concepts are not fixed. They travel – between disciplines, between individual scholars, between historical periods, and between geographically dispersed academic communities. Between disciplines, their meaning, reach, and operational value differ. These processes of differing need to be assessed before, during, and after each 'trip.' (Bal 2011, 12; on traveling concepts see Neumann and Nünning 2012)

Working with concepts in the study of culture more generally and in the study of migration more specifically thus requires a continued reflection on the commonalities and differences of these concepts as they change across time, across scholarly communities, and across the boundaries of academia. The contributions to this section focus on such changes and individual contexts while reflecting on the analytical tools in the field as central elements of our knowledge construction.

In that vein, Evangelos Karagiannis and Shalini Randeria scrutinize the concepts 'migrant' and 'culture' and their specific employments that shape the discourse on migration in academia as well as in politics. To draw attention to the nuances, changes, and political contexts of such concepts, the authors call for a socio-cultural anthropological approach to migration as a societal phenomenon. The authors demonstrate the importance of a nuanced contextualization of an analytical category like 'migrants' that allows us to trace semantic shifts over time in allegedly timeless concepts. It also reveals how these seemingly neutral descriptive or analytical terms are situated in a concrete socio-political discourse. Increasing skepticism towards the analytical fruitfulness of the concept of culture, on the one hand, and its parallel ascent in political discourses of migration, on the other, led anthropologists to treat 'culture' not as an analytical term but as an emic category – that is, as one belonging to the semantics of the societies under study. The increasing identification of the category of migrants with Muslims can be seen in the context of the recent practice of coupling the discourse on migration with Orientalism, i.e., the figuration of the 'Undesirable' as the quintessential 'Other.'

Offering a sociological perspective on migration, Heike Greschke problematizes the proximity of conceptualizations of migration in the social sciences to nation-state notions of normality in society and argues that this proximity is also reflected in the prioritization of questions regarding social, or more specifically, cultural integration or segregation. In her approach to migration in the case of (transnational) migrant families, Greschke departs from 'methodological nationalism' and provides a two-fold, shifting perspective on transnational migration processes and practices. Greschke employs a specific actor-centered methodology that combines a micro and a macro approach and sheds light on the concept 'migration' in its interrelation with the concept 'integration.' In her attempt to disentangle the analytical concepts from the often emotionally charged terms used in the political arena, Greschke restricts the concept 'integration' to a systematic definition and thus differentiates 'migration' into distinct modes of mobility.

Changing concepts and new contexts bear great potential to create new perspectives for the study of migration particularly when these changing concepts go along with the development of new 'operative' analytical lenses. Doris Bachmann-Medick's contribution to this volume is a case in point. She invites us to approach various forms of migration with the analytical lens of 'translation' in order to find new angles from which to reevaluate existing and inform future case studies of migration that have been presented in the disciplines of history, sociology, or cultural anthropology. If embedded in a wider multidisciplinary approach, a translational perspective can shed light on the decisive pivot points in migration processes: their 'pre-translations,' context shifts, negotiation procedures, and processes of inclusion and exclusion, as well as multiple linguistic affiliations and conflicts of the migrants themselves. Such an employment of the concept 'translation' in the context of migration may also help us to be more attentive to the ambivalent role of migrants as agents of translation on the one hand and their day-to-day struggles as translated individuals and groups on the other.

In a similar vein, it is a principal concern of this volume to put forward a perspective on concepts that not only regards them as academic tools but also attends to their important role in political discourses as well as in people's personal experience. To explore the translatability of these different dimensions in the work of concepts is thus one of the fundamental challenges in the study of culture and migration.

References

Amelina, Anna, Devrimsel D. Nergiz, Thomas Faist, and Nina Glick Schiller. *Beyond Methodological Nationalism: Research Methodologies for Cross-Border Studies*. New York/Abingdon: Routledge, 2012.

Bal, Mieke. "From Cultural Studies to Cultural Analysis." *Kritische Berichte* 2 (2007): 33–44.

Bal, Mieke. "Working with Concepts." *European Journal of English Studies* 13.1 (2009): 13–23.

Bal, Mieke. "Interdisciplinarity: Working with Concepts." *Filolog* III (2011): 11–28.

Baldassar, Loretta, and Laura Merla, eds. *Transnational Families, Migration and the Circulation of Care: Understanding Mobility and Absence in Family Life*. New York/London: Routledge, 2014.

Bartram, David, Maritsa V. Poros, and Pierre Monforte. *Key Concepts in Migration*. Los Angeles/London/New Delhi/Singapore/Washington, DC: Sage, 2014.

Brockmeyer, Bettina, and Levke Harders. "Questions of Belonging: Some Introductory Remarks." *InterDisciplines* 1 (2016): 1–7.

Brubaker, Rogers. *Nationalism Reframed: Nationhood and the National Question in the New Europe*. Cambridge/New York: Cambridge University Press, 1996.

Butler, Judith. *Frames of War: When Is Life Grievable*. London/New York: Verso, 2009.

Casas-Cortes, Maribel, Sebastian Cobarrubias, Nicholas De Genova, Glenda Garelli, Giorgio Grappi, Charles Heller, Sabine Hess, Bernd Kasparek, Sandro Mezzadra, Brett Neilson, Irene Peano, Lorenzo Pezzani, John Pickles, Federico Rahola, Lisa Riedner, Stephan Scheel, and Martina Tazzioli. "New Keywords: Migration and Borders." *Cultural Studies* (2014): 1–33 (article "Border Regime" 15–16).

Cassarino, Jean-Pierre. "Theorising Return Migration: The Conceptual Approach to Return Migrants Revisited." *International Journal on Multicultural Societies (IJMS)* 6.2 (2004): 253–279. <http://unesdoc.unesco.org/images/0013/001385/138592E.pdf#page=60> [accessed: 15 March 2018].

Glick Schiller, Nina. "Beyond the Nation State and its Units of Analysis: Towards a New Research Agenda for Migration Studies." *Concepts and Methods in Migration Research: Conference Reader*. Ed. Karin Schittenhelm. 2007. 39–72. <http://sowi-serv2.sowi.uni-due.de/cultural-capital/> [accessed: 15 March 2018].

Glick Schiller, Nina, Linda Basch, and Cristina Blanc-Szanton. "Transnationalism: A New Analytic Framework for Understanding Migration." *Annals of the New York Academy of Sciences* 645.1 (1992): 1–24.

Goffman, Erving. *Frame Analysis: An Essay on the Organization of Experience* (1974). Boston: Northeastern University Press, 1986.

Hess, Sabine. "De-Naturalising Transit Migration: Theory and Methods of an Ethnographic Regime Analysis." *Population, Space and Place* 18.4 (2012): 428–440.

Hoerder, Dirk. "From Immigration to Migration Systems: New Concepts in Migration History." *OAH: Magazine of History* (Fall 1999): 5–11.

Horvath, Kenneth, Anna Amelina, and Karin Peters. "Re-thinking the Politics of Migration: On the Uses and Challenges of Regime Perspectives for Migration Research." *Migration Studies* 5.3 (2017) (special issue: *Migration Regimes in the Spotlight: The Politics of Migration in an Age of Uncertainty*): 301–314.

Lazăr, Andreea. "Transnational Migration Studies: Reframing Sociological Imagination and Research." *Journal of Research in Anthropology and Sociology* 2.2 (2011): 69–83.

Lässig, Simone, and Swen Steinberg. "Knowledge on the Move: New Approaches Toward a History of Migrant Knowledge." *Geschichte und Gesellschaft* 43.3 (2017): 313–346.

Le Goff, Jacques. *Phantasie und Realität des Mittelalters* (1985). Stuttgart: Klett-Cotta, 1990.

Levitt, Peggy. "What's Wrong with Migration Scholarship? A Critique and a Way Forward." *Identities: Global Studies in Culture and Power* 19.4 (2012): 493–500.

Mayer, Benoît. *The Concept of Climate Migration: Advocacy and its Prospects.* Cheltenham/ Northampton: Edward Elgar Publishing, 2016.

Neumann, Birgit, and Ansgar Nünning, eds. *Travelling Concepts for the Study of Culture.* Berlin/ Boston: De Gruyter, 2012.

Pfaff-Czarnecka, Joanna. *Multiple Belonging and the Challenges to Biographic Navigation.* MMG Working Paper 13-05. Göttingen: Max Planck Institute for the Study of Religious and Ethnic Diversity, 2013.

Sassen, Saskia. "A Massive Loss of Habitat: New Drivers for Migration." *Sociology of Development* 2.2 (2016): 204–233.

Sayad, Abdelmalek. *The Suffering of the Immigrant.* Preface by Pierre Bourdieu. Cambridge: Polity, 2004.

Schittenhelm, Karin, ed. *Concepts and Methods in Migration Research: Conference Reader.* 2007. <http://sowi-serv2.sowi.uni-due.de/cultural-capital/reader/Concepts-and-Methods.pdf> [accessed: 15 March 2018].

Scott, James C. *Domination and the Arts of Resistance: Hidden Transcripts.* New Haven/London: Yale University Press, 1990.

Stoler, Ann Laura. *Duress: Imperial Durabilities in Our Times.* Durham, NC/London: Duke University Press, 2016.

Tamas, Kristof, and Joakim Palme, eds. *Globalizing Migration Regimes: New Challenges to Transnational Cooperation* (2006). London/New York: Routledge, 2016.

Tazzioli, Martina, and Nicholas De Genova, eds. "Europe/Crisis: Introducing New Keywords of 'The Crisis' in and of 'Europe.'" New Keywords Collective. *Zone Books Near Futures Online.* <http://nearfuturesonline.org/europecrisis-new-keywords-of-crisis-in-and-of-europe/> [accessed: 15 March 2018].

World Migration Report 2018. International Organization for Migration (IOM)/The UN Migration Agency. Geneva: IOM, 2017. <https://publications.iom.int/system/files/pdf/wmr_2018_en.pdf> [accessed: 15 March 2018].

I: The Power of Images and the Imaginary

Christine Bischoff

Migration and the Regime of the Gaze

A Critical Perspective on Concepts and Practices of Visibility and Visualization

1 Introduction

Culture today is a visual culture: nothing should escape visibility. Our knowledge, our ideas and experiences are shaped by images that do not merely illustrate or reproduce them, but also structure, change, and highlight them. Accordingly, our perceptions and interpretations of migration are also determined by images; they play a decisive role in the construction of "imagined communities" (Anderson 2016).

Whether and how we see migration is a sociocultural process: we see what has been made visible to us.[1] The (in)visibility of things is not a manifest quality, but, as Peter Geimer put it, "designed in studios and laboratories" (Geimer 2002, 7). For example, something can become visible at the price of something else leaving the spotlight. Visibility is much more than visual perceivability. The invisible can become visible, as illustrated in the example of video surveillance (see Holert 2000, 19). But it is not only whether and how we see migration that is determined by images but also how we remember and compare migration discourses and events, since they are our paradigmatic memory reservoir. Aleida Assmann emphasizes that memory is often no longer linked primarily to stories and characters but to free-floating images (see Assmann 2001, 114).

The context in which an image appears is decisive for its analysis because the question is in fact how a multi-layered phenomenon such as migration can be made visible at all. Migration is merely an idiom behind which a multitude of sociocultural realities and economic and political structures are hidden, which in turn are bundled and abstracted in it. Behind the idiom are individual fates with very different backgrounds, with obvious and intimated changes, the actual or supposed causes of which are seen in migration. Images can for the most part only show very specific things, people, and situations. If, for example, one looks at press pictures on the topic of migration and the contexts accompanying them, this bundled and abstracted web of narratives has to again be dismantled – "into

1 On the term visibility in its political, historical, and cultural contexts see Holert 2000, 19.

https://doi.org/9783110600483-003

images of reality that, although they cannot define them, can question presupposed concepts" (Jaeggi 2006, 12).

The issue of migration in the mass media has not suffered from a lack of visibility in the last few years. However, in the context of media it is primarily linked to the concept of "otherness," which in turn is understood primarily as a cultural difference. The migration scholar Mark Terkessidis argues that Otherness is often used as a backdrop for discussing globalization processes and carrying out the consequent conflicts:

> In fact, the [...] visual habits in terms of cultural difference must change radically – because difference does not appear there where the majority thinks it sees it. In order to understand the process of globalization in the field of culture, it is necessary to perceive its disquieting difference or the alterity of identity. (Terkessidis 2006, 312)

That said, it is clear that visual narratives in the media have already taken up contemporary concepts that emphasize the decenteredness, brittleness, incoherence and non-linearity of human lives and represent life in terms of transitions with different spatial and boundary experiences (see Bischoff 2016, 14–16). Changes of perspective can also be found in these media narratives, despite the frequent accusations that they represent migration as out of the ordinary or a crisis or only consider economic perspectives (see Projekt Migration, 6–7). Both migration processes and developments in the media landscape are non-linear and characterized by breaks and non-simultaneous simultaneities. The latter are characterized by different (visual) representation strategies that cross, meet, intersect, compete, and are subject to fluctuation and changing relevance.

Despite the growing importance of globalization processes, national perspectives remain dominant socially and politically. The dominant visual regimes attach to the phenomena of migration a touch of something problematic that needs to be overcome in one way or another. The viewers are placed in imagined centers from which the changing periphery of the migrant is looked upon. The migration scholar and author Aytac Eryilmaz understands visual regimes above all as the perspectives of the nation that creates others out of people crossing borders:

> Foreigners who need to be studied and understood, defended against and controlled, used and integrated. Whether with empathy, economic pragmatism or racist exclusion, the nation uses the other to position itself at the centre. Thus emerges the narrative of the majority and its minorities. (Eryilmaz 2006, 16)

The visual regime determines what we see and how we see it. Seeing and perceiving are thus socioculturally constructed processes. It is necessary to reveal accustomed visual regimes and to challenge the claimed self-evidence of the dominant (national) gaze. And this is not possible without focussing especially on those who are decisive in the production, mediation, and reception of the visual regimes relevant to migration: the media. Medial representations in word and image manage and popularize the ideas and the knowledge of people, objects and their movements and thus the "iconography of the other." But images of migration themselves are in motion. They are often even preceeding the actual migration processes. They are a technique and a practice of migration, as the art historian W. J. T. Mitchell puts it:

> Images are 'imitations of life', and they turn out to be, in a number of important senses, very much like living things themselves. It makes sense, therefore, to speak of a 'migration of images' of images themselves as moving from one environment to another, sometimes taking root, sometimes infecting an entire population, sometimes moving onward like rootless nomads. [...] Images 'go before' the immigrant in the sense that, before the immigrant arrives, his or her image comes first, in the form of stereotypes, search templates, tables of classification, and patterns of recognition. At the moment of first encounter, the immigrant arrives as an image-text, whose documents go before him or her at the moment of crossing the border. (Mitchell 2010, 13–14)

In the case of images, there is always a variety of possible interpretations or multiple encodings marked by an interpretative flexibility (see Leimgruber, Andris, and Bischoff 2013, 257). There is no guarantee whatsoever that an image will have only 'one true' meaning. And there is no law saying that meanings will not change over time. It makes little sense to engage in debates about 'right' and 'wrong' meanings, but about equally plausible, as well as competing and contested meanings and interpretations. "One soon discovers that meaning is not straightforward or transparent, and does not survive intact the passage through presentation. It is a slippery customer, changing and shifting with context, usage and historical circumstances. It is therefore never finally fixed" (Hall 1997, 9). Images are not puzzles that are solved according to rote rules; they are ambiguous and unstable. The ability to see and decipher specific codes can change, decrease, or expand among both image producers and image recipients. The key is to continually deal with the empirical material and to look again at the concrete examples.

In today's media and information societies, only what has circulated in the media or what has been captured in an image appears to have taken place or existed at all. What is not captured in an image does not exist. To be excluded is above all to be unable to make oneself visible (see Bischoff 2016, 17). Inclusion and exclusion are not absolute concepts but they are always related. Particular attention must be paid to medial mechanisms and strategies that lead to the creation of inclusion and exclusion. One cannot exist without the other. In this respect, images are used to highlight both relations and the construction of difference. They mark conceptions of identity by visualizing ideas of 'the own,' which always involve presentations of 'the foreign Other.' Just as there is no definition without distinction, representation of the self is not possible without addressing the Other. In order to make definitions of borders medially intelligible, internal social cohesion and a difference to what is outside must be portrayed as a plausible fact, as historically given and immediately apparent in everyday life:

> All these images and narratives thus try to suggest this plausibility. In fact, it is especially the everyday and cognitively verifiable images that are largely 'constructivist' stereotypes that are only plausible when they are ideologically and mnemotechnically prepared, that is, when people see rose-coloured images through appropriately coloured glasses. (Kaschuba 2008, 297–298)

In images of migration, there is no 'objective reality' to be found, they are not 1:1 images of an immediate, experienced reality. Synthetic image production does not go so far as to replace experienced reality.[2]

The following chapter will not discuss images of migration in the sense of them being an 'iconography of the Other' and question how appropriate or inappropriate they may be in terms of their reflection of the 'reality' of what is pictured. The visual narratives are less intended to discuss what is represented than they are an examination of processes of representation and the accompanying conscious and unconscious intentions and desires of both the producers and recipients. The analysis will be concerned with both the "image of the other" and the gaze of the invisible subjects constituted in the definition of the "deviating other." The "constitutive outside" (see Butler 1990) points to this function of the image narrative that only discriminates secondarily as "fine images," and "primarily serves the constitution of collective and individual identities supported by the signifiers of alterity" (Schmidt-Linsenhoff 1997, 12).

In order to examine more closely the complex, interrelated relationships between the visible and the speakable visible, the article addresses two main questions: first, it will discuss why the concept of visual regimes might be fruitful in the analysis of 'images of migration' and the 'migration of images' and how it relates to image discourses understood as everyday media practice. This will

2 The 'iconic turn' postulated in the humanities and social sciences as a cultural turn marking the shift from a spoken and written culture to an image culture in the course of the institutionalization and universalization of photography, film, and electronic media "includes not only the clearly increasing quantity of the visualisations; the core of this process of change is the domination of subjective perception through the images" (Müller-Doohm 1993, 439).

be followed by a case study that will demonstrate how the implementation and application of visual regimes work through specific processes of image production, mediation, and reception by constructing specific image narratives, providing selected versions and controlling or completely blotting others out. The focus here will be on images of the religious in the context of migration, since these have a prominent proxy function in image discourses about 'the Other.'

2 Visual Regimes in Relation to 'Images of Migration' and the 'Migration of Images' in Everyday Media Practice

2.1 The Figure of the Other

The Other is, despite the human desire and at times the human compulsion to small and large scale mobility and thus to permanent reorientation, still perceived primarily as a problem. The Other arises in the mind of the beholder: images do not simply show 'the Other,' they at the same time conceal the photographer, the author and the addressee of these media products. The representation of the Other, of migrants, in the media thus does not only say something about the medial representation of the Other, but also about the constitution of the respective society itself. This mechanism of reprojection is guided by unconscious interests and projections. Images of migration, therefore, are not only codes of individual perceptions but of entire social discourses in their respective specific historical social forms of production, mediation, and reception. In the discussion of the Other, the discussion of the social and national self is always an inherent part; one cannot be thought or represented without the other. In order to perceive and declare the Other as such, it must already be part of an imagined collective. To visualize migration – even with the intention of understanding, of comprehending and in doing so normalizing it – is always a form of Othering, a process of transformation, design, and creation.

Since the 1980s, the study of culture has become increasingly aware of the fact that culture-based differences between people are less a result of actual empirical facts than they are the result of specific forms of representation. Cultures are not objects; they are artefacts of academic and other discourses that work with cultural models.

> In scholarship, it is no longer a question of what marks 'culture x' and distinguishes it from 'culture y'. Instead, the question has become how societal conflicts create cultures as representations, for example by designing specific images of the self and the other, what actors are involved in this process and how these representations of cultures affect social actions or become reflected in social structures. (Sökefeld 2007, 47)

2.2 Visual Discourses of 'the Self' and 'the Other'

The process of globalization impacts permanently and intensively models of national autonomy. The concept of the national perspective nevertheless persists in all its facets in medial, political, and even scholarly discourses. Migration thus retains a flavor of the marginal and the problematic that needs to be coped with, tamed and, when seen as menacing, controlled or pushed back. Migration is always associated with Otherness, which in turn is seen as cultural difference, and usually also as inferiority. Mark Terkessidis sees this as a legacy of the Enlightenment and Idealism as these movements expressed themselves in the German-speaking countries in particular:

> In contrast to the lip service and the ubiquitous rhetoric of postmodernity, the ideals of Johann Gottfried Herder can be heard implicitly in all debates: cultures are [...] still regarded as independent, spherical structures in which the externally visible characteristics of individuals (appearance, clothing, customs, etc.) appear as embodiments of an invisible, substantial cultural commonality – an identity. (Terkessidis 2006, 311)

The Other does not suffer from a lack of visibility, quite the contrary: the Other's outer appearance is instrumentalized in the visualization of aspirations of integration, demarcation, and distinction. Medially mediated images in particular are used to define identity-enhancing patterns that in turn can be used for the construction of the self and the line separating that self from the Other. At the same time, the increasing complexity of clear associations of self and Other, belonging and not belonging, is often met in medial representations by simple umbrella concepts that suggest clear behavioral markers (religious, traditional, etc.) and a reduction to a manageable number of clear external identification markers (skin color, clothing, headscarf, etc.). Images serve both as identity marker and as a resource for self-constructions of the migrants themselves. Images of the self and of the Other are causally linked.

The results are impressive narrative models and catchy metaphors. Gerhard Schulze argues that the return of categories such as "western Christianity," "Christian values" or "core Europe" to public discourse exposes a normative search for belonging in a West that feels exposed to growing threats and fragmentation (Schulze 2008, 139). In the representation of national identities and models of society, both older and new discursive and visual motifs are used again and again, since national consciousness and national identity constructions are orientation models mediated in particular by oral and visual images:

> In different times and contexts, different motifs are activated, old images are discarded or forgotten and new ones are added [...]. Some of these motifs are, on the other hand, enduring, have become canonised since the period of 'nation-building' in the nineteenth century as virtual national symbols and have repeatedly been charged with changing meanings. (Götz 2005, 188)

These media products with all their prejudices and stereotypes are not suitable for linguistic or visual witnesses of one's own and other cultures. Rather, their meaning can be found in the documentation of cultural challenges, confrontations, and rapproachments; they reveal how cultures and groups react to changes, uncertainties, and challenges.

In the 1990s, the concept of the "pictorial turn" (Mitchell 1997) or the "iconic turn" (Boehm 2006) emerged to identify the increasing importance of the visual. These terms signify that visuality has become a paradigm of information and knowledge transfer. However, this is not to be understood simply as an increase in imagery, that is, an ever-growing tide of images,[3] but a change in consciousness that attributes the visual a much more important role in thought processes and knowledge acquisition than has hitherto been the case (see Bruhn 2003, 22).[4]

---

3 The topos of the flood of images and informationlessness was first introduced by Siegfried Kracauer in reference to the success of illustrated magazines in the 1920s. Kracauer suggested that the "flood of images" would lead to an increasing decline in rational-argumentative language culture (see Kracauer 1963, 33).

4 That said, the terms "pictorial turn" and "iconic turn," which have themselves become topoi in the scholarly and more general discourse, have also been criticized. It was not the identification of the shift from the word to the image and the resulting dramatic changes in knowledge (acquisition) that was called into question but that the claim (especially by historians) that a "visual revolution" had taken place was being overdramatic. The publicist Karl Pawek suggested in 1963 in his *Das optische Zeitalter* (*The Optical Age*) that not the 'visual' age was the historical exception but the 'written' age (see Pawek 1963). The "pictorial turn" or "iconic turn" could thus in no case be understood as a linear progression from a writing to an image culture. The media scholar Elke Grittmann also considered the visual age to be the rule and not the exception: "Images have always characterised the perception of an epoch. It was once art that explained political world-views to an elite circle, today it is press photography that contributes so decisively to the emergence of a visual public. The traditions are unmistakable and yet everything is different. The images of press photography are now created in the context of a functional system, as journalism, and no longer at the behest of rulers" (Grittmann 2007, 13).

Images have emerged in relationship to text – and this distinguishes the use of the image in recent years and decades from past centuries – as having their very own news value (see Müller and Geise 2003, 91).

2.3 The Logics of Visual Meaning-Making

Migration discourses conveyed through images are more than a mere form of visual language. Beyond its visual core, what is shown becomes engrained as social and cultural practice.[5] "They are not expressions or images of materialities, but are themselves materialities *sui generis*. They contain elements of consciousness and thus transport and form consciousness" (Jäger 2006, 330–331). Image discourses and therein generated visual regimes of migration structure how certain things are thought and spoken about and how they are visualized and how we act again and again on the basis of this thinking, speaking, and visualizing. They form our knowledge of migration, our understanding, and the impulses that lead to action. This is always about the enforcement of 'legitimate' definitions of migration and the determination of prevailing categories of perception about it. The process of naming and imaging is crucial because the power of conceptual and visual categorization, that is, the power to make something public and explicit, is an extremely significant form of power. The naming and imaging of migration is ultimately also the foundation of political power (see Landwehr 2001, 92).

Visual regimes of migration generated via images have, for example, the power to represent them as normal or deviant, to define migrants as a 'race' or 'ethnic group' and ground them negatively or positively. The producers and mediators of images, such as photographers, photo editors, and the illustrated press act as authoritative presenters, moving within historically specific and regulated modes of thought, language, and visualization that dominate the discourse in specific sociocultural contexts and periods. The resulting visual formations follow a certain logic and can be understood as interpretation models, in which ideals and images are articulated.

Sociocultural life can be understood as a construct marked by media made up of the ideas that people have about life and the practices that evolve from these ideas. This means that the medial image formations are not a transparent

5 Michel Foucault's work on historical, psychiatric, and medical discourses in particular points to one's perception of the mechanisms of inclusion and exclusion that form and discipline subjects in their modes of thought and action (see Foucault 1974).

window for the recipient but always a specific interpretation of the world; in part they become autonomous simulations and a substitute for reality. The media theorist Stefan Weber uses the term "medial turn" to illustrate these upheavals in the fundamentals of knowledge mediation through media (see Weber 1999). Doris Bachmann-Medick in turn reveals the inner "differentiation impulses" of the so-called cultural turns, revealing their different perspectives as concepts visible and accessible for the analysis of social and cultural interconnections (see Bachmann-Medick 2016). Visibility becomes

> a critical category of cultural and social analysis. Visibility refers not only to the possibilities of social self-portrayal and the new sensibilities regarding social staging and forms of surveillance, but also to strategies of social power and exclusion that are aimed at concealment and invisibility (of poverty, inequality, disease, etc.). However, any attempt to render visible such concealed phenomena presupposes a complex context of visualization. For this context and many others, the question of the text-image relationship, the interaction between images, their tense relationship in the media and the need for them to be commented on through writing and texts is of eminent importance. (Bachmann-Medick 2016, 270)

Behind the medial representations are concealed worlds of imagination that draw their logic and meaning-making from (historical) imaginations. They reflect, on the one hand, social desires and longings, and on the other, collective anxieties and repressions. In terms of migration and the Other, they often manifest themselves in a juxtaposition of idealization and depreciation. In this context, processes of interconnection and concentration are constant, especially through the visual, and have significant influence on the medialities of the social appropriation of reality.

Visual regimes of migration become pointed in how they approach the religious. Medial visualization practices show that perceptions and understandings of 'the Other' or of the 'especially Other' are (once again) increasingly being linked to questions of religiosity or religious belonging.

3 The Religious and its Proxy Function in the Image Discourse on Migration

3.1 Visible in Invisibility

At the latest since the events of September 11, 2001, the headscarf has become an icon of Otherness and the associated, so-called integration problems in the medial migration discourse. This hypothesis is common in the scholarly literature (see Tarlo 2010; Terkessidis 2006; Farrokhzad 2006; Räthzel 2005). The headscarf is discussed not only in the media but in the scholarly debate as a symbol of difference. It is, in its various forms, a visual marker of being an 'Other,' of being different and of deviating norms and values. It is a central emotive force in discussions about Muslim migration processes in European countries. As a visually and discursively mediated potential threat narrative, it has a transnational scope that European media like to access particularly when contrasting the 'Self' and the 'Other' or the incongruences between Europe and countries marked by Islam. While the media illustrate and construct differences using the headscarf, scholarly studies emphasize how such visual and discursive narratives are condensed into apparently constitutive images that impact an individual's perceived knowledge of the 'Self' and the 'Other' (see Beck-Gernsheim 2004, 51–57).[6] The headscarf can be (mis-)used as a marker of the oppression of women. The headscarf becomes an alibi. The image-rhetorical operation that takes place here has been described by Roland Barthes in his *Myths of Everyday Life*: any random sign from our everyday lives, so Barthes, can be transformed into a myth of the present. The sign is deprived of its concrete history and transformed into a timeless gesture (see Barthes 1964, 4–6). The headscarf becomes a symbol of an Islamic threat, a code for oppression, and a "gesture of conquest by the other." The headscarf is thus present in media debates, even if it is visually absent. It is a visible and invisible pars pro toto when migration is discussed in the media as a source of conflict, both religious and non-religious, and in the confrontations of conscience and worldview. As the cultural scholar Nanna Heidenreich says, Western "habits of seeing" make it possible for us to see the headscarf where it is and where it is not (Heidenreich 2003, 316). While in the European context the headscarf has become an automatic visual code for a premodern tradition, in the Islamic context, Heidenreich points out, it is seen in a much more differentiated manner:

> Wearing carsaf (the Turkish version of a chador) and/or a turban, that is, a headscarf worn in an Islamic-style, marked upwards social mobility, urbanity and education in opposition to the mostly rural traditional headscarf worn by the lower and middle social classes that was not measured by how much it covered. (Heidenreich 2003, 316)

6 Contrary to its alleged omnipresence, the role of typical 'headscarf images' as a visual marker in the media is often quantitatively overrated. That, at least, was the result of an empirical study of Swiss print media. It is more so that the associative efficacy of the 'headscarf' is very strong because this genre includes many formal and stylistic similarities. These similarities lead the viewer to immediately draw similar conclusions to the appearance of a headscarf that remain particularly vivid in his image memory (see Bischoff 2016, 237–239).

In the Anglo-American context, unlike in the German-speaking countries, the headscarf is increasingly being considered under other aspects that transcend one-dimensional religio-political connotations. Emma Tarlo consciously addresses the increasingly aesthetic and fashionable aspects of the headscarf in her attempt to find a more relaxed approach to this piece of clothing:

> The topic of Muslim women's appearances is by no means a blank canvas. Rather, it has the quality of a familiar painting, so often reproduced that representation gets confused for reality and we fail to see what might have been left out of the picture or how things could have been painted differently. Representations of Muslim women are dominated by one single all-consuming image, word and concept – *the veil*. [...] Not only does it have the effect of marginalizing those Muslim women who do not wear visible indicators of faith, but it also reduces those who do to a single coherent and monolithic category as if they speak and think with one voice. [...] In the case of images of women, hijabs, jilbabs and niqabs function both to indicate (and by implication, confirm) the idea that Muslim cultural and religious values and behaviour are somehow alien and different. (Tarlo 2010, 2, 5, 9–10)

3.2 Perspectives of Ethnification

The position that the headscarf should not simply be considered a piece of clothing associated with religious behavior norms in addition to its fashion, practical, and socio-conventional aspects is the reason for the confrontation it provokes. Instead, the interpretation that wearing a headscarf is a sign of the "stubborn and exclusive behaviour of an other who does not want to adapt to the respective national or leading culture" (Berghahn 2009, 34). The image of Muslim migrant women is thus reduced in media debates to their 'cultural' affiliation, irrespective of their social status or their life circumstances. A migrant with a headscarf is constructed as the symbol of a cultural 'Other' within an imagined homogeneous entity marked by national borders, inside of which it is stylized as a threat to peaceful coexistence. This form of media migration discourse marks different boundaries: in addition to nation-state boundaries, there are inner boundaries, namely those of cultural differences that emerge in the country of migration and lead to conflicts. Etienne Balibar coined the phrase neo-racism as "racism without races" at the core of which no longer stood, so his argument, biological difference but the "postulate of insurmountable cultural differences" (Balibar and Wallerstein 1992, 28–30). This form of racism suggests that different ways of life and traditions are not really compatible and thus lead to dismissive attitudes. Differences were no longer created and ontologized by nature but by culture. The so-called cultural difference, combined with the concept of ethnicity, becomes the new concept of "racial difference" (see Balibar and Wallerstein 1992, 29–30).

The headscarf becomes a cultural and visual marker that initiates such construction processes of difference and can be (mis-)used as an instrument to modernize racism discursively (see Bischoff 2016, 246).

Finally, boundaries are drawn between and within the sexes: these boundaries are created by a form of 'ethnic stratification' that shows itself in the significant differences between the ways of life of indigenous and immigrant women. With the increasing delegation of reproductive work (housework, cleaning, etc.) to migrant women, inequality between women in a society increases. At the same time, the continuing structural gender conflicts in the respective country of immigration are made invisible, since the reproductive work remains at its core 'women's work.' This contributes to the idea that within this migration subdiscourse, 'Western' women can be represented as emancipated, independent, and competent, and thus as a counterpoint to migrant women, especially Muslim, headscarf-wearing migrant women. The 'Western' man is in turn not considered patriarchal, but modern, open-minded, and cooperative. The problems Muslim migrant women have with regard to language, education, professional opportunities, and so on are not discussed in the context of discriminatory institutional structures but as family structures primarily defined by (backward) traditions and (religious) fanaticism. The essence of this discourse is the headscarf and burka debate, which unifies all these images. "The social conflict that results out of the hegemony of the dominant culture is transformed in this prevailing discourse into an inner-ethnic conflict and reduced to the private sphere of gender relations" (Weber 2007, 99).

3.3 Faith Politicized

Visual regimes of migration are increasingly determined by the question of the role of religion. What is the proper relationship between faith and the state? What is the best model for the culturally diverse societies of the 21st century? Islam has become an icon of the radical other in media debates on migration processes, associated with danger and violence. This becomes all too clear in contemporary portrayals of terrorism, homeland security and crime illustrated with praying Muslims, bearded men and headscarf- or burka-wearing women. Migration processes are stylized in the media as the origin of sources of religious conflict and vice versa. This medialized perspective on religion in the context of migration primarily as potential conflict is a development of the last ca. 15 years (see Bischoff 2016, 249–251). It is pointed out again and again, especially with regard to Islam, that religion is not merely a form of faith but a way of life that infiltrates the whole human being and all of society in a manner that is not necessarily compatible with modern European norms. Elisabeth Beck-Gernsheim stresses that the meaning of religiosity undergoes a clear and visible transformation in migration. Religious services in the diaspora, for example, become a social event in addition to being a religious one. They become a meeting point where personal networks and contacts can be developed, with the congregation as a helpful institution that can provide concrete information and advice (e.g., for finding work) and moral support:

> Here one can meet people with a similar history and a similar fate; here one can share memories and catch up on current news; here one can talk with like-minded people about dreams and disappointments, about privations and future visions. Here one can find help, support, community – not only in the spiritual sense, in relation to God, but also in a very profane sense, in relation to people and especially to other migrants. (Beck-Gernsheim 2004, 30)

In migration, religion gains in presence because it emerges as a feature of one's own group belonging and identity. Religion becomes a symbolic space of community, a space for symbolically charged practices, for example wearing a headscarf or beard. However, these symbols are not always to be understood as primarily religious. For Beck-Gernsheim they are "signs of cultural difference" and social markers of a form of cultural Islamic identity (Beck-Gernsheim 2004, 33).

3.4 Religion as an Anchor of Identity

The medial attribution that religion is often the most important anchor of identity among migrants finds its pinnacle in pictorial representations of the phenomenon of conversion. One example can be found in the Swiss *Tagesanzeiger-Magazin* under the title "Die Überläufer. Wie aus einem Schweizer Katholiken ein Taliban wurde und aus einer pakistanischen Muslimin eine Christin"[7] in which two images are placed centrally that are typical for the visual as well as the textual narrative on the subject (Fig. 1 and Fig. 2).

Figure 1 shows a young woman who migrated with her family from Pakistan to Austria when she was a young girl. She converted from Islam to Catholicism in her early 20s, was cast out by her family and has had to live in hiding since then for fear of being found out and killed. With the publication of her book *Vom Islam*

7 "The Converts. How a Swiss Catholic Became a Taliban and a Pakistani Muslim Woman a Christian" (my translation).

Fig. 1: The figure shows a Pakistani woman living in Austria who converted from Islam to Christianity. The article was published in the Swiss *Tagesanzeiger-Magazin* under the headline "The Converts. How a Swiss Catholic became a Taliban and a Pakistani Muslim Woman a Christian." The caption reads: "More freedom, less fight: Sabatina at Vienna's St. Stephen's Cathedral" (*Tagesanzeiger-Magazin* 2004, 8).

Fig. 2: Figure of a young Swiss man who converted to Islam, portrayed in the same article. The capture reads: "The Quran is the absolute truth. Jassin in his apartment in Zurich" (*Tagesanzeiger-Magazin* 2004, 9).

zum Christentum – ein Todesurteil[8] under an alias (see James 2007) she gained significant media attention in Austria. Figure 2 on the other hand shows a young Swiss man who grew up in a village and after a gang career was introduced to Islam by fellow university students who were Turkish. He converted and, so he states, feels committed to the ideology of the radical Taliban.

At first glance, both images have strongly religious connotations: the young woman is embedded in Christian iconography, standing behind rows of devotional candles and the reference to St. Stephen's Cathedral in Vienna. The image is staged in such a way that it can be deciphered by the viewer, regardless of

8 "From Islam to Christianity – A Death Sentence" (my translation).

whether one is aware of the Christian pictorial traditions. The young Muslim in turn is shown in a Muslim context, in particular through the mentioning of the Quran and his dress. He is wearing a Kufiya, a scarf wrapped around his head and face into a turban as is often worn by men in the Arab world. Simple visual markers – candles and clothing – are sufficient for the recipients to properly contextualize and decode the images.

Images such as these make a broad range of interpretations possible. But they also press political and social events and issues into specific interpretive frameworks by means of their creation contexts, means of mediation, addressee-appropriate 'presentation,' captions and contexts. The subject of conversion is conveyed to the addressees specifically by means of visual accentuations and attributions of predefined characteristics. These 'frames' (Goffman 1986) are emotionally and normatively predetermined and draw on explicit and implicit knowledge and ideas. The framing influences the perception and the reaction of the addressees. While recipients of Muslim heritage might not necessarily associate the clothing and the covered face of the portrayed convert with danger, these attributes can initiate an association chain of "converted Muslim – danger – radicalization – islamist terror" for other recipients, especially when seen in contrast to the depicted Christian context – the converted Christian is shown in a madonna-like pose in a sea of candles. In this way, a sublime, peace-loving Christianity and a radical, terrorist Islam are juxtaposed, without this contrast being addressed explicitly in the accompanying text.

An example of the complexity and inconsistency of such immanent visual sensual structures are the photographs that accompany a different *Tagesanzeiger-Magazin* article on another convert.[9] In the article, the Swiss convert Ahmed Huber is portrayed as a dangerous anti-Semite, a right-wing extremist and an Islamist who acts as a "networker and pioneer of Islamist fascism" (*Tagesanzeiger-Magazin*, 3). The menace of Muslim converts even graces the magazine's cover through the headline "Allah and Hitler are Great. Confessions of the Islamist Ahmed Huber," this time without an accompanying photograph. Instead, the cover image shows an older Muslim man in prayer. In the background a television can be seen, a stereo, a shelf with colorful videocassettes, power cables lying around and colorful oriental pictures hanging from the wall (Fig. 3).

The suggestion of potential danger is built up gradually in the images of the reportage. First, Mr. Huber is shown sitting next to a stack of books, the title "Anecdotes about Hitler" quite visible at the edge of the photograph. The next

9 A portrait of Ahmed Huber, who converted to Islam several decades ago; see *Tagesanzeiger-Magazin* (22 May 2004): 20–29.

Fig. 3: Front cover of a photo reportage with the headline: "Allah and Hitler are Great. Confessions of the Islamist Ahmed Huber" (*Tagesanzeiger-Magazin* 2004, 1).

Fig. 4: Figure in the same article shows Ahmed Huber in his study in front of his library with his favorite books (*Tagesanzeiger-Magazin* 2004, 22).

photograph is of Mr. Huber standing in front of a bookcase, followed by numerous portraits (Fig. 4). In this gallery of familiar faces, the usual 'right-wing' and 'Islamist' suspects from different epochs meet, from Richard Wagner and King Ludwig II to Hitler and Jean-Marie Le Pen, the Ayatollah Khomeini and Gamal Abdel Nasser, the Turkish prime minister Necmettin Erbakan and Cosima Wagner. Ahmed Huber is portrayed as a 'showman-Islamist' of whom the reader does not know whether he is merely ridiculous or in fact dangerous.

In contrast to Erwin Panofsky's iconography model, a cultural studies image analysis cannot assume that a universal code exists to decipher images. Images are not puzzles where the pieces always fit together in the same way; they are ambiguous and unstable. The ability to read certain codes can decrease, disappear or expand among both image producers and image recipients.

The photo reportage on the Swiss convert Ahmed Huber works with coded picture puzzles that demand a good deal of previous (image) knowledge, since it cannot be assumed that all recipients at all times will to an equal degree be able to completely decipher the transported image messages. The recipients are indirectly prompted by such puzzle pictures (Fig. 4) to find the corresponding symbols and allegories that verify the narrative of Islamist and anti-semitic danger emanating from such radical converts in the text. In such images, easy-to-decode symbols exist next to more subtle references.

Not all recipients can necessarily encrypt all image messages equally well. Peter Burke points out that recipients may no longer be able to decipher a whole "picture programm," but only fragments of it (see Burke 2010, 41–42). For him, such images bear witness "to the stereotypical but gradually changing viewpoints that individuals or groups have of the social world, including the world of their own imaginations" (Burke 2010, 211).

In this process the meaning of representation also shifts. Representations can no longer only be defined as a form of visibility but as (im-)possibilities of participation that manifest themselves in manifold ways (see Rogoff 1999, 98–99). For example, it is no longer a question of merely analyzing the representation of ethnic minorities, but of analyzing the relations and references between image producers, visual material, and recipients in the public, which can no longer be reduced to a simple active-passive paradigm. Except: the position of the external observer does not exist, since the observing recipient must also be thought of as active and changeable.

The cited examples show that the media discourse is quick to essentialize religion. Religion per se or the 'Other' religion is often demonized as a destructive force threatening the liberal nation-state. Or it is presented as generally 'innocent' and instrumentalized appropriately. Both visualization strategies ignore the role of religion in the development of these very same liberal societies. The jurist and sociologist Felix Ekardt emphasizes, for example, that liberal democracy has its origins not only in the Enlightenment, but also in religious currents such as Calvinism:

> This is true not only for philosophers such as Thomas Hobbes, John Locke, Immanuel Kant, Francis Bacon, Johannes Althusius, and early natural scientists and politicians. This Calvinist basis is all the more important for much of the population, or at least the bourgeois elite in the countries of Western Europe and North America, in which classical liberal ideas came to prevail in philosophy, economics and law during the Enlightenment. This is crucial because it marks the process characteristic of the West: religious moral and political concepts are gradually secularised, and religion itself becomes the catalyst of a liberal, pluralistic state. (Ekardt 2009, 298)

Instead of continually asking headscarf-wearing women about their loyalties vis-à-vis the liberal nation-state, it would make more sense to discuss religion as lived practice in these media portrayals, and in doing so make the wearing of a headscarf comprehensible as a natural aspect of religious practice. The to date unresolved challenge reflected in the medially staged images and debates about the headscarf and burka is the question of which visible public spaces liberal democracies provide for their religious citizens. The headscarf and/or the burka are not just “shifting images” (Oesterreich 2005, 41) perceived as a provocation in image or media reception because they are understood as a double rebellion – as a rebellion against the discriminatory majority society (the formation of a conscious counteridentity with the help of the headscarf) and rebellion against discriminatory traditions (refusal to recognize the preeminence of the man by women socialized in the West, for example by calling for the reconciliation of wage-labor and religiosity). The wearing of a headscarf is by no means only a visible sign and expression of an act that is “a mixture of ultra-conservatism and rebellion. Ultra-conservatism in religious clothing style and rebellion against a society that demands conformity” (Oesterreich 2005, 42). Rather, the headscarf can also be a visual marker of whether liberalism, pluralism, and tolerance truly prevail in liberal democracies and have established themselves in the consciousness and practice of these societies and are thus reflected in the media and other institutions. These forms of contemporary religious pluralism on the global market of world views reveal themselves in surprising but ever more frequent alliances and coalitions, for example in the case of headscarf-wearing Muslim women who because of a headscarf-ban in state schools found asylum in Catholic private schools.

4 Concluding Remarks

The genre of press photography shows most clearly that the oft-cited boundless possibilities to interpret visual material (see Boehm 2015) are in fact limited. The visual regimes in this genre create normative and ritualized narrative and representation forms (see Elia-Borer, Sieber, and Tholen 2011). Space for the interpretation of images is restricted in the print media by predefined ‘plausible’ and accepted interpretation horizons. Images of migration are characterized by their persistence. This applies to text components as well as to, for example, photo reportages, but these are marked, in contrast, by more and broader developments and changes in terms of language use, the disappearance of old terms and the emergence of new, as well as different narrative perspectives (see Bischoff 2016, 286–287). The simple oppositions found in images on the other hand persist between monocultural (marked by linearity and continuity) and transcultural (determined by multilocality and disparity) visual regimes. Dealing with this opposition contributes to feelings of identity.

In public media discourses, this often leads to a discrepancy between monocultural and transcultural perspectives. But it is only when both are thought of and formulated together that sociocultural lifeworlds of migrant groups can be understood and the misunderstanding and mystification that frequently characterizes medial debates on migrations be prevented. Images published in the context of public debates on migration initiate a process of ethnification in everyday cultural communication. Some groups find prominence and become identifiable in visual and narrative genres in the form of ethnic figurations. Conceptually, the underlying paradigm of cultural difference defines the 'Self' and the 'Other,' which is prerequisite for migrants being identified as an independent, albeit problematic and special, social category. The division into 'Self-images' and 'Other-images' as a basic associative pattern transcends the media as a sociocultural organization principle. The purpose of 'Self-images' as a paradigmatic concept is that a symbolic self-identification emerges in collective images in which groups form socioculturally and confront one another. These images and narratives are an essential part of the construction and constitution of the imagined 'Self' and 'Other,' which then becomes effective in the different sociocultural representations of individuals, groups, generations, genders, and religions. That said, this 'imaginary' quality has very 'real' and 'lasting' political consequences because it can be transformed into symbolic social and cultural practices, especially in everyday patterns of inward and outward looking identity politics (see Kaschuba 2008, 297).

For the history and present of migration processes, the question arises whether these should continue to be presented in images that are ethnically marked, to thus monumentalize a form of ethnification governed by eternal repetitions and variations, or whether it might ever be possible to imagine something new that "takes into account the contradictions and problems of every representation, as well as the contingency of the image itself" (von Osten 2005, 29). This seems all the more difficult because the concept of ethnicity is becoming ever more amorphous as a result of constantly changing sociocultural living circumstances. Arjun Appadurai, who is critical of the term, but cannot avoid it completely either, argues instead to use the term "ethnoscape" (Appadurai 2006, 50),[10] writing:

---

10 Appadurai defines ethnoscapes as "the landscape of persons who make up the shifting world in which we live: tourists, immigrants, refugees, exiles, guestworkers, and other moving groups and person constitute an essential feature of the world and appear to affect the politics of and between nations to a hitherto unprecedented degree [...]. But it is to say that the warp of these stabilities is everywhere shot through with the woof of human motion, as more persons and

> The 'ethno' in ethnography takes on a slippery, nonlocalized quality, which the descriptive practices of anthropology will have to respond to. The landscapes of group identity – the ethnoscapes – around the world are no longer familiar anthropological objects, insofar as groups are no longer tightly territorialized, spatially bounded, historically unselfconscious, or culturally homogenous. (Appadurai 2006, 50)

Thus, the point is primarily that the 'we' lexicalized in the media to defend imagined majority positions against 'the Other' must be challenged. The meaning of transitions, in-between-spaces, and processes of mixing – and not only among people identified as 'migrants' but in general – must be visualized and narrated while mystifying ideas of any original 'purity' of cultures need to be abandoned – medially and especially visually.

References

Anderson, Benedict. *Imagined Communities: Reflections on the Origin and Spread of Nationalism.* London/New York: Verso, 2016.

Appadurai, Arjun. "The Power of the Imagination." *Projekt Migration. Ein Initiativprojekt der Kulturstiftung des Bundes.* Eds. Kölnischer Kunstverein, Dokumentationszentrum und Museum über die Migration in Deutschland, Köln, Institut für Kulturanthropologie und Europäische Ethnologie der Johann Wolfgang Goethe Universität Frankfurt a.M., and Institut für Theorie der Gestaltung und Kunst. Zurich/Cologne: Dumont, 2006. 50–53.

Assmann, Aleida. "Wie wahr sind Erinnerungen?" *Das soziale Gedächtnis. Geschichte, Erinnerung, Tradierung.* Ed. Harald Welzer. Hamburg: Hamburger Edition, 2001. 103–122.

Bachmann-Medick, Doris. *Cultural Turns: New Orientations in the Study of Culture.* Berlin/Boston: De Gruyter, 2016.

Balibar, Etienne, and Immanuel Wallerstein. *Rasse, Klasse, Nation. Ambivalente Identitäten.* Hamburg: Argument-Verlag, 1992.

Barthes, Roland. *Mythen des Alltags.* Frankfurt a.M.: Suhrkamp, 1964.

Beck-Gernsheim, Elisabeth. *Wir und die Anderen. Vom Blick der Deutschen auf Migranten und Minderheiten.* Frankfurt a.M.: Suhrkamp, 2004.

Berghahn, Sabine. "Deutschlands konfrontativer Umgang mit dem Kopftuch der Lehrerin." *Der Stoff, aus dem Konflikte sind. Debatten um das Kopftuch in Deutschland, Österreich und der Schweiz.* Eds. Sabine Berghahn and Petra Rostock. Bielefeld: transcript, 2009. 33–73.

Bischoff, Christine. *Blickregime der Migration. Images und Imaginationen des Fremden in Schweizer Printmedien.* Münster/New York: Waxmann, 2016.

Boehm, Gottfried. *Was ist ein Bild?* Paderborn: Fink, 2006.

Boehm, Gottfried. *Wie Bilder Sinn erzeugen. Die Macht des Zeigens.* Berlin: Berlin University Press, 2015.

groups deal with the realities of having to move or the fantasies of wanting to move" (Appadurai 2006, 50).

Bruhn, Matthias. *Bildwirtschaft. Verwaltung und Verwertung von Sichtbarkeit.* Weimar: VDG, 2003.

Burke, Peter. *Augenzeugenschaft. Bilder als historische Quellen.* Berlin: Wagenbach, 2010.

Butler, Judith. *Gender Trouble: Feminism and the Subversion of Identity.* New York/London: Routledge, 1990.

Ekardt, Felix. "Pluralismus, Multikulturalität und der 'Kopftuchstreit'. Politik und Religion in liberalen Demokratien." *Der Stoff, aus dem Konflikte sind. Debatten um das Kopftuch in Deutschland, Österreich und der Schweiz.* Eds. Sabine Berghahn and Petra Rostock. Bielefeld: transcript, 2009. 297–313.

Elia-Borer, Nadja, Samuel Sieber, and Georg Christoph Tholen, eds. *Blickregime und Dispositive audiovisueller Medien.* Bielefeld: transcript, 2011.

Eryilmaz, Aytac. "Vorwort zum Projekt Migration." *Projekt Migration. Ein Initiativprojekt der Kulturstiftung des Bundes.* Eds. Kölnischer Kunstverein, Dokumentationszentrum und Museum über die Migration in Deutschland, Köln, Institut für Kulturanthropologie und Europäische Ethnologie der Johann Wolfgang Goethe Universität Frankfurt a. M., and Institut für Theorie der Gestaltung und Kunst. Zurich/Cologne: Dumont, 2006. 14–25.

Farrokhzad, Schahrzad. "Exotin, Unterdrückte und Fundamentalistin. Konstruktionen der 'fremden Frau' in deutschen Medien." *Massenmedien, Migration und Integration. Herausforderungen für Journalismus und politische Bildung.* Eds. Christoph Butterwegge and Gudrun Hentges. Wiesbaden: Springer VS, 2006. 53–84.

Foucault, Michel. *Die Ordnung der Dinge. Eine Archäologie der Humanwissenschaften.* Frankfurt a. M.: Suhrkamp, 1974.

Geimer, Peter. "Was ist kein Bild? Zur 'Störung der Verweisung.'" *Ordnungen der Sichtbarkeit. Fotografie in Wissenschaft, Kunst und Technologie.* Ed. Harald Welzer. Frankfurt a. M.: Suhrkamp, 2002. 313–341.

Götz, Irene. "Zur Wirkmacht inszenierter Bilder im Medienzeitalter." *Der Bilderalltag. Perspektiven einer volkskundlichen Bildwissenschaft.* Eds. Helge Gerndt and Michaela Haibl. Münster: Waxmann, 2005. 187–198.

Goffman, Erving. *Frame Analysis: An Essay on the Organization of Experience.* New Hampshire: Northeastern University Press, 1986.

Grittmann, Elke. *Das politische Bild. Fotojournalismus und Pressefotografie in Theorie und Empirie.* Cologne: Herbert von Halem Verlag, 2007.

Hall, Stuart, ed. *Representation: Cultural Representations and Signifying Practices.* London/Thousand Oaks/New Delhi: Sage, 1997.

Heidenreich, Nanna. "'Deutsche' (Un-)Sichtbarkeiten." *Fremdes Begehren. Transkulturelle Beziehungen in Literatur, Kunst und Medien.* Eds. Eva Lezzi and Monika Ehlers. Cologne/Weimar/Vienna: Böhlau, 2003. 307–319.

Holert, Tom. "Bildfähigkeiten. Visuelle Kultur, Repräsentationskritik und Politik der Sichtbarkeit." *Imagineering. Visuelle Kultur und Politik der Sichtbarkeit.* Eds. Tom Holert and Mark Terkessidis. Cologne: Oktagon, 2000. 14–33.

Jäger, Siegfried. "Zwischen den Kulturen. Diskursanalytische Grenzgänge." *Kultur – Medien – Macht. Cultural Studies und Medienanalyse.* Eds. Andreas Hepp and Rainer Winter. Wiesbaden: Springer VS, 2006. 327–351.

Jaeggi, Martin. "Migration im Bild – Menschen und Räume." *Migration im Bild. Ein Inventar.* Ed. Tiberio Cardu. Baden: Hier und Jetzt, 2006. 12–16.

James, Sabatina. *Vom Islam zum Christentum – ein Todesurteil.* Munich: A & M, 2007.

Kaschuba, Wolfgang. "Deutsche Wir-Bilder nach 1945. Ethnischer Patriotismus als kollektives Gedächtnis?" *Selbstbilder und Fremdbilder. Repräsentationen sozialer Ordnungen im Wandel.* Eds. Jörg Baberowski, Hartmut Kaelble, and Jürgen Schriewer. Frankfurt a.M./New York: Campus, 2008. 295–329.

Kracauer, Siegfried. "Die Photographie." *Das Ornament der Masse. Essays.* Frankfurt a.M.: Suhrkamp, 1963. 21–39.

Landwehr, Achim. *Geschichte des Sagbaren. Einführung in die Historische Diskursanalyse.* Tübingen: edition diskord, 2001.

Leimgruber, Walter, Silke Andris, and Christine Bischoff. "Visuelle Anthropologie. Bilder machen, analysieren, deuten und präsentieren." *Europäisch-ethnologisches Forschen. Neue Methoden und Konzepte.* Eds. Sabine Hess, Johannes Moser, and Maria Schwertl. Berlin: Reimer, 2013. 247–281.

Mitchell, W. J. T. "Der Pictorial Turn." *Privileg Blick. Kritik der visuellen Kultur.* Ed. Christian Kravagna. Berlin/Dresden: Id-Verlag, 1997. 15–40.

Mitchell, W. J. T. "Migration, Law, and the Image: Beyond the Veil of Ignorance." *Images of Illegalized Immigration: Towards a Critical Iconology of Politics.* Eds. Christine Bischoff, Francesca Falk, and Sylvia Kafehsy. Bielefeld: transcript, 2010. 13–30.

Müller, Marion G., and Stephanie Geise. *Grundlagen der visuellen Kommunikation. Theorieansätze und Analysemethoden.* Konstanz: UTB, 2003.

Müller-Doohm, Stefan. "Visuelles Verstehen – Konzepte kultursoziologischer Bildhermeneutik." *"Wirklichkeit" im Deutungsprozess. Verstehen und Methoden in den Kultur- und Sozialwissenschaften.* Eds. Thomas Jung and Stefan Müller-Doohm. Frankfurt a.M.: Suhrkamp, 1993. 438–457.

Münkler, Herfried, and Bernd Ladwig. "Dimensionen der Fremdheit." *Furcht und Faszination. Facetten der Fremdheit.* Eds. Herfried Münkler and Bernd Ladwig. Berlin: Akademie Verlag, 1997. 11–44.

Oesterreich, Heide. "Das Kopftuch als Kippfigur." *Politik ums Kopftuch.* Eds. Frigga Haug and Katrin Reimer. Hamburg: Argument Verlag, 2005. 41–46.

Osten, Marion von. "Die Fahrbahn ist ein graues Band. Überlegungen zu einigen privaten Fotos mit Automobilen." *31 – Das Magazin des Instituts für Theorie der Gestaltung und Kunst* 6/7 (2005): 25–33.

Pawek, Karl. *Das optische Zeitalter. Grundzüge einer Epoche.* Olten/Freiburg i. Br.: Walter, 1963.

Projekt Migration. Ein Initiativprojekt der Kulturstiftung des Bundes. Eds. Kölnischer Kunstverein, Dokumentationszentrum und Museum über die Migration in Deutschland, Köln, Institut für Kulturanthropologie und Europäische Ethnologie der Johann Wolfgang Goethe Universität Frankfurt a.M., and Institut für Theorie der Gestaltung und Kunst. Zurich/Cologne: Dumont, 2006.

Räthzel, Nora. "Begegnungen mit dem Kopftuch." *Politik ums Kopftuch.* Eds. Frigga Haug and Katrin Reimer. Hamburg: Argument Verlag, 2005. 112–119.

Rogoff, Irit. "Wegschauen. Partizipation in der Visuellen Kultur." *Texte zur Kunst* 9 (1999): 98–112.

Schmidt-Linsenhoff, Viktoria. "Einleitung." *Projektionen. Rassismus und Sexismus in der Visuellen Kultur.* Eds. Annegret Friedrich et al. Marburg: Jonas, 1997. 8–14.

Schulze, Gerhard. *Die Sünde. Das schöne Leben und seine Feinde.* Munich/Vienna: Fischer 2008.

Sökefeld, Martin. "Zum Paradigma kultureller Differenz." *Europa und seine Fremden. Die Gestaltung kultureller Vielfalt als Herausforderung.* Eds. Reinhard Johler, Ansgar Thiel, Josef Schmid, and Rainer Treptow. Bielefeld: transcript, 2007. 41–57.

Tagesanzeiger-Magazin (16 October 2004).

Tarlo, Emma. *Visibly Muslim: Fashion, Politics, Faith.* Oxford/New York: Bloomsbury, 2010.

Terkessidis, Mark. "Globale Kultur in Deutschland: Der lange Abschied von der Fremdheit." *Kultur – Medien – Macht. Cultural Studies und Medienanalyse.* Eds. Andreas Hepp and Rainer Winter. Wiesbaden: Springer VS, 2006. 311–325.

Weber, Martina. "'Das sind Welten.' Intrageschlechtliche Differenzierungen im Schulalltag." *Eva ist emanzipiert, Mehmet ist ein Macho. Zuschreibung, Ausgrenzung, Lebensbewältigung im Kontext von Migration und Geschlecht.* Eds. Chantal Munsch, Marion Gemende, and Steffi Weber-Unger-Rotino. Weinheim/Munich: Juventa, 2007. 91–101.

Weber, Stefan. "Die Welt als Medienpoiesis. Basistheorien für den 'Medial Turn.'" *Medien Journal. Zeitschrift für Kommunikationskultur* 23.1 (1999): 3–8.

Heidrun Friese

Framing Mobility

Refugees and the Social Imagination

1 Mapping the Field

Following the deaths of more than three hundred people close to the shores of the southern Italian island of Lampedusa on October 3, 2013, increased media hype and global attention have sought to address mobility in the Mediterranean, and the highly controversial negotiation of hospitality therein. As a result, since then, the name 'Lampedusa' has become an empty signifier.[1]

Media images of overcrowded and flimsy vessels, shipwrecks and death, and catastrophe and tragic loss fuel the notion of a tragic border regime. At the same time, racialized images of 'black masses' evoke an 'invasion,' an uncontrollable 'flood.' Dramatic pictures of debilitated people arriving on the island reinforce a social imagination that views so-called 'undocumented mobility' as a humanitarian catastrophe, or as a threat to Europe's welfare and national identities that asks for drastic and robust measures against 'traffickers' as well as for a permanent 'state of exception,' that became a normal paradigm of government as the camp became a nomos of modernity (Agamben 2003).[2] Critical stances, humanitarian impetuses of rescue, and fantasies of invasion intersect in media representations of this situation. In June 2011, Angelina Jolie-Pitt, American actress and Goodwill Ambassador of the UN High Commissioner for Refugees, arrived on Lampedusa; stating that she "was moved" by the boatpeople. Soon thereafter, the former head of the French *Front National* Marine Le Pen, the former Prime Minister of Italy Silvio Berlusconi, and Pope Francis personalized competing socio-political discourses by emphasizing either humanitarian or security-oriented aspects of mobility, demarcating common conceptions of the "limits of hospitality" (Friese 2014). Then, following the shipwreck in October 2013, global media and national

1 A discussion of the notions of 'empty signifier' or 'floating signifier' and its various theoretical implications, reaching from Claude Lévi-Strauss to Roland Barthes, Jean Baudrillard, Ernesto Laclau, and Jacques Derrida, is not intended. Furthermore, the differences between master signifier and hegemonic signifier in Slavoj Žižek and Ernesto Laclau cannot be discussed here. The following remarks instead take up an argument developed in Friese 2017, 2017a, 2017b.

2 On the relations between mobility, 'aesthetics, ethics, and politics,' see Musarò and Parmiggiani 2014.

https://doi.org/9783110600483-004

politicians gathered around coffins of the victims to join in the humanitarian spectacle. And a few weeks later, the mission *Mare Nostrum* was launched, which not only pursued humanitarian Search and Rescue (SAR) missions, but also served in policing the Channel of Sicily next to the Libyan coast.[3]

As an empty signifier, 'Lampedusa' became mobile, and reached Europe's capital cities. On Berlin's Oranienplatz, a media-saturated *Lampedusa Village* accommodated local activists from the Kreuzberg neighborhood, who declared: *Lampedusa is everywhere*. This transfer between tropics allowed the historical and geographical site of Lampedusa to now be 'located' literally any- and everywhere, interchanging center and periphery, and proving the power of what the signifier could activate and mobilize. It allowed for different political positions – *populist, humanitarian*, and *critical* – as well as a perspective that imagines the foreigner as a hero, a substitute for a lost revolutionary subject. Alongside inherited images of the foreigner as enemy or victim, the old image of the foreigner as liberator could thus be activated.

The global mediatization of mobility and its critical-antagonistic discourses about security, humanitarianism, and activism led to an astonishing media prominence for the island and its six thousand inhabitants. In interviews, locals commented rather skeptically on the onslaught of aid-workers, activists, and scientists, with statements that consistently echoed the following sentiment: "They arrive and depart but nobody keeps an interest in local problems." Additionally, locals voiced fears that global media coverage might damage the image of the island as an uncontaminated paradise for leisure and tourists, its powerful production of images consuming Lampedusa as a historical place. Previously known only to a limited number of tourists, Lampedusa has to deal with global media attention, scientific discourse, and artists' representations, becoming part of the global social imagination and its signifying processes. For example: Maria Iorio's and Raphaël Cuomo's film *Sudeuropa* (2005–07) explores visibility and invisibility with reference to tourism and the camp on Lampedusa; and Zakaria Mohammed Ali's film *To Whom it May Concern* (2008), Jakob Brossmann's images *Lampedusa im Winter* (2015), and Gianfranco Rosi's *Fuocammare* (2016) depict varying perspectives on the island as well. Morgan Knibbe's *Shipwreck* (2014) offers a narration of the shipwreck of October 3, 2013; Enrico Chiarugi recites the names of the perished; and *Havarie* (2016), by Philip Scheffner, points its camera on a flimsy refugee boat. Giuseppe Di Bernardo, in *Viaggio a Lampedusa* (2010);

3 *Mare Nostrum* has been replaced by the Frontex lead mission *Triton* and EU NAVFOR MED operation *Sophia*, launched in June 2015, with the aim of countering the activities of human smugglers and traffickers in the Mediterranean.

Costanza Quatriglio, in *LampeduSani* (2014); and Davide Camarrone, in *Lampaduza* (2014) document their trips to the island. Migrants' belongings and parts of boats arriving on Lampedusa are used in artwork and shown in museums; the composer Ennio Morricone sets the voices of the drowned to music. Lampedusa is presented on stage by Henning Mankell, in *Lampedusa* (2006); by the Théâtre Senza, in *Miraculi* (2014); and by the Italo-Tunisian theater Teatro dell'Argine/Eclosion D'Artistes, in *Lampedusa Mirrors* (2015). Lina Prosa offers a trilogy of shipwrecks (2013), and the writer Maylis de Keregal reflects on the Mediterranean tragedy (2015). By no means exhaustive, this list indicates the glut of artistic production and reproduction of emotion designed to move the spectator, and to inscribe the migrant's destiny onto a powerful narration of misery, pity, indignation, and victimhood.

The making of borders and the governmentality of mobility comprises an ensemble of multiple and even contradictory actors, practices, models, and economic calculation (Foucault 2006). These historically shifting constellations – no longer ordered by a centralized site of power – are shaped by (military) technologies of policing and surveillance, administrative acts and logics, pastoral practices and orientations, humanitarian impetuses and scientific discourse. On one hand, the dominant discourse of security brings forth scenarios of threat. It insists on the defense of national sovereignty or, in its culturalist version, on an alleged national identity, opting for the increasing surveillance of borders if not the outright shielding of national territories against an 'invasion.' On the other hand, humanitarian and critical discourses address (religious) prescriptions, ethics, and values; or (antagonistic) political mobilization. The social imagination and images of mobile people are part of governmentality, its techniques, and its production of truth and power. Images are certainly not innocent, nor are processes of signification. Mixing tragic fate with guilt and pathos, anguish and suffering, the social imagination and the production of specific images and figures of mobile people as enemies, victims, or heroes are part of such powerful constellations.

If we follow Cornelius Castoriadis's understanding of tragedy as political drama displaying conflicts that make up *the political* (1997, 284), then the border regime is to be considered tragic: not because human beings are left to die at the borders,[4] but because the border regime reveals and elaborates the founding

4 Following the Report of the United Nations High Commissioner for Refugees, in 2015 at least 3,770 people died or were reported as missing in the Mediterranean (UNHCR 2015, 32). In 2016, 5,079 people drowned. The death toll of these voyages therefore amounted to 14 dead people per day. (<https://missingmigrants.iom.int/mediterranean> [accessed: 18 March 2018]). In 2015,

tensions of democracy.[5] The central feature of democracy is autonomy: the autonomy of a *demos*, the *polis*, to determine its own laws. Democracy is founded on its own autonomous will, and it is the explicit political form of an autonomous political community. As no (divine) law is "above" the *demos* as autonomous subject of legislation, it grounds itself in autonomous will. Such circularity – a *demos* decides who is to be the *demos* – opens up the paradox of the original and arbitrary moment of foundation, which decides about membership, citizenship, and borders. At the same moment, it is therefore inclusive and exclusive. (Undocumented) mobility challenges democracy as it points towards the democratic paradox of the constitution of the *demos*, a political community that is founded on the exclusion of those who do not belong. Discourse on mobility cannot escape the basic paradox of democracy, which always already sets the limits of hospitality and marks the figure of the stranger.

Media hype, the pathos of the event-society, and the "economy of attention" (Citton 2014) make the tragic border regime work. Much as the routes of migrants from Turkey to the Aegean islands of Kos and Lesbos have attracted massive media coverage (in early September 2015 the picture of Alan al-Kurdi's body, washed ashore on the holiday beach of Kos, became a media icon and "a symbol of refugee misery"),[6] Lampedusa is likewise a prominent part of the popular media spectacle, an eternal backdrop of 'tragic' and 'catastrophic' events. The island became the *terra santa* of pilgrimage, the veneration of innocent victims,

---

1,015,078 refugees crossed the Mediterranean; and around 850,000 crossed the Agean Sea. According to the Italian Minister of the Interior, 181,436 people reached Italy in 2016. <http://www.libertaciviliimmigrazione.dlci.interno.gov.it/sites/default/files/allegati/cruscotto_statistico_giornaliero_31_dicembre.pdf> [accessed: 01 December 2017].

5 Furthermore, it reveals "self-limitation," and, "more than that, tragedy shows not only that we are not masters of the consequences of our actions but that we are not even masters of their meaning" (Castoriadis 1997, 284).

6 "Refugee Crisis: Following the Tragic Journey of Aylan Kurdi's Family from Syria to Kos: The Death of Aylan al-Kurdi en Route to the Holiday Island has Made it a Symbol of Refugee Misery." *The Independent* (5 September 2015). <http://www.independent.co.uk/news/world/europe/refugee-crisis-following-the-tragic-journey-of-aylan-kurdis-family-from-syria-to-kos-10488358.html> [accessed: 18 March 2018]. Regarding contemporary art, I am referring to the pictures taken of Ai Weiwei at the shores of Lesbos, their display of tragic misery, and their intent to transmit authenticity while resting within the simulacrum of repetition. <http://www.repstatic.it/content/nazionale/img/2016/01/31/215336129-fb4b9d87-1c56-46ca-be4f-64e82f2b2929.jpg> [accessed: 17 April 2017]

and the repository of 'authentic' feelings of post-humanitarian catharsis. Images of indignation and pity are constantly produced, staged, and replayed.[7]

The following section addresses the social imagination and the organization of visibility and invisibility in public discourse. Against this background, the staging of mobile people as *enemy*, *victim*, or *hero* will be analyzed, in an effort to show how the social imagination is connected to the border regime, the migration industry, and current governance, which will be shown to legitimize a continuum of political positions, taking *populist*, *humanitarian*, or *critical* perspectives.

2 The Social Imagination

"Each society," Cornelius Castoriadis points out, "is a construction, a constitution, a creation of a world, of its own world" (Castoriadis 1987, 3). Social imagination however, cannot be considered as a "mere copy, a reflection of the outside world." Rather, the imagination *allows for* human relations, intersubjectivity, and the "social-historical world" (Elliot 2012, 355). The social imagination thus, does not merely call forth visible images, symbols, and iconographies, but is "a fundamental creation of the social imaginary, [wherein] the gods or rules of behavior are neither visible nor even audible but *signifiable*" (Castoriadis 1997, 182–183, emphasis mine). Social imagination relates "an aesthetics of imagination [and] the primary institutions of society (language, norms, customs and law)." As such, it is to be understood as a "form of relation through which individuals and collectivities come to relate to such objects of representational and affective investment" (Elliot 2012, 356). The social imagination therefore not only creates and allows for intersubjectivity, but society is moved by "the work of creative imagination, the eruption of a radically new that did not exist in any prior form" (Elliot 2012, 355). Therefore, the social imagination is not a mere reproduction of society, a secondary or 'second order' product, but one that plays a decisive and vital role in the creation, reshaping, and reorganization of social life.

7 In contrast, the social imagination and the images produced by the North-African *harragas* (as people who "burn the borders" are called throughout the Maghreb) challenge the dominant imaginaries of 'undocumented' mobility (for such an account, see Friese 2013, 2017b). The *harraga* experience of people from the Maghreb, however, is very different from experiences that sub-Saharan people encounter on their journeys, as they often have to face racism, brutal exploitation, and imprisonment in Libya.

In this sense, a space of negotiation is enabled by the production of images of the 'stranger' as one who does not belong to a political community, who is not endowed with the same rights as a European citizen, and whose exclusion from the political community nonetheless creates the political community. As part of the social imagination's processes of signification, these images constitute society, and are endowed with the creative power to constantly reshape the social and political order. The social imagination thus constitutes and shapes the space of the political, and it organizes presence and absence, visibility and invisibility.

2.1 Visibility – Invisibility

The social imagination and its processes of signification create those who are excluded, pointing toward global asymmetries and the repressed traces of colonialism, and carrying a postcolonial framing. Media attention and strategies of visualization certainly do not escape this ordering, but instead bring forth an astonishing movement between visibility and invisibility. Despite massive media coverage, those who arrive on Lampedusa are invisible, as they are usually immediately transferred to a reception center situated at the periphery of the island's small village (*Centro di primo soccorso e accoglienza*, *Cpsa*, a so-called 'hot spot' for identification and registration). The migrants are then not permitted to leave the center, nor is the public permitted to visit the camp.

In such a way, the status of these *clandestini*, and the semantic vicinity of the stranger to this secret and potential threat, is repeated and enforced. *Clandestini*, the Italian term for illegalized mobile people, points towards 'that which is kept secret or is done secretly' because it is prohibited. Etymologically, the word's root, *clan*, signifies 'hidden' (*di nascosto*), from *kal/cal* as that which is hidden from the light of the day (*dies*); that which is 'hidden from light or sheds the light of the day.' The "secret," as Jacques Derrida notes (Derrida 2002, 198), is the "separated," which marks as well the double presence of the *clandestini*: present and absent, invisible and visible.

Images mobilize, and visibility is part of political strategy. Therefore, the *clandestini*, the invisibles, must become visible if they are to be marked as enemy, as a threat that legitimizes the practices of the border regime, its governance, and its eminent economic interests. The media's gaze, the production of visibility, the staging of the invasion of 'black masses,' and the conjured 'exodus of biblical dimensions' serve the production of illegality and become part of the tragic border regime. In the same vein, visibility is a vital feature of humanitarian and critical discourse. Suffering has to be made visible in order to move the spectator and to raise indignation and eventually action. In order to understand the common features connecting humanitarian to critical discourse, we must take a short historical detour: "History tells us that it is by no means a matter of course for the spectacle of misery to move men to pity," Hannah Arendt remarks (1963, 70). The poor person's misery is due to the fact that while "he is not disapproved, censured, or reproached; *he is only not seen* [...]. *To be wholly overlooked, and to know it,* [is] *intolerable*" (Arendt 1963, 69, emphasis mine). Revolutionary pathos and

> the personal legitimacy of those who represented the people and were convinced that all legitimate power must derive from them, could reside only in [...] the capacity to suffer with the "immense class of the poor," accompanied by the will to raise compassion to *the rank of the supreme political passion and of the highest political virtue.* (Arendt 1963, 75, emphasis mine)

The visibility of the poor in the course of the French Revolution was related to a public compassion with the *malheureux* (Arendt 1963, 94). Visibility, thus, is bound on one hand to compassion, and to the (political) structures of "recognition," the becoming visible in the public sphere, on the other (Honneth 2003).

Luc Boltanski's *Distant Suffering* (1999) takes up Hannah Arendt's analysis in examining at modes and practices of compassion that are not directed towards the immediate environment or the concrete and immediate sight of suffering, but rather to distant concerns and mediated suffering. Boltanski's account concerns ethics and moral philosophy as well as the question of how indignation and political action in the face of human suffering are developed. In order to address these questions, he develops a triple topic: the "topic of denunciation," the "topic of sentiment" and the "aesthetic topic" (Boltanski 1999). One may add to this the topic of knowledge, as social science may contribute to political mobilization (even if individual suffering is not at the center of its attention). These topics display tensions between the universal and the particular, between mass and singular suffering:

> In fact, while it is easier to integrate general forms of presentation into the logic of political programs [...], nonetheless it is necessary to go into particular cases, that is go into details, in order to arouse pity, involve the spectator and call on him to act without delay. (Boltanski 1999, 33)

Such a *dispositif*, however, has been modified, not least with the day-to-day use of digital media. Pity, commotion, and solidarity were once tied to religious prescription; moral-philosophic universalism; or 'grands récits' that distanced suffering through insight, critical judgement, and knowledge. Today, however, suffering is imprinted into the subjectivity of the spectator. Universality has been replaced by irony, and subjectivity and reason have been replaced by authenticity. Solid principles have been converted into consumerism, and rational judgment into subjective opinion. Solidarity has transformed into a feature of the lifestyle industry and the "ironic spectator" and "post-humanitarian imagination" identify with celebrities as Chouliaraki (2014, 24, 26) pointed out.

The post-humanitarian imagination and the adjoined "economy of attention" (Citton 2014) are characterized by several tensions: the tension between suffering and its branding; the tension between actor and victim(ization); the tension between consumer and the political (Chouliaraki 2014, 82–83); and the tension between pathos, distance, and prudence. The demand to render the invisible visible belongs to these tensions, and thus, the quest for visibility of the marginalized may reiterate stereotypes and even promote further marginalization (Schaffer 2008). In such a vein, Susan Sontag has argued that the (postcolonial) gaze on the suffering of others in art and contemporary photography adds a double message to the "unforgettable" and "ubiquitous images" of "large-eyed victims": on one hand, they expose appalling misery; on the other hand, they ontologize suffering and withdraw historical circumstances from change (Sontag 2003, 56).

The visibility of suffering not only generates victims, it also allows one to imagine the Other as threat and enemy; thus, strategies of visibilization might solidify what they seek to combat. Additionally, we should not forget that imagination is connected to massive economic interest; and that images are created to legitimize and justify the current border regime, the migration industry, and current forms of governance that depict mobile people as victim or foe.

Current policies of expanding European borders to its former colonies via bilateral or multilateral accords – such as the recent Italian accords with Libya, and European efforts to cooperate with African dictators – are intended to close the central Mediterranean route, to keep mobile people 'in their place,' and at the same time to render them invisible. Images of death are no longer produced close to European shores; death is no longer visible for the European public: it occurs out of sight in the desert.

3 Framing Mobile People

3.1 Enemy

"The overcrowded boat is a common visual representation of threatening immigration to the West. In the European context, it is usually a flimsy-looking craft filled with black Africans," Gilligan und Marley remark (2010). Such images confirm a racialized, biopolitical security-*dispositif*, they call for the repulsion of 'invaders,' the 'catastrophic flood' and the 'parasites' attacking the imagined body of the nation and its prosperity. Such images legitimize the permanent "state of exception" (Agamben 2003).

Indeed, the Italian Civil Defense Department (*Protezione Civile*), generally charged with coping with natural disasters, has been present on Lampedusa for decades, becoming an integral part of the tragic border regime and its economic interests. In an effort to allow this body to circumvent common rules for public tender and to promote clientelist networks, in July 2008 the government of Silvio Berlusconi declared a state of "emergency migrants."

In February 2011 a "state of humanitarian emergency in the territory of North Africa" (sic) was again asserted "in order to allow efficient measures hindering the exceptional influx of aliens on national territory" (Presidenza del Consiglio dei Ministri 2011). Part of this "emergency" was a fiscal moratorium for local tourist operators who complained about the decline of their sector, which they attributed to the recent migratory fluctuations and extensive media coverage that damaged the island's image as an 'uncontaminated' paradise. Indeed, the dominant media-gaze fostered the social *imaginaire* of assault. The production of such images was part of a political strategy by former Minister of the Interior, Roberto Maroni (*Lega Nord*), designed to press EU for financial aid. The arriving Tunisian *harraga* (migrants) were not immediately transported to mainland Sicily, but were forced to camp in the streets of the small Lampedusan village. For weeks, the island resembled an open-air television studio hosting journalists, photographers, artists, and anthropologists from all over the world who coproduced the media's 'humanitarian crisis.' The moratorium imposed by Berlusconi (who visited the island during his own politically produced 'state of emergency' to promise inhabitants milk and honey) has been extended by successive governments until 2017. This now-famous 'state of emergency' has indeed become the island's 'new normal,' fostering the economic interests of local and national actors.[8]

8 The newly elected major of Lampedusa, a hotel owner who defeated the famous Giusi Nicolini for that role, is pressing for yet another prolongment. For an account of the (local) migration industry and clientelistic networks, see Friese 2012. In late 2014, Italy was shaken by the scandal *Mafia Capitale*. Through a network of links with politicians and businessmen, a criminal gang held an influence over public tender for cooperatives that offer services for migrants and Roma. The web of corrupt links between city hall officials, neo-fascist militants, and mobsters included Salvatore Buzzi, who was jailed for murder in 1984. He had been alleged by prosecutors to have been the 'entrepreneurial right hand' of Massimo Carminati, the top mobster and a former member of the NAR neo-fascist terrorist group. "His organisation is claimed to have bribed officials

The notion of catastrophe – from the Greek *kata* 'down' (*cata-*) + *strephein* turn'; *katastrephein,*'to overturn, turn down' denotes the 'reversal of what is expected' (especially a fatal turning point in a drama) – has been extended to 'sudden disaster' and is part of the (tragic) border regime, the political imagination, and the permanent state of exception.[9] The permanent state of emergency and the normalization of the exception are promoted by highly symbolic visualizations of undocumented mobility. The *imaginaire* of mobile people as a sudden, abrupt catastrophe penetrating civil society generates a public-siege mentality against 'invasion,' which in turn has to be calmed by efficient, well-calculated measures of border management. Dominant mass-media images, policies of border management, and the state of exception are interrelated: images reiterate the exception, as the exception needs its images in order to command public and political legitimacy. At the same time, the imaginaire of catastrophy fosters (local) economic interests and the migration industry. However, not only the security industry refers to catastrophe. The scenario of catastrophe and tragedy is employed by humanitarian entrepreneurship and the need to promote and

to win contracts, including for the management of migrant-holding facilities and Roma camps. Evidence submitted by police in support of their application for warrants includes a wiretapped phone conversation in which one speaker is claimed to be 59-year-old Buzzi. 'Do you know how much you earn from immigrants?' the speaker asks. 'Drug trafficking earns less.' More than 100 other people have been formally placed under investigation, including Gianni Alemanno, the mayor of Rome from 2008 to 2013, and a former member of the neo-fascist Italian Social Movement (MSI)" (Hooper and Scammell 2014). Connected to investigations regarding *Mafia Capitale*, the links between politicians and migration entrepreneurs in Sicily came under critical scrutiny. The largest camp for asylum seekers in Europe is the CARA in Mineo/Sicily, which has been converted from a decommissioned US military base in Sigonella. The mayor of Mineo, Anna Aloisio, member of the *Ncd* (Nuovo Centro Destra/New Center-Right) party of Minister of the Interior Angelino Alfano (from the Sicilian province of Agrigento) is head of the *Consorzio calatino terra di accoglienza*. A member of the consortium, the coop *La Cascina* won a public tender worth € 100 million and has been accused by investigators of *Mafia Capitale* because he had bribed the member of the commission with € 10,000 per month. Additionally, the undersecretary Giuseppe Castiglione (*Ncd*), Angelino Alfano's right-hand man, had nominated Luca Odevaine, a man of the mentioned Salvatore Buzzi as consultant of the CARA and as member of the National Roundtable on Immigration. Odevaine confessed that he was 'compensated' by *La Cascina*: "They gave me € 10,000 per month as ... let's say, 'contribution.'" (*La Repubblica*, 03 December 2014). <http://palermo.repubblica.it/cronaca/2015/03/12/news/cara_di_min..._illegittima_otto_indagati_tra_cui_un_sottosegretario-109315938/> [accessed: 15 March 2015]. In July 2017, Massimo Carminati was convicted to 20 years, and Salvatore Buzzi to 19 years of imprisonment. The court did not charge the defendants with belonging to a Mafia organization.

9 'First recorded 1748,' see <http://www.etymonline.com/index.php?term=catastrophe> [accessed: 18 March 2018].

sell humanitarian aid and services on a competing market.[10] Private SAR missions – as they have been carried out until recently – such as *Moas, Jugend Rettet, Save the Children, SOS Mediterranée, Proactiva Open Arms*, and *Sea Eye* – must promote their activities with drastic images of capsizing boats, drowning people, and dead black bodies.

The border regime also fosters an expanding branch of competing IGOs and NGOs (such as *International Catholic Migration Committee*, the *International Organization for Migration, Save the Children*, and *Doctors Without Borders*), public-private enterprises, knowledge-generating research institutes, and various other infrastructures that manage mobilities and make up a dense transnational fabric of actors, practices, policies, and powerful images of human mobility. Images of flow, invasion, crisis, and emergency are an integral part of such entrepreneurship as the media gaze produces and disseminates dominant views of undocumented mobility. This dense fabric of organizations, the security and military industry, and the European border agency (*Frontex*) make up the highly dynamic postcolonial border regime that connects policing, surveillance, the state of exception, and the imagination of threat that must be kept under control. Sarcastically, the militarization of borders is justified with the humanitarian fight against 'human trafficking.'

The oft-repeated image of invasion inverts the postcolonial situation: it relies on forgetting, withholding, and the repression of European colonialism, and installing instead historical revisionism. These images invoke the invasion of black masses, which take what has been taken from them; they show the phantasms of colonialism and the ghastly return of the colonized on the shores of the colonizer: the threat of a revenant.

Georg Simmel has already noted that the wanderer provokes "latent or open antagonism (*Gegnerschaft*)," and becomes an "irreconcilable enemy" and a "parasite of the sedentary elements of society" (Simmel 1992, 760). The provocation, the affront, is the other who menaces pollution, an intruded heterogeneity that contaminates and infects the sane body of the people, the interior of the nation, and its welfare. The intruder, the parasite, is to be detected at the borders; he is to be neutralized, and eliminated symbolically or physically by 'letting him die' (Foucault 2001, 291), thereby employing modern biopolitics: which is not to govern a territory but the 'own' population. As Francesca Falk notes: "[T]errito-

10 For an account of the 'humanitarian industry,' see Mesnard who delineates a "doubled typology: the first one corresponds to the specific visibility of victims and the second regards both the visibility of the NGO and their *own victims* and the relations to rival organisations" (Mesnard 2004, 17).

rial borders are superimposed on the boundaries of the body; migration appears at the same time as an assault upon the integrity of one's own body and that of Europe," and a "discursive and iconic connection between infection and immigration" is established (Falk 2010, 90, 89).

Dominant media representations evoke catastrophe and contamination. They also repeat juridical classifications and distinguish between different categories of mobile people. At one pole exists the miserable refugee who escapes from war and annihilation; at the other pole there is the economic migrant who flees economic circumstances on the grounds of autonomous decision and will. It seems that the demonstration of individual decision and autonomy cannot be pardoned, it is exactly that demonstration that distinguishes the figure of the victim – who is not responsible for his decisions – from the enemy, the figure of threatening autonomy, decision, and will. The migrant is not only a parasite of the nation's body and welfare but an autonomous agent. Paradoxically, autonomy and will both characterize the modern subject. A helpless, needy and dependent person can be pardoned for that which in reverse marks the Other as a threat: namely autonomy, choice, and decision.

3.2 Victim

"The victim is [...] the hero of our times," Daniele Gigliolo boldly states (2016, 9). The refugee as victim is depicted as the innocent object of appalling circumstances, a victim of warfare and violence who seeks only a safer life, and who is ready to take risks, even deadly risks, in order to do so. In this, the terrifying, threatening stranger is transformed into a ward and a supplicant, a passive, needy object of temporary care who in turn must be grateful for the offering of shelter and asylum. Using the "topic of emotion, sentiment and aesthetic" (Boltanski 1993, 10), the social imagination recognizes the victim in the figure of the miserable, and turns the needy into an icon (Falk 2010, 86).

Usually, the dominant image of suffering and the wide-open eyes of victims capture the post-humanitarian spectator, as Susan Sontag confirms in her remarks on the dominant "shots" and the "full frontal views" of suffering people in distant countries described in *Regarding the Pain of Others* (2003, 56). The spectator can hardly evade the direct gaze of suffering that calls for pity, and for donations. In fact, all NGOs employ these images to that end.

The post-humanitarian imagination is apolitical and ahistorical. Replacing one image of disaster with another, images are distributed around the globe with unprecedented speed, suggesting a fictitious copresence. In its need to attract funding, emotion, and authenticity, post-humanitarian discourse does not name responsibilities; it does not ask for causes or consequences; it does not have a referent other than the conveying of authenticity: in short, it is self-referential. Violence, ferocity, war, destruction, mobility, blind and merciless nature, pitiless fatality, and catastrophe are indiscriminately and uncritically comingled in an effort to attract attention and an immediate subjective response.

> Another salient attraction of images of victims is that their closeness to physical pain creates a semblance of authenticity resulting from the unreflecting crudeness of the representation. [...] The representation of pain in images of victims does not tell a personal story, it rather generates attention by means of its event character. It suggests evidences that are not meant to inform, but to create the impression that similar events have taken place since times immemorial. By means of their irrational perspective and the persistent repetition of their motifs, they achieve a narcotic effect. From this perspective, they could be compared to spiritual imagery and its language of ritual. Mostly, representations of suffering – as a kind of *visual humanism* – are deployed as a means to propagate proper ethical values. In scenarios generated by PR agencies, they become mere clichés of an aesthetics of pity geared to sell *issues*. The overstimulation of this emotional approach to ethics numbs the viewer until he becomes used to shocks. (Sontag 2003, 56)

By exposing human plight, such a discourse resembles mythical-religious narratives, erasing and overwriting time, history, and post-colonial entanglements.

Within the accelerated economy of attention, the commodification of *pietà*, *misericordia*, solidarity, and the selling of humanitarian issues via personal story-telling intends to render deplorable fate even more catchy for the consumer's subjectivity. Images of humanitarian rescue and protection against blind fate and sudden catastrophe are the only possible offer on the highly competitive market.

Identification with the victim in its visibility allows for subjectivity and ultimately, as Renata Salecl states, for the deviation of anxiety:

> There seems to be an attempt in today's society to find a cure for anxiety by constantly exposing the disturbing objects that might incite it (even in contemporary arts, for example, we try to figure out what is anxiety-provoking in death by exposing cadavers). (Salecl 2004, 15)

This post-humanitarian discourse can be labeled as tragic because the actor must bear responsibility for his acts despite being entangled in fatal circumstances. From their own comfortable and secure shores, spectators bear witness to shipwreck, death, and destruction; in so doing, they become part of the commenting media chorus and the border regime that in one tragic breath assembles threats, victims, and heroes.

3.3 Hero

The symbolism of critical discourse demanding indignation and protest is by no means autonomous; instead, it is well integrated into the current 'economy of attention.' It has to produce spectacular and tragic images in order to incite emotion, subjectivity, identification, and solidarity. Compassion is no longer tied to universal religious and moral claims (*pietá, misericordia*) or to the promise of salvation; and solidarity is no longer connected to critical analysis and its *grand récit* of revolution, but to individual feeling and the mediatized authenticity of the Other's suffering. Thus the "ironic spectator" and the call for solidarity in the "era of post-humanism" (Chouliaraki 2014, 24, 26) are to be kindled by dramatic, 'tragic' scenarios such as shipwreck, disaster, and death, and not by the discreet and 'invisible' day-to-day existence of *sans-papiers* in search of a normal life.

The image of the refugee mobilizes. Compassion, in the sense that Hannah Arendt describes it, is elevated "to the rank of the supreme political passion" and "the highest political virtue" (Arendt 1963, 75) without being embedded in a universalistic framework. Next to the figure of the refugee as *victim* is produced the figure of the refugee as *hero*, a hero who is acknowledged and legitimized by his dangerous, heroic voyage. In this symbolic order, the hand of the drowning victim desperately gasping for help becomes the sheet anchor of the (red) fist and battle. Compassion is converted to solidarity with refugees and, even more so, with the 'just fight of refugees.'

Alongside the topics of emotion and aesthetic, the social imagination of the refugee as hero reworks the topic of denunciation and indignation (Boltanski 1993, 10). This symbolic transformation elevates those who are excluded from 'legal' mobility (but who mobilize nonetheless) into modern heroes of an autonomous fight for opportunity and for the realization of autonomous will. In short, the excluded become heroes of modern subjectivity.

Such representations do not refer to the humanitarian discourse and its alleged neutrality. However, paradoxically, this discourse is apolitical even as it seeks to open a political space. It is apolitical because the figure of the hero is situated between the innocent victim of capitalism and neoliberal governance on one hand, and, on the other, the figure of threat, the enemy who menaces the existing order. Whereas at one time the attention of activists was directed to the movements of liberation in the 'Third World,' and to the hope that revolutionary élan of the periphery might pressure the capitalist center, the hero of liberation has been replaced by the fighting refugee. He is to take up the fight against power and capitalism, and yet he is assigned to an asymmetric position that allows the activist to celebrate the (missed) triumph of the revolution, ensuring for himself the position of protector or guardian angel of a just and humanitarian cause.

The "Declaration of Solidarity" of the committee "Lampedusa is Everywhere" states: "The inhumane events of Lampedusa during October of 2013 are only one example of how the 'fortress Europe' claims countless human lives every day. We are *shocked, angry and mournful* that the defence (sic) of western wealth is indiscriminately violating human rights."[11] Even if critical discourse establishes analytical connections to European policies and the border regime, it falls in with the pathos of humanitarian discourses of catastrophe and victimhood, declining a scale of possible distress by the (alleged) authenticity of suffering in one's own shock, anger, and mourning. By depicting victims whose autonomous actions end in tragic fatality, and even *have to* end in tragic fatality for its legitimizing power to be established, the tension of the tragic-humanitarian discourse is repeated. Taking up the topic of denunciation, such a discourse laments the defense of affluence on one hand. On the other hand, the compassionately suffering activists participate in western prosperity, even if they seek to exorcize such participation through the displayed authenticity of emotion.

Additionally, if power works through "dispersed, heteromorphic and local procedures" which are "adopted, enforced and transformed by strategies, resistance, inertia" and therefore do not lead to "massive domination" or a binary structure of domination and subordination (Foucault 1994, 425, my translation), the fighting hero is positioned within a binary structure. For some strands of activism, the fighting hero/non-citizen is to act on behalf of those whose revolutionary practices did not succeed. Therefore, such representations must equip mobile people with revolutionary courage. In such a vein, Antonello Mangano describes the revolt of illegalized day-laborers in the Calabrian town of Rosarno in 2010 as follows: "[P]art of the Senegalese community cordoned off the villa of a capomafia [...] no one ever dared to do such a thing. This is why Africans will save us" (Mangano 2010, 135). Against the superior might of circumstances, the foreigner/victim/sacrificed is transformed into a redeemer, a *Salvatore* of the enslaved and oppressed: he is the one who acts for others and speaks on behalf of those without voices of their own. The hero redeems from voicelessness and acquits from responsibility; the image of the hero who ceaselessly fights injustice, power, and the state not only allows the spectator to celebrate victory, but allows even the hero to celebrate himself as an incessant and victorious fighter for the just cause. However, the foreigner as imagined hero, as figure of salvation, and as icon of struggle is advised to arrange that which political circumstances deny.

---

11 "Refugee Struggle for Freedom" <http://refugeestruggle.org/en/solidarity/lampedusa-everywhere> [accessed: 18 March 2018], emphasis mine.

The cult of heroes blocks claims to a life according to wishes and dreams. Mobile people cannot be inscribed into an imagined, imaginary collective subject.

The stranger, as Georg Simmel has remarked, is one who is "not tied down in his action by habit, piety, and precedent," but is "bound by no commitments which could prejudice his perception, understanding, and evaluation of the given," allowing thus for 'objectivity' and freedom (Simmel 1992, 766–767, my translation). As Bonnie Honig demonstrates, the figure of the stranger as hero and/or liberator reflects that of the "foreign founder" who inaugurates a political community, a figure well known in political thought. Rousseau's contract theory, for example, assures "a foreign founder's foreignness [...] the distance and impartiality needed to animate and guarantee a general Will that can neither animate nor guarantee itself" (Honig 2001, 21). The figure of the stranger, the foreigner as founder and lawmaker, or as immigrant, reveals the problematic nature of democracy and its aforementioned paradox. Additionally, as a homogeneous people or nation is nothing but a fiction, the foreigner always already constitutes a (political) community.

Mobility challenges democracy; it incessantly points towards the paradox of democracy and to its inclusive-exclusive structure. In fact, current national-populist "excitable speech" (Butler 1997) and its emphasis on borders and national sovereignty – the sovereignty of an imagined homogeneous, identitarian *demos* to decide who belongs and who is excluded from the community – exhibits the paradox. The call for open borders or humanitarian impetuses also cannot escape this tension. Mobility challenges efforts to contain the excluded. The signifying processes of the social imagination that produce the racialized figures of mobile people as a threat, as victims, or as heroes reiterate and elaborate these tensions as well. Mobile people are none of these.

References

Agamben, Giorgio. *Stato di eccezione. Homo sacer II, 1.* Torino: Bollati Boringhieri, 2003 (engl. *State of Exception*. Chicago: The University of Chicago Press, 2005).

Arendt, Hannah. *On Revolution*. London: Penguin, 1963.

Boltanski, Luc. *La souffrance à distance: Morale humanitaire, médias et politique*. Paris: Métailié, 1993 (engl. *Distant Suffering: Morality, Media and Politics*. Cambridge: Cambridge University Press, 1999).

Butler, Judith. *Excitable Speech: A Politics of the Performative*. New York: Routledge, 1997.

Castoriadis, Cornelius. *The Imaginary Institution of Society*. Cambridge, MA: Polity, 1987.

Castoriadis, Cornelius. *The Greek* Polis *and the Creation of Democracy: The Castoriadis Reader*. Oxford: Blackwell, 1997.

Chouliaraki, Lilie. *Lo spettatore ironico: La solidarietà nell'epoca del post-umanitarismo (a cura die Pierluigi Musarò)*. Udine: Mimesis, 2014.
Citton, Yves, ed. *L'économie de l'attention: Nouvel horizon du capitalisme?* Paris: La Découverte, 2014.
Derrida, Jacques. *Politik der Freundschaft*. Frankfurt a.M.: Suhrkamp, 2002.
Elliott, Anthony. "New Individualist Configurations and the Social Imaginary: Castoriadis and Kristeva." *European Journal of Social Theory* 15.3 (2012): 349–365.
Falk, Francesca. "Invasion, Infection, Invisibility: An Iconology of Illegalized Immigration." *Images of Illegalized Immigration: Towards a Critical Iconology of Politics*. Eds. Christine Bischoff, Francesca Falk, and Sylvia Kafehsy. Bielefeld: transcript, 2010. 83–100.
Foucault, Michel. "Pouvoirs et stratégies." *Dits et écrits*, Vol. II, 1970–1975. Eds. Daniel Defert and Francois Ewald. Paris: Gallimard, 1994 (1977). 418–428.
Foucault, Michel. *In Verteidigung der Gesellschaft. Vorlesungen am Collège de France (1975–1976)*. Frankfurt a.M.: Suhrkamp, 2001.
Foucault, Michel. *Sicherheit, Territorium, Bevölkerung. Geschichte der Gouvernementalität I: Vorlesungen am Collège de France (1977–1978)*. Frankfurt a.M.: Suhrkamp, 2006.
Friese, Heidrun. "Border Economies: Lampedusa and the Nascent Migration Industry." *Shima: The International Journal of Research Into Island Cultures* 6.2 (2012) (special issue: *Detention Islands*. Eds. Alison Mountz and Linda Briskman): 66–84.
Friese, Heidrun. "'Ya l'babour, ya mon amour' – Raï, Rap and the Desire to Escape." *Music, Longing and Belonging: Articulations of the Self and the Other in the Musical Realm*. Ed. Magdalena Waligorska. Newcastle: Cambridge Scholars Publishing, 2013. 176–201.
Friese, Heidrun. *Grenzen der Gastfreundschaft. Die Bootsflüchtlinge von Lampedusa und die europäische Frage*. Bielefeld: transcript, 2014.
Friese, Heidrun. *Flüchtlinge: Opfer – Bedrohung – Helden. Zur politischen Imagination des Fremden*. Bielefeld: transcript, 2017.
Friese, Heidrun. "Representations of Gendered Mobility and the Tragic Border Regime in the Mediterranean." *Journal of Balkan and Near Eastern Studies* 19.5 (2017a) (special issue: *Women in the Mediterranean*): 541–556.
Friese, Heidrun. "Repräsentationen illegalisierter Mobilität. Lampedusa als tragisches Grenzregime." *Migration ein Bild geben. Visuelle Aushandlungen von Diversität*. Eds. Christoph Rass and Melanie Ulz. Wiesbaden: Springer VS, 2017b. 269–296.
Gigliolo, Daniele. *Die Opferfalle. Wie die Vergangenheit die Zukunft fesselt*. Berlin: Matthes & Seitz, 2016 (orig. *Critica della vittima*. Roma: nottetempo, 2014).
Gilligan, Chris, and Carol Marley. "Migration and Divisions: Thoughts on (Anti-)Narrativity in Visual Representations of Mobile People." *Forum Qualitative Social Research* 11.2 (2010): article 32.
Honig, Bonnie. *Democracy and the Foreigner*. Princeton: Princeton University Press, 2001.
Honneth, Axel. *Unsichtbarkeit. Stationen einer Theorie der Intersubjektivität*. Frankfurt a.M.: Suhrkamp, 2003.
Hooper, John, and Rosie Scammell. "Rome's 29 June Co-operative Alleged to be Base of Mafia-style Gang." *e Guardian* (7 December 2014). <http://www.theguardian.com/world/2014/dec/07/june-29-co-operative-italy-rome> [accessed: 18 March 2018].
Mangano, Antonello. *Gli africani salveranno l'Italia*. Milano: BUR, 2010.
Mesnard, Philippe. *Attualità della vittima: La rappresentazione umanitaria della sofferenza*. Verona: ombre corte, 2004.

Musarò, Pierluigi, and Paola Parmiggiani, eds. *Media e migrazioni: Etica, estetica e politica del discorso umanitario*. Milano: Franco Angeli, 2014.
Presidenza del Consiglio dei Ministri. "Dichiarazione dello stato di emergenza umanitaria nel territorio del Nord Africa per consentire un efficace contrasto all'eccezionale afflusso di cittadini extracomunitari nel territorio nazionale." *Gazzetta Ufficiale* n. 83 dell'11 (April 2011). <http://www.protezionecivile.gov.it/jcms/it/view_prov.wp?toptab=2&contentlc=LEG24032#top-content> [accessed: 21 September 2017].
Salecl, Renata. *On Anxiety: Thinking in Action*. London: Routledge, 2004.
Schaffer, Johanna. *Ambivalenzen der Sichtbarkeit. Über die visuellen Strukturen der Anerkennung*. Bielefeld: transcript, 2008.
Simmel, Georg. "Exkurs über den Fremden." *Soziologie. Untersuchung über die Formen der Vergesellschaftung*. Ed. Otthein Rammstedt. Frankfurt a.M.: Suhrkamp, 1992 [1908]. 764–771 (engl. *The Sociology of Georg Simmel*. Transl. Kurt Wolff. New York: Free Press, 1950. 402–408).
Sontag, Susan. *Regarding the Pain of Others*. New York: Picador, 2003. <http://monoskop.org/images/a/a6/Sontag_Susan_2003_Regarding_the_Pain_of_Others.pdf> [accessed: 18 March 2018].
UNHCR. *Global Trends: Forced Displacement in 2015*. 2015. <www.unhcr.org/576408cd7> [accessed: 18 March 2018].

Werner Schiffauer

Migration and the Structure of the Imaginary

Migration theory is characterized by a big divide. One side is populated by postcolonial and postmodern authors who discuss hybridity, fragmentation, creolization, localization, and similar phenomena. The other side is occupied by political economy approaches that concentrate on issues such as the globalization of capital or the construction of transnational networks and examine factors that determine migration decisions or the flow of savings. The former focus on the symbolic realm – formation of subjectivities, racism, heterogeneity, counter publics – while the latter analyze the 'hard stuff' – migration routes, opportunity structures, and so on. Both approaches exist side by side and hardly take note of each other. If anything, there is strong dislike between the two. While the latter find the former a lofty lot, the former consider the latter a boring bunch.

As an anthropologist I find myself caught in the middle. Working on field sites, and with case studies, we see the 'real' and the 'symbolic' interpenetrate all the time. In order to grasp this interpenetration more systematically I want to introduce the term 'imaginary opportunity space.' As a composite it connects (objectively given) opportunities with (subjective) imaginations. By introducing it I want to direct attention to three facts.

First: It has, of course, been widely recognized that objectives, such as push-and-pull factors and networks, may exist as such. But we must analyze their subjective side in order to understand why and how they become effective. They need to be known, and thus perceived and recognized. Analysis has focused on this cognitive aspect. But I want to show that this is not all. There is also a strong affective component: objectives give rise to desires, daydreams, anxieties, and hopes. They are felt. By using the term 'imaginary' as a qualifier, I want to draw attention to the emotional life of opportunity structures.

Second: While there is an imaginary aspect of objectives, there is also an objective side of imaginations. I propose that 'imaginations' are not as arbitrary as the term might suggest. 'One can imagine everything,' but not every imagination becomes symbolically powerful. In order to become powerful, an idea must relate to something 'real.' Various theoreticians of the imaginary have drawn our attention to this point. Wolfgang Iser (1993) introduced the figure of the imaginary in order to determine the relation of the fictitious to the real in literary texts. Jean Paul Sartre (1943/1962) defined the imaginary as the negation of the real – it is what reality not yet is, but could be. For him the imaginary is defined by a co-existence of presence and absence. Robert Musil (1978) saw the imaginary as

https://doi.org/9783110600483-005

the sense of the potential (*Möglichkeitssinn*), as opposed to the sense of reality (which is confined to the given).

I would like to suggest that imaginations become powerful when they relate to opportunities: hence the term 'imaginary *opportunities*.' An example may help to establish this concept. As European intellectuals, we might have received abstract media information about the lifestyle and work conditions in the Gulf States. However, we will start to relate to this abstract information in a new way when we meet somebody like us (a colleague of the same age and discipline) who has actually been there. Because we could picture ourselves to have been in his or her place, all of a sudden the abstract information means something. We can relate to it and it affects us in a new way. It becomes interesting in the literal sense of *Interesse*, by establishing a connection between us and the place. The abstract imagination turns into an imaginary reality as soon as we can say we could have been there.

This happens worldwide and on a daily basis. Mediascapes, as Arjun Appadurai has pointed out, compose widespread knowledge about lifestyles and life worlds:

> Mediascapes [...] tend to be image-centered, narrative-based accounts of strips of reality, and what they offer to those who experience and transform them is a series of elements (such as characters, plots, and textual forms) out of which scripts can be formed of imagined lives, their own as well as those of others living in other places. These scripts can and do get disaggregated into complex sets of metaphors by which people live [...] as they help to constitute narratives of the Other and protonarratives of possible lives, fantasies that could become prolegomena to the desire for acquisition and movement. (Appadurai 1998, 35–36)

These imaginations, one might add, are concretized and condensed in fashionable consumer goods such as *Mercedes* or *Apple*. But this is by far not enough of a motive for migration. It becomes a motive, and a powerful one, when someone like oneself (an uncle, a cousin) shows that he has access to these goods. That is frequently the case when a migrant comes home and displays fancy consumer goods that represent lifestyles. All of a sudden one can imagine oneself in his or her place. Suddenly there seems to be a way in which dreams might become reality.

Third: By talking about imaginary opportunity *spaces* I suggest that we are dealing with a structured entity. It is helpful to turn to grammar to understand what is happening. Imaginary opportunities are usually expressed in the form of unreal conditional sentences. Taking up an idea formulated by Nelson Goodman (1955) who sees in unreal conditional sentences articulations of laws that exist regardless of their actual occurrence, I want to add an additional twist. I want to show that unreal conditional sentences are characterized by a complex relationship between four statements referring to a law, a condition, a category, and a moral statement, respectively. The complex relationship between these four statements highlights the relationship between the 'real' and 'unreal,' or rather, between a realm one would agree exists beyond our individual or collective imagination and another one would agree exists "only in our imagination."[1]

1 The Interdigitation of Reality and Imagination in Unreal Conditional Sentences

To show the structure of an unreal conditional sentence, let us use a basic example: 'If the glass had fallen on the floor, it would have definitely/surely/probably shattered.'

a) First, such sentences state a causal relationship that takes the form of a premise, a rule, or a probability – in other words, a statement of reality. Hence unreal conditional sentences are thought experiments. This can be taken a step further, as any premise, rule or probability is most adequately expressed by an unreal conditional sentence. Laws allow prediction – by definition they are independent from whether something is actually the case, or not.
b) In addition to a premise, unreal conditional sentences contain a particular reference and thus state a category. The particular glass I am referring belongs to the category of glasses and shares with other glasses a characteristic that is relevant to the situation – in this case, fragility.
c) Furthermore, a statement is also made concerning condition: in this way, we differentiate between true and hypothetical, unreal conditional sentences (Goodman 1955). True unreal conditional sentences are statements about something that can indeed take place. Accordingly, the sentence only makes sense within a context in which the glass could indeed have fallen. It would make no sense at all if the object were, say, immovable. Likewise, the sentence, 'if I had hurried, I would have caught up to him,' can only make sense if haste would have made an actual difference. There are of course hypothetical unreal conditional sentences ('if I had lived in the 17th century,' 'if tri-

1 I regard the concept of the real and unreal as a frame analysis according to Goffman (1993). Even strict adherents of constructivism cannot get around making this distinction. There is a difference between play and seriousness, between responsibility and imagination. Modality is one of the methods of combining them.

angles were squares'), but these are clearly distinguishable from true unreal conditional sentences in everyday speech by markers that emphasize their special meaning. Such sentences commonly refer to dreams, identification with literary figures, fairy tales, etc. In these sentences it is merely the condition that is unreal – not the premise. Further below, we will refer back to the difference between true and hypothetical unreal conditional sentences.

d) Finally, unreal conditional sentences can only make sense as long as the situation has not been realized yet but lies in the future.[2] For instance, if the subject's reality is (for the time being) different, the thought experiment is just that, an experiment in thought. Therefore, temporality is implied, which opens a space for *moral judgment*. It can be either good or bad if the glass fell, or that it did not fall at all.

In short: unreal conditional sentences, although seemingly simple at first glance, represent a complex relationship of premise – category – condition – moral judgment. These sentences are decisive for the constitution of what I term the 'realm of potential.' The realm of potential refers to events that are possible, or not; that could have been or were possible, but did not occur. The realm of potential is the space of missed opportunities but also of avoided catastrophes. It is the space against which we measure and judge 'chances taken' and 'real developments.' The realm of potential is also the backdrop against which statements of bad luck, fortune, destiny, or tragedy start to make sense. It is hence constitutive for our self-conception, for our place in the world and therefore for our identity.

Condition

Premise Moral Judgment

Category

Let us briefly return to the migrant (in biographical reports it is usually an admired uncle) coming home and displaying an *Apple* computer. The imaginary opportunity space of a younger relative meeting him consists of four statements: he was successful – so it is possible to be successful (premise). As my relative I am like him: if he made it I can make it (category). He left at the age I am now (con-

2 There are some complications when we encounter irrational conditional sentences of the past ('if I had migrated, I would be better off today'). I will return to this point.

dition) – so now I should give migration a serious thought: should I or shouldn't I (moral judgment).

In what follows I want to show how engaging with these categories can contribute to a better understanding of three key questions of migration theory: the migratory decision, transnational communities/diasporas, and irregular migration.

2 Dreaming about Migration

During the 1960s and early 1970s, a drastic change of outlook with regard to migration took place in the rural regions of Turkey.[3] Mass migration from the Black Sea Region to the urban centers of Ankara, Istanbul, and Izmir started in the 1950s. Initially only young men from very poor and large families migrated. Their lot was pitied: migration was interpreted as a strategy to combat poverty. In terms of our analytical quadrangle they belonged to the rural underclass rather than to the well-established. With the exception of individual adventurers no one among the better-off entertained the idea. Rather, the life in *gurbet* was associated with hardship and relative isolation. The idea of having drawn the better lot if one stayed home was reinforced by the fact that migration eased the pressure on land – more land per capita was available than before.

The situation changed radically when the first generation of migrants turned out to be successful, a success that was structurally related to the boom years. When the migrants strolled through the villages in nice clothes during vacation time, feelings of envy surged. What happened was a categorical restructuring. If these subalterns were able to make it, one could certainly have done so as well – and even better. What was regarded as a privilege (namely, to stay in the village) in the 1950s, loomed as a relative failure by the late 1960s – a missed opportunity. This completely changed the outlook on the village: although the living situation was better than ever, life in the village lost its appeal. This was expressed by sayings like 'the village has no future' or 'there is no life in the village.'

Numbers were a significant part of this dynamic. There had been adventurers before, but here, a collective rather than an individual had migrated. Individual adventurers are considered exceptions; they are simply 'different.' No law can be deduced, as they do not belong to the same category as oneself. The choices of

3 In what follows I return to the fieldwork I carried out in rural Turkey during the 1970s and among migrants in Germany. The results are published in Schiffauer 1987 and 1991.

such isolated village expats provided material for good stories but had no relevance for one's self-perception. It was only when many migrated that the impression took hold that one could also have been in their position. Hence the rule of 'you can get it if you really want' was established. The validity of this law of migration had been established first with regard to internal migration in Turkey and, some years later, for international migration.

The rule did not apply to all categories though, as age set a conditionality: it was believed that only men below 40 would stand a chance to succeed in migrating to Izmir, Ankara or Istanbul, and the age limit was even lower for those migrating to Germany. The rule also applied only to men. The only categorical exception was made for young women who had a professional education (mostly nurses and teachers).

Once rule, category, and conditionality were established, moral evaluation started. The decision to migrate had to be weighed against familial responsibilities: problems arose when the parents were elderly and had to be cared for; when there were young children, or when issues of honor required someone's presence in the village. All these moral questions were more pressing with regard to migration to Europe than with regard to internal migration: Germany was far away and considered to be the country of infidels, where sexual libertinage, alcohol, and consumption of pork were epidemic. And it was clear that, unlike with internal migration, young men would be outside of social control. While it seemed to be established that migration to Germany was a pathway to economic success, a discussion of the moral implications arose.

Young men faced a situation of moral ambivalence: on the one hand they experienced increasing moral pressure to migrate, exerted primarily by the peer group with whom they compared themselves. One ran the risk of being regarded as a failure if one stayed behind. But there was also moral pressure by older family members who insisted on the fulfillment of familial duties. According to them, migration would only be okay if certain conditions were fulfilled, i.e. after the death of the parents. This conditionality, however, contradicted the conditionality set by the imperatives of the labor market: to wait until the parents passed away would mean to risk that the window of opportunity might close. Moral negotiations led to compromises. The most favored one was to let the elder sons migrate and force the youngest one to stay behind. The older brothers would support the youngest and thus compensate him for having to stay in the village. This compromise failed after it turned out that this obligation of support often ended with the death of the parents. Once this became apparent, the moral evaluation changed. It started to be considered irresponsible for parents to insist that their youngest son would stay; rather, they were pressured to allow him to move to the city. In the 1970s, only a few parents gave in; by the 1990s, the new rule was firmly established. Only a few 'stubborn' parents insisted on what had before been considered the normal way.

3 Imaginaries of the Transnational Space

The term "imagined communities" is probably one of the most successful terminological coinages of recent years: originally applied to the imagined community of the nation by Benedict Anderson in 1983, imagined communities now also describe diasporas, social and nativist movements. The term refers to (commonly strong) feelings of community and solidarity among groups of otherwise strangers who share the same national space (Anderson 1991, 6), or to feelings of kinship by persons who are of common descent, but beyond that hardly have anything in common (diasporic communities). I want to put forward the claim that imagined communities are based on imaginary spaces.

Benedict Anderson draws our attention to two interrelated phenomena that are constitutive for the phenomenon of imagined communities. The first, and most commonly referred to, is the media thesis: besides novels, daily newspapers were instrumental in developing a nationally imagined space. Day after day, according to Anderson, newspapers represent a space in which certain events play out: "The idea of a social organism moving calendrically through homogeneous, empty time is a precise analogue of the idea of the nation, which also is conceived as a solid community moving steadily down (or up) history" (Anderson 1991, 26). The second thesis could be called the career thesis. Anderson attempted to explain why the creole communities of the Americas developed a national consciousness well before most of Europe. He put forward the plausible argument that a national space was revealed to bureaucrats through "administrative pilgrimages" (Anderson 1991, 140). In the case of creole bureaucrats, these pilgrimages were restricted not just horizontally (promotion to a different administrative unit in the colonies was generally barred) but also vertically (the highest positions of the imperial administration invariably went to functionaries from the mother country). It was this circumstance alone that turned the reference to an administrative unit into one's home territory. The idea of the imaginary opportunity structure relates these two hypotheses to each other: the nation as a *realm of potential* was revealed to creole functionaries since other locations within the respective colonial administrative unit were locations they *could have been sent to* or gladly *would have been transferred to*. The mother country, on the other hand, did not belong to the realm of potential. Real/unreal space was limited to the colonial administrative unit: it had the status of a special legal space. That these were also represented in the form of newspapers certainly helped: but the newspapers alone would have been insufficient to develop strong national feelings. As Michael Warner (2002) has accurately pointed out, newspapers constitute *publics*. Such publics are sociologically different from communities and evolve from reading a particular product. They have symbolic power – but this power unfolds much differently from collectives. Publics can structure one's conception of the world, but do not account for communities of solidarity.[4]

What, then, when it comes to "transnational imagined communities"? Again I would argue that there is more to transnationalism than just networks, a feeling related to the fact that networks do not only represent ongoing transactions but also constitute a realm of the possible, a *Möglichkeitsraum* (Musil 1978), which allows for hopes and dreams, a space where one compares oneself with others. In this realm, unreal conditional sentences of the present ('if I moved to another country within this space I would fare like my cousin there') or of the past ('If my parent had moved / had not moved (let's say) to Germany I would be in a different position today') can be constructed.

Let me draw attention to the fact that transnational imaginary spaces, by definition, cross borders. This, in fact, distinguishes them from national imaginary spaces and defines their specific dynamics. Whereas national imagined communities are surrounded by borders (and literally defined by them), transnational imagined communities are internally structured by national borders that cut through them. Some members of a transnational community live in the home country, others in one of the immigration countries. With regard to imaginary opportunity spaces this creates complexity.

National opportunity spaces are also status regimes. The term status regime refers to ideas about the normal (i.e. both in the sense of normative and the regular) correlation of education, profession, income, gender, race, and social background – or, to put it in Pierre Bourdieu's terms – to the conversion rates of the social, economic, and cultural capital (Bourdieu 1982, 193ff.). Boris Nieswand (2011) coined the term "status paradox" in order to grasp how migration upsets this normal order of things. When labor migrants accept the hardships of migration they do so because it is profitable. For those who stayed at home this means

4 It should lastly be mentioned that the idea of the imaginary opportunity structure connects quite easily with Benedict Anderson's (1991) and Ernest Gellner's (1983) theories on nationalism. The imaginary space of the nation is, according to Gellner, also a realm of potential: he implies that one could say with some realism that within its space, one could work in another location – possibly because the diplomas are recognized, because one speaks the common language, because one has the necessary documents.

that the migrants circumvented the rules that regulate ascendancy within the nation-state. Referring to Ghana, he states:

> [The problem is that] global inequalities in wages and buying power [...] provide them with an opportunity structure to gain status in Ghana by transferring resources acquired through their work in Germany. The visibility of the *Burgers'* houses, consumer goods, dependents and money in Ghana causes irritation and envy, in particular among those who feel illegitimately overtaken [...] Both admiration for their success and moral condemnation for the means by which it is achieved can be observed. (Nieswand 2011, 137)

I observed similar reactions among Turks in Turkey who coined the derogatory term *Alamanci* for the migrants in Germany. The associations of the terms *Alamanci* or *Burger* resemble those about nouveau riches: they have no manners; they gave up the Turkish (resp. Ghanaian) way of life; they dress funny; they make unreasonable claims. Esin Bozkurt quotes a Turkish migrant to Germany who complains about these stereotypes:

> They find us ignorant, unmannered, crude peasants. We are great when it comes to pay the bill, but after that we become mountain Turks crude peasants [...] they call us Germanized. They say we are no real Turks. We earned money, gained financially, but we lost people's respect. They do not take us serious anymore. They call us diluted Turks. (Bozkurt 2009, 102–103)

The concept of an imaginary opportunity space provides a clue to this widespread phenomenon. Persons in the same category as oneself pose challenges for the self because one relates to them in the mode of unreal conditional sentences ('If I were in his/her place,' 'if I acted in the way he or she does'). In a way, persons in my category constitute the parallel universes of my own life. They show what I could have reached (or where I could have ended up) in life. This implies feelings of superiority or inferiority and can be a source of pride or depression. The often not very fair judgments about the persons living in the other part(s) of the transnational space derive from this dynamic. When they bypass the normal order of things, they do not commit an abstract violation of the rules of the game: it is very personal. Category, law, and condition – everything is equal, what remains is the question of morality. And this is the stuff from which the venom is brewed. There are a variety of reasons why one did not go or claims not to have gone (because one had to be loyal to the parents, one's country, one's traditions), implying – of course – that the other did not care so much about these issues. The moral question is the realm to which those who are worse off (usually in the home country) resort to: the others might be richer – but I am morally superior.

The quotation above makes an interesting point, stating: "We are great when it comes to pay the bill." There is widespread complaint among the *Alamancis* that they are charged higher prices or cheated. There seems to be a claim to distributive justice associated with the structure of imaginary opportunity spaces. There seems to be widespread feeling that one is somewhat entitled to a fair share.

While those who stayed on one side of the divide are challenged by the display of wealth, migrants often have different problems. As their new country usually does not recognize their degrees, they accept a devaluation of their social status in exchange for monetary gains. This poses a particular problem when the economic turnout is smaller than expected, e.g. in case of unemployment. One of my interlocutors expressed strong self-doubts when he realized that all his former classmates had taken high administrative positions in Turkey, while he was still desperately fighting for earning money as a worker in Germany. Interestingly, he accused his former classmates of arrogance and of "having forgotten where they came from."[5]

Morality plays an even greater role when violence comes into play. The tension between migrants and those left behind is particularly pronounced in cases of flight from civil war or conflict zones, as the *irrealis* (had I only stayed / had I only left) takes on existential dimensions in instances of physical violence. As a result, the prejudice leveled against refugees returning home is particularly bitter. Those who stayed behind often considered refugees who fled Bosnia and Herzegovina during the civil war of the 1990s as 'traitors,' since they did not take part in defending the home territory. This was especially prevalent in Sarajevo, a city that was under siege for two years. The situation was exacerbated by widespread and false rumors about the luxury the refugees allegedly experienced during their stay in Europe (Black and Koser 1999).

Feelings of guilt experienced by emigrants frequently coincide with accusations of treason. Such feelings of guilt can be exploited for the purpose of collecting money. In the case of Eritrea, a 'voluntary' tax of 2% was imposed on migrants living in Europe (Black and Koser 1999). In the Berlin Eritrean community, lists of donors were posted, resulting in more pressure on those who had so far avoided payment. Similar actions have reportedly been undertaken on behalf of the Kurdish PKK and the Sri Lankan Tamils. This rationale also demonstrates an interdigitation of the real and unreal: since the migrants live in another country and are physically and materially better off, the structure of the *irrealis* can fully manifest itself. This helps explain an initially confusing set of circumstances: a secured residence permit and relative economic success in the country

5 The statement was made by Memed Kaynar whom I portrayed in *Die Migranten aus Subay* (Schiffauer 1991, 263–291); his despair is described in 281, 282.

of immigration seldom lead to a weakening of the ties to the homeland but often increase the readiness to donate (Black and Koser 1999).

The relevance of opportunity structures for maintaining transnational imagined communities has been recognized by nation-states. During the 1990s and early 2000s most nation-states revised policies that previously aimed at bringing migrants back home, instead encouraging them to stay where they were while at the same time reinforcing their ties to their home country. This is usually accomplished by granting citizenship rights to émigrés, thus turning them into 'expats.' The state of Israel, for example, grants a right to citizenship to all Jews, wherever they are, and in this way effectively creates ties to those who live in other countries. Thus the sentence: 'if I were there I also would be affected by (let's say) acts of terrorism' becomes a real unconditional sentence. When Hungary passed a status law in 2001 that granted citizenship rights to Hungarians living in Romania, Serbia or Slovakia in order to further their national sentiments, it came as no surprise that Romania protested this attempt of extension of the imaginary community. By favoring national sentiments both the governments of Israel and Hungary also try to build up pressure groups in the other countries.

The same holds true for attempts to accord migrants special status with regard to the right to return. It would be interesting to compare states that grant such opportunities to migrants with those that do not. During the wars of independence in former Yugoslavia, the Croatian diaspora, which enjoyed such rights, was much more active than the Serbian diaspora, which at the time did not. This culminated in the electoral success of Franjo Tudjman, with donations of $ 4 million collected from members of the American and Canadian diaspora.

It is the possibility to form unreal conditional sentences which creates a space 'one can relate to' or 'which matters.' Transnational media do have an impact on this process, but it seems to be secondary. I would argue that it is the possibility to construct irrational conditional sentences that constitute interest in transnational media, a prerequisite for the creation of transnational ties. It is the interplay of categorical equality and difference that is crucial in this process. Rather than shared norms, relations set by the unreal conditional sentence structure an imaginary community. In this vein, an observation by Asu Aksoy and Kevin Robins about Turkish television viewers in London might be of interest: media convergence has not produced higher levels of identification with the home country among these viewers, but rather more distance. The disgust or anger experienced by Anglo-Turkish viewers at television reports exposing the failure of Turkish institutions during national crises (e.g. after an earthquake) have led to sentiments like 'good that we don't live there anymore' (Aksoy and Robins 2000).

What matters here, however, is not just the *irrealis* of the present, but also that of the past. The migration decision of parents or grandparents is structuring diasporic awareness. Sabine Mannitz described this awareness among students at a high school in Berlin. During a group discussion the young participants complained about limited mental horizons and parochial values they encountered while on vacation in the respective homeland villages of their parents. "Not having a limited horizon like the relatives 'there' made the biggest difference and was marked by a mixture of pride and fortunate coincidence. *After all, they could have easily become like them*, had their parents not emigrated [...]" (Mannitz 2002, 267). A time dimension seems to be involved here: in an analysis of three generations of Turkish migrants roughly ten years later, Bozkurt (2009) managed to show how a growing temporal distance led to an increasing mystification of the home-country. While children of the 'Gastarbeiter' in Mannitz's study had rather ambivalent feelings about the home country, their grandchildren started to romanticize it.

> Turkey is perceived as a space of embracement, hope and belonging as opposed to Germany, which is illustrated as the refusing, exclusionary place of residence [...]. This gaze finds its expression in the way youngsters depict Turkey as seen in their emphasis on its history and beauty that sound like taken from a travel catalogue (like 'warm-blooded and hospitable people', 'wonderful nature' and 'ever present sun'). The way they refer to Turkey and the extent to which they mystify their homeland depends on imaginary extensions of short summer holidays and their retrospective reconstruction in Germany. (Bozkurt 2009, 154)

Again, this awareness is often greatly intensified when violence was involved. It is especially pronounced among descendants of Polish and Russian Jews who immigrated to the United States near the end of the 19th and at the beginning of the 20th century. The sentence: 'Had my grandfather not decided to emigrate, my parents would have perished in a concentration camp and I would not be alive today,' expresses an existential experience. The incomprehension many Jewish intellectuals expressed over European attitudes after September 11 followed this pattern.

An interesting case is given by the Armenian diaspora. After the fiftieth anniversary of the genocide, a series of terrorist attacks started against representatives of the Turkish state. It seemed that the public staging of the collective memory had led to an actualization of the unreal conditional sentence, and had reinvigorated hatred. If the unreal conditional sentence determines the structure of imaginary space, then the history of the migration movement must be decisive for the expression of this imagination. It determines who is comparing him or herself to whom. I have the impression that the Holocaust plays a different role in Ashkenazi and Sephardic Judaism – mainly because the former (due to the structure of the unreal conditional sentence) was more strongly affected than the latter. Or, to put it another way: the above-mentioned sentence comes from an Ashkenazi Jew. Abhorrence of the Holocaust is of course widespread among all Jews, but the extent of that concern is dependent on the spatial modality.

To sum up: what looks like a network from a bird's eye view resembles an imaginary opportunity space from an actor's point of view. Networks certainly allow access and open opportunities (and this is what has been studied), but networks and transnational communities have a collective psyche. The concept of the imaginary opportunity space allows us to access it.

4 Imaginaries of Irregular Migration

What was said in regards to premise in our imaginary quadrangle has to be modified by the aspect of risk. The risks of migration can be arranged on a continuum starting with low risk/low profit on one end and high risk/high profit at the other. Usually, low risk/low profit is the rule in regular migration whereas high risk/high profit is the rule in irregular migration.

Marta Kindler conducted an excellent study of the risk analysis Ukrainian women performed when engaging in irregular migration to Poland. They distinguished between low risk/low profit countries and high risk/high profit countries. Poland was low risk because languages were similar; add to that cultural similarity and spatial proximity. Migration was irregular but the difference to legal migration was minor. Risk, in other words, was calculable. Migration to Italy, on the other hand, was considered to be high risk: young women in particular might be forced into prostitution; earned wages might be withheld and in case of illness it might be difficult to find support. But it was also high profit – one could make four times more money than in Poland (Kindler 2011, 80, 84). It was clearly Italy that sparked fantasies. The stories of successes, missed opportunities and failures almost had fairy-tale qualities. Italy was an adventure; Poland very clearly was not. In terms of our imaginary quadrangle, high risk/high profit versus low risk/low profit refers to the category of law. It has implications for the other dimensions as well:

	Poland	Italy
Law	low risk/low profit	high risk/high profit
Category	broad	small
Condition	unlimited	limited
Moral Issue	low	high

Low risk/low profit entails broad categories: almost everybody can do it. High risk/high profit is much more selective. It usually attracts young singles (or, in the case of Ukraine, women in their forties whose children can already take care of themselves), while parents of small children tend to avoid it. Condition works in a similar fashion: the cost for high-risk border crossings usually correlates with travel costs. And, while moral issues associated with low risk/low profit are also low, as decisions are reversible, high-risk migration choices are heavily debated. Kindler mentioned warnings issued in Ukraine, cautioning against the dangers of falling into prostitution (Kindler 2011, 87), as well as articles pointing to neglected children ("Italian orphans") in the home country (Kindler 2011, 93).

A prominent example of high profit/high risk migration occurs from Sub-Saharan Africa to Europe. In this case, it seems that the daring and dangerous voyage to Europe is conceived as a challenge and often as a test of manliness. "It is the possibility to fail [...] which emphasizes the 'courage,' the 'persistence,' the 'energy,' if you want the 'entrepreneurial spirit' demanded as a condition for success" (Hoffmann 2016, 96). Heidrun Friese similarly refers to rap songs about illegal border crossings: "To accept the challenge, the ability to overcome dangers and risks on the way you had decided for; to assert oneself, to prove oneself, to succeed – all these are dimensions of male subjectivity, male self-confidence and dominant representations of manliness" (Friese 2012, 235).

In the hothouses of Almeria in Southern Spain, such exaggerated fantasies clash with a harsh reality.[6] The risk of finding no job, or only an extremely badly paid one, tends to be dramatically higher than expected. A vicious circle sets in. Difficult access means there are reality checks on the Jimmy Cliff-law-of-migration ("you can get it if you really want, but you must try, try and try"). Those who stay behind usually have no idea of the real working conditions of the migrants. They are not aware of the fact that failure, not success, is the rule. The validity of the law or its consequences are not questioned: rather, what is said is, "You did not get it – so you either did not really want it or you did not try hard enough." As a consequence, the virtue of those who report the truth risks to be questioned. An informant of Boris Nieswand told him that "only if he is able to adequately perform his gain of status which is related to his stay in Europe, will it render him respectable. This leads to the fact that when migrants contact their families and

6 See detailed description by Felix Hoffmann 2016: The situation of illegal workers is characterized by irregular labor conditions and low wages. Especially during the off-seasons workers have to live from their savings, "We don't work for money we work for papers." There is no hope for savings before they win a permanent permission to stay (which, if everything goes well, takes at least seven years after entering the country).

friends in the home country, they usually paint an embellished picture. When returning to the home country, they are under pressure to give material evidence of having been to Europe. Again people in Ghana "could question his moral integrity if he refused to perform his migrant's status or might even doubt he had physically been to Europe" (Nieswand 2011, 141). Nieswand describes migrants who postponed a visit to their home country because they were not in a position to make these statements (142).

Factual information counts little when it is confronted with dreams. One of Hoffmann's informants in Almeria recalled his own childhood when he told him: "'I would not believe it – because some our houses are close to each other in Africa – so every month I saw that they were sending back money to their parents. So when they told me: 'Do not come here, it's too hard, I would tell them: 'It is bullshit boy! You don't tell me that! Because you are sending money to your parents! And you expect me to stay here and you tell me that place is hard?'" (Hoffmann 2016, 101). It is very clear that the selection of information is governed by desire.

High risk/high profit makes for good stories. It seems to be especially relevant that this type of information circulates in the form of narratives. As Kindler notes they are told and re-told again, and again (Kindler 2011, 81). In this process they acquire a dramatic form: exposition, rising action, culmination, falling action and denouement follow each other. It seems to me that this literary form reinforces longings and desires. In this vein, Frank Pieke (2002), who studied illegal migrants from the Province of Fujian in East China, has demonstrated how a culture of migration as such develops:

> Discourses on social mobility that prescribe what constitutes success and how to attain it, while proscribing (often simply by remaining silent about them) others that locally quite literally are not considered an option. In the source areas of Fujianese mass migration a *culture of migration* has taken root that prepares all able-bodied men and women for their eventual departure. In the Fujian home communities, the culture of emigration stigmatizes local alternatives to emigration as second rate or even a sign of failure. (Pieke 2002, 32)

These narratives seem to play a crucial role in arousing collective obsession over the process of migration.

5 Conclusion

The concept of the imaginary space allows us to take a fresh look at the migration process. It gives us access to the emotional life of opportunity structures, social and economic networks. What from above can be mapped as a web of social relations connecting actors in different countries actually looks like an assemblage of possible worlds from below. A network consists to a large part of people like me. It is people like me to whom I compare myself. They show through their very existence what my life could look like, or could have looked like, if I just had made a different decision. They thus stand for alternatives of the present and possible futures. These alternatives incite hopes, doubts, fears, desires, and dreams, as well as the jealousies and irritations associated with the migration process. The emotional dimension also allows us to understand seemingly irrational decisions.

By enabling us to analyze the emotional side of networks, the concept of imaginary opportunity structures allows us to close the gap that exists between postmodern and postcolonial approaches on one hand, and sociological and political economy approaches on the other. By analyzing the degree to which networks form imaginations and imaginations relate to 'real' structures, we can approach the material basis of imaginations without reducing the latter to the former.

References

Aksoy, Asu, and Kevin Robins. "Thinking across Spaces: Transnational Television from Turkey." *European Journal of Cultural Studies* 3.3 (2000): 343–365.

Anderson, Benedict. *Imagined Communities: Reflections on the Origin and Spread of Nationalism* (1983). 2nd rev. ed. London/New York: Verso, 1991.

Appadurai, Arjun. "Disjuncture and Difference in the Global Cultural Economy." *Modernity at Large: Cultural Dimensions of Globalization*. Minneapolis: University of Minnesota Press, 1998. 27–47.

Black, Richard, and Khalid Koser. *The Mobilisation and Participation of Transnational Exile Communities in Post-Conflict Reconstruction: A Comparison of Bosnia and Eritrea* (Transnational Communities Programme Working Papers). 1999. <http://www.transcomm.ox.ac.uk/wwwroot/black.htm> [accessed: 13 March 2018].

Bourdieu, Pierre. *Die feinen Unterschiede. Kritik der gesellschaftlichen Urteilskraft*. Frankfurt a.M.: Suhrkamp, 1982.

Bozkurt, Esin. *Conceptualising "Home": The Question of Belonging Among Turkish Families in Germany*. Frankfurt a.M./New York: Campus, 2009.

Clifford, James. *Routes: Travel and Translation in the Late Twentieth Century*. Cambridge, MA: Harvard University Press, 1997.

Friese, Heidrun. "Y'al babour, y'a mon amour. Rai-Rap und undokumentierte Mobilität." *Deutscher Gangsta-Rap. Sozial- und kulturwissenschaftliche Beiträge zu einem Pop-Phänomen*. Eds. Marc Dietrich and Martin Seeliger. Bielefeld: transcript, 2012. 231–284.

Gellner, Ernest. *Nations and Nationalism*. Oxford: Blackwell, 1983.

Goffman, Erving. *Frame Analysis: An Essay on the Organization of Experience* (1974). 3rd ed. Boston: Northeastern University Press, 1993.
Goodman, Nelson. *Fact, Fiction, and Forecast.* 4th ed. Cambridge, MA: Harvard University Press, 1955.
Hoffmann, Felix. "We don't work for money – We work for papers! Zur Normalisierung des 'Illegalen' auf dem agroindustriellen Legalisierungsmarkt Almerias. Frankfurt (Oder): Kulturwissenschaftliche Fakultät Europa-Universität Viadrina, PhD-thesis, 2016 (published under the title: *Zur kommerziellen Normalisierung illegaler Migration. Akteure in der Agrarindustrie von Almería, Spanien.* Bielefeld: transcript, 2017).
Iser, Wolfgang. *The Fictive and the Imaginary: Charting Literary Anthropology.* Baltimore/ London: Johns Hopkins University Press, 1993.
Kindler, Marta. *A Risky Business: Ukrainian Migrant Women in Warsaw's Domestic Work Sector.* Amsterdam: Amsterdam University Press, 2011.
Mannitz, Sabine. "Auffassungen von kultureller Differenz." *Staat-Schule-Ethnizität.* Eds. Werner Schiffauer, Gerd Baumann, Riva Kastoryano, and Steven Vertovec. Münster: Waxmann, 2002.
Musil, Robert. *Der Mann ohne Eigenschaften.* Reinbek bei Hamburg: Rowohlt, 1978.
Nieswand, Boris. *Theorising Transnational Migration: The Status Paradox of Migration.* New York/Abingdon: Routledge, 2011.
Pattie, Susan P. *Longing and Belonging: Issues of Homeland in the Armenian Diaspora.* (Transnational Communities Programme Working Paper Series), 1999.
Pieke, Frank N. *Recent Trends in Chinese Migration to Europe: Fujianese Migration in Perspective.* Geneva: IOM International Organization for Migration, 2002.
Sartre, Jean Paul. *Das Sein und das Nichts* (1943). Reinbek bei Hamburg: Rowohlt, 1962.
Schiffauer, Werner. *Die Bauern von Subay. Das Leben in einem türkischen Dorf.* Stuttgart: Klett-Cotta, 1987.
Schiffauer, Werner. *Die Migranten aus Subay. Türken in Deutschland: Eine Ethnographie.* Stuttgart: Klett-Cotta, 1991.
Warner, Michael. "Publics and Counterpublics." *Public Culture* 14.1 (2002): 49–90.

II: (Border) Regimes

Sabine Hess

Border as Conflict Zone

Critical Approaches on the Border and Migration Nexus

"Far from disappearing, many borders are being reasserted and remade through ambitious and innovative state efforts to regulate the transnational movement of people." This conclusion by Peter Andreas and Timothy Snyder in their seminal work *The Wall around the West* (Andreas and Snyder 2000, 2) never seemed as up-to-date as today, when we consider the US president's announcement of his intent to construct a nine-meter-high wall on the US-Mexican border, or when we look at recent developments across the European continent in the wake of the so-called 'summer of migration' of 2015, when unexpectedly high numbers of refugee-migrants managed to make their paths to western European countries by – quite literally – overrunning the various border technologies and infrastructures installed by European Union nation-states and supra-national organizations over the last decade. Since then, we have been witnessing an unexpected reemergence of national and regional border apparatuses on the European continent,[1] in their very material shape of fences, ditches, dogs, and watch-towers such as along the Macedonian-Greek, Hungarian-Serbian, and Hungarian-Croatian land borders. Even prior to 2015, fences as border-control mechanisms have been reemerging: around the Spanish enclaves Ceuta and Melilla, for example, and along the Greek-Turkish and Bulgarian-Turkish land borders separating Europe from "the rest," as Stuart Hall described this post-colonial division of the world (Hall 1992). In addition to these obvious and manifest acts and architectures of re-bordering, a large number of rather technological border apparatuses have been installed, even long before 2015, which tend to be more or less 'invisible': radar- and computer-controlled, digital, intelligent border surveillance technologies that establish network-like security spaces (integrating satellites, drones, and radar systems with big data banks) such as Spain's Integrated System for External Vigilance (*Sistema Integrado de Vigilancia Exterior*, SIVE), introduced in 2002; the Maritime Surveillance (MARSUR) system, introduced in 2005; and the European Border Surveillance (EUROSUR) system, introduced by the EU in 2013 alongside big databases like the fingerprint data bank *Eurodac*, the Schengen Information

1 Rebordering efforts, in their very material manifestations, can be studied not only in Europe, but also in fence constructions in Israel, around the Spanish enclaves Ceuta and Melilla in Morocco, and in the fence between Mexico and the US.

https://doi.org/9783110600483-006

System (SIS), and the Visa Information System (VIS). Funds amounting to millions are flowing into research and development of this type of technology, while civil and military protagonists are becoming increasingly fused. In relation to the control of the Mediterranean, Sergio Carrera and Leonhard den Hertog, from the Center for European Policy Studies, have described these developments as a "surveillance race" (2015, 16), which is producing a new spatialization and digitalization of borders. This extension and multiplication of the border from what used to be a recognized line around the territory of nation-states led Etienne Balibar to speak of the "ubiquity of borders" already at the beginning of the new century (2002, 84).

The hope for a "borderless world," associated with the end of the Cold War, with the enlargement and harmonization of the European Union – especially with its creation of "Schengenland" as an "area of freedom, security and justice" alongside the internal market (Walters and Haahr 2004) – as well as in the general context of the ongoing economic, cultural, and political globalization processes seem to belong to the past (see Newmann 2006; Donnan and Wilson 2010, 2). During the high phase of the globalization debate, in social and cultural sciences and in the public alike, there were already dozens of academic reports and approaches warning against an overly simplistic and overly euphoric assessment of globalization (see Ong 1999; Smith and Guarnizo 1998), proposing instead medium-range concepts like glocalization (Robertson 1998) or transnationalism (Glick Schiller, Basch, and Szanton Blanc 1995). Nevertheless, the social- and cultural-science debate at that time was dominated by metaphors of 'flow' and 'network' (see Castells 2000; Sassen 2002); and by the proclamation of paradigm shifts that favored mobility, fluidity, and hybridity over seditiousness, fixity, and homogeneity, as in James Clifford's volume *Routes: Travel and Translation in the Late Twentieth Century* (1997) or John Urry's concept of 'mobility turn' (2000).

Now, on the contrary, one could speak of a 'border turn' or border paradigm. In this regard, in his 2006 overview on concepts and approaches of/in the interdisciplinary field of border studies, David Newman was already emphasizing that the "contemporary studies of borders has become a major growth industry" (Newman 2006, 144). Even if this does not hold true for the German-speaking academic context – as, to this day, no research center or professorship with such a denomination exists – we can nevertheless observe a certain kind of "explosion" of studies and research projects on borders, as Hastings Donnan and Thomas M. Wilson described in their 2010 compendium *Borderlands: Ethnographic Approaches to Security, Power, and Identity*. But not only in border studies, also in the much wider field of migration and mobility studies, the new and growing significance of the "border" can be noted. For example, in their recent article "Regimes of Mobility across the Globe" (2013), Nina Glick Schiller and Noel

B. Salazar also centrally address this issue. Glick Schiller and Salazar note an accelerated return of national borders and lines of ethnic separation in the midst of global economic crises, following Ronen Shamir's observation that a "single global mobility regime" is arising that is "oriented to closure and to the blocking of access, premised not only on 'old' national or local grounds but on a principle of perceived universal dangerous personhoods [...] In practice, this means that local, national, and regional boundaries are now being rebuilt and consolidated" (Glick Schiller and Salazar 2013, 199).

However, what is meant in the different texts by the notion of 'border' or 'boundary'? And how can we conceptualize the relation between borders and movements, especially the movement that is politically labeled as 'international migration'? In the following, I sketch out recent debates in border studies on how to conceptualize 'the border' and how to make sense of its role and function, illustrating that conceptualizations that are customary in international border studies also view this as a relationship with a top-down structure. On the one hand, this produces a representational regime in which migrants appear as the dependent variable, as structurally powerless and as 'victims.' On the other hand, it also leads – even in critical research work – to a reification of the controlling power of the border regime. But how can the cross-border practices of migration be incorporated into a theorization of the border that would do justice to the agency of migration, while at the same time being capable of analytically examining the expansion of the border regime?

This question, which I attempt to address in the final section, brings me back to the starting point for my first interdisciplinary research project, "Transit Migration," which considered the construction of the European border regime in 2002. In the early 2000s, this project led me, with an interdisciplinary group of seven other researchers, to Southeastern Europe, which at that time was already being seen as a 'migration hot spot' and a target of attention of European media and politicians (Transit Migration Forschungsgruppe 2007). In the face of the migrants' compelling and evidently unrestrainable desire to reach a safer and better life in Europe – as they described it to us at that time in the transit- and buffer-zones in Southeastern Europe – we developed a conception of Europe's external border as representing a "zone of conflict and interaction" (Karakayali and Tsianos 2007; Hess 2012). We therefore referred to the 'border regime' as a dynamic ensemble consisting of various agents, discourses, and practices. This enabled us to formulate our approach as "ethnographic border regime analysis" that incorporates migration as a powerful force into the way in which the border itself was theorized, rather than regarding migration merely as being an object of the border (Karakayali and Tsianos 2007; Tsianos and Hess 2010). The developments that took place in the summer and fall 2015 – which were astonishing for me as a longtime observer – put the force of migration (that we had then already noted) back on the international agenda, in a breathtaking and sometimes disturbing way.

1 Border as Barrier, Gate, and Transformation Regime?

David Newman stresses that there is no single theory of borders (2006, 145). However, a notion of 'border' prevails across all disciplines: this still focuses on the territorial "state border,"[2] despite widely recognized and profound changes in regard to its shape, territoriality, and spatialization; its production and governance; as well as its character, role, and function (Donnan and Wilson 2010). In this regard, following Etienne Balibar's quote on the ubiquity of borders, it is 'common sense' in international border studies that state borders can no longer be conceptualized as static lines or demarcations of sovereignty and nation-state power encompassing a national territory that ontologically possess an 'essence,' as the manifesto "Lines in the Sand: Towards an Agenda for Critical Border Studies" (Parker and Vaughan-Williams 2009) also firmly stresses.

This shift induced not only a geographical refocusing of research away from the level of the state – down to regions, municipalities, and even neighborhoods (e.g., gated communities); and up to transnational, regional, or global settings (such as the EU or NAFTA) – but also a methodological reorientation with a focus on bordering processes and practices, "rather than [on] the border per se" (Newman 2006, 144). Nevertheless, most studies primarily conceptualize "the border" as a "barrier of exclusion and protection" even as concrete research, like that of Hastings Donnan and Thomas M. Wilson, focuses on borderlands and the daily border-crossing practices of "borderlanders" (Donnan and Wilson 2010, 11). Borders in this respect are seen as ordering technology – in a Foucauldian sense – that differentiates 'us' from 'them,' 'here' from 'there,' and 'inside' from 'outside,' as Henk van Houtum and Ton van Naerssen have also spelled out in their article

---

2 For "geographers," as David Newman puts it, "we have traditionally understood borders (or boundaries) as constituting the physical and highly visible lines of separation between political, social and economic spaces" (2006, 144); also Donnan and Wilson refer to the state border and "geopolitical enclosures" in their conceptualization of "borderlands" as the main object of research for cultural anthropological border studies that would focus "on those local people and communities who live, work and cross borders" (2010, 8).

"Bordering, Ordering and Othering" (2002).[3] This thinking of borders as boundary-drawing technologies invites a wide metaphorical usage of the notion of 'border' in a de-territorial and rather social sense that comes very close to the concept of boundary drawing spelled out by Fredrik Barth (1998).

The territorial state border was, in this regard, always more than simply a territorial line on the ground manifested or enforced by a material infrastructure at that location. Borders always required the performance of further social and cultural ingredients and investments, as historical research clearly indicates how difficult it was to establish borders and border practices (François, Seifarth, and Struck 2007). Borders are not only "meaning making and meaning carrying" (Donnan and Wilson 2010, 4), they also need to have a place in our mental maps, cultural imaginaries, and moral judgments; they additionally require laws and re-presentations to be enacted by various people and institutions in the first place. In his seminal paper "Mapping Schengenland: Denaturalizing the Border," William Walters outlines three historical typologies of borders: the "geopolitical," the "national," and the "biopolitical border." He asserts that even the "geopolitical border," as the expression of the 18th and 19th century understanding of nation-states as territorially defined and fixed units – when the border was the potential line where armed forces are arrayed – has to be thought of as an "assemblage." He states: "There is a whole apparatus connected with the geopolitical border – not just a police and military system, but cartographic, diplomatic, legal, geological, and geographical knowledges and practices" (Walters 2002, 563). Walters speaks of borders as an "art of government," and during the colonial period, border-making was especially an "art of international government" (Walters 2002, 564).

But the border is not only a 'barrier of exclusion,' and perhaps, as I outline below, it is decreasingly thought of as a spatially employed barrier. Rather, the border was always also a gate, and, in this respect, an institution of mobility,[4] "designed to break up and manage the flow of items and personnel into and out of the state" (Donnan and Wilson 2010). Borders differ considerably in regard to this flow-management capacity, not only if flows are slowed or quickened, but rather what kind of flows are treated, and how. In this respect, borders are experienced very differently by different people – a fact that the European Union devel-

3 Newman: "For all disciplines, borders determine the nature of group (in some cases defined territorially) belonging, affiliation and membership, and the way in which the processes of inclusion and exclusion are institutionalized" (2006, 147).

4 Migration research on "suitcase tourists" and commuting migrants, for example, show that it is the border and currency differences across borders that motivate mobility in the first place.

oped into a distinct and globally unique feature via its differentiation between free internal circulation for the newly constructed entity of "European citizens," and its "external border" presented to the so-called "third country nationals" (Walters and Haahr 2004; Hess and Tsianos 2007). The US cultural anthropologist Michael Kearney referred to this capacity of borders quite early in his research on the long-militarized US-Mexican border in the following way:

> Rhetoric aside, and as noted above, the *de facto* immigration policy of the unitedstatesian government is *not* to make the U.S.-Mexican border impermeable to the passage of 'illegal' entrants, but rather to regulate their 'flow', while at the same time maintaining the official distinctions between [...] kinds of people, that is to constitute classes of peoples. (Kearney 1991, 58)

Kearney thus goes a step beyond just referring to the "filtering" function of border control "that separates out the unwanted from the wanted cross-border flows" (Anderson 2000, 4). In this regard, Peter Andreas already broke with a widely held societal 'wisdom' that nourishes metaphors like "fortress Europe" by pointing to the fact that "the wall around the west [...] is by design highly permeable" (Andreas 2000, 4). Chris Rumford describes borders in this respect as "asymmetric membranes" (2008, 3) and William Walters uses the metaphor of a "firewall" that hits and selects on a very differentiated grid (Walters 2006, 197).

However, what Kearney is hinting at with his formulation, "constitutes classes of people," delineates the border not as a repressive instrument that selectively deters, but rather as a "productive" mechanism, in the sense of Foucault's notion of bio-power as opposed to the traditional conceptualization of power following a repression hypothesis. The border, in this understanding, can be seen as a transformational regime of rights and statuses – questioning and sometimes even removing the civic status that people had at the moment of their crossing by delegitimizing the cross-border mobility as illegal or irregular, and hence ascribing to them a status of being 'undocumented' or 'unlawful,' without citizenship rights. In this sense, the border is a gigantic transformational regime, producing new hierarchies of people by categorizing and processing the unchecked mobilities as different 'migration' categories. This leads to a "differential inclusion" (Mezzadra and Neilson 2013, 7) and "civic stratification" (Morris 2002), and to what Sandro Mezzadra and Brett Neilson termed the "multiplication of labor" (2013, 20). William Walters sums up this role and function of the border as a "locale where power is produced" in his concept of the "biopolitical border":

> The biopolitization of the border is signaled by the political concerns, events, and means by which the border will become a privileged instrument in the systematic regulation of

national and transnational populations – their movements, health, and security. (Walters 2002, 571)

As such, Walters shows that nation-state immigration policies only started to "become in any way systematic" around the turn of the last century, in the wake of "economic downturns in many countries" and heightened national security concerns with the introduction of passports, visas, foreigner police, and respective foreigner laws. Even the US only started to federally regulate immigration in the 1880s. If already the "geopolitical border" cannot be reduced to a line in the sand, as I have just outlined, then the biopolitical border has to be imagined rather as the "machine" that Walters describes, or as a regime "with an assortment of technologies, simple and complex, old and new. These include passports, visas, healthy certificates, invitation papers, transit passes, identity cards, watchtowers, disembarkation areas, holding zones, laws, regulations, customs and excise officials, medical and immigration authorities" (Walters 2002, 572).

Walters points here to an additional function of the border that is distinct from understandings that focus on its exclusionary aspects. Rather, Walters is stressing the fact that borders are a "privileged site" where political authorities can acquire biopolitical knowledge about populations: "In a sense, then, the border actually contributes to the production of population as a knowable, governable entity" (Walters 2002, 573).

2 The EU as a New Border Laboratory

A common denominator of border studies is an emphasis on the transformation of the border from a demarcation line surrounding national territory to an ubiquitous, techno-social, deterritorialized apparatus or regime producing geographically stretched borderscapes.[5] This holds especially true for the European Union – as Bernd Kasparek and I argue in the recent article, "De- and Restabilising Schengen: The European Border Regime after the Summer of Migration" (Hess and Kasparek 2017) – which can be regarded as a "laboratory" of said transformation. With the Schengen agreement of 1985, the European project heralded the creation of a continental border regime, with the newly created notion of an 'external border' as the pivotal mechanism and space for migration control. Even despite being initially outside the formal EC/EU framework, this globally unique process of regionalization and of supra-national harmonization was a

5 Chris Rumford speaks of a "condition of postterritoriality" (2006, 159).

driving force towards an accelerated and deepened process of Europeanization, culminating in both the Treaty of Amsterdam (1999) and later the Treaty of Lisbon (2009). The process resulted in the creation of an "area of freedom, security and justice" through the Treaty of Amsterdam and the parallel construction of the European border regime as a fluid, multi-scalar assemblage. This assemblage involves European Union agencies like FRONTEX (the European border and coast guard agency), bodies of European law (like the Common European Asylum System), processes of standardizations and harmonizations especially in the field of border management (called "Integrated Border Management"), and a growing military-industrial-academic complex largely funded by the EU, alongside more traditional political national apparatuses of migration control that had evolved since the 1970s and a flexible involvement of IGOs (international and intergovernmental organizations such as the UNHCR or the IOM).

If there is one central rationale at the core of the European border regime, it is driven by what Gallya Lahav and Virginie Guiraudon have called the fundamental "control dilemma" (2000). With the creation of the EU internal market at its peak, this dilemma refers to the question how to reconcile a neoliberal economic paradigm of the (preferably global) free circulation of goods, services, and capital, with the continued biopolitical desire to control the movements of people. While the EU upholds these four freedoms internally, towards outside entities, the EU is merely committed to the first three of these freedoms. There is no commitment to a global freedom of movement for people; rather, many authors of border studies and European studies have pointed to the fact that the creation of the single EU market opened the door to a wide field of security actors and led to an intensified securitization of questions of mobility (Huysmans 2000; Bigo and Guild 2005). William Walters and Jens Henrik Haahr argue in this respect: "Schengenland can be seen as having certain acts of securitisation as its conditions of possibility" (2004, 95).

In regard to the border regime, the main practical answer to the control dilemma was, according to Lahav and Guiraudon's intriguing title of the paper, to move border controls "away from the border and outside the states" (Lahav and Guiraudon 2000), leading to a new spatialization and geographical expansion. In this, the European Commission's vision was a "smart," techno-scientific, invisible yet selective border that is able to distinguish between bonafide travelers and unwanted migrants (Commission of the European Communities 2008).

To this end, broadly speaking, four paradigms were enacted within the European border regime. The first was a paradigm of "remote control" and externalization that led to an unforeseen multiplication and diversification of actors from state to non-state, from international actors to very local ones (Lavenex 2004; Lahav and Guiraudon 2000; Zolberg 2006; Hess and Tsianos 2007; Bialasiewicz 2012). The second, as already indicated, was a paradigm of a fortified, yet smart external border through technology, digitalization, and biometrization (Koslowski 2005; Broeders 2007; Dijstelbloem, Meijer, and Besters 2011; Kuster and Tsianos 2014). These two dimensions have been analyzed extensively by border studies, yet a third also exists, namely, an internal regime steeped in the institution of asylum and put into practice through the Dublin/*Eurodac* regulations, which aims at the immobilization of migrant populations within the European territory (Schuster 2011; Kasparek 2016; Borri and Fontanari 2015).

The fourth and final paradigm, evident especially in recent years, is an increasing humanitarization[6] of the border, described by Walters as the "birth of the humanitarian border" (2011). This accelerated in the context of the growing number of deadly shipwrecks and tragedies in the Mediterranean Sea in recent years, as the crossing of the border became, very obviously, a "matter of life and death." However, the humanitarian discourse dates further back to the "White Paper" of former British Prime Minister Tony Blair from the year 2002, entitled "Secure Border, Safe Haven." It is mostly read as a founding document for externalization. But it is more than that as the paper also used and – I would say - instrumentalized a strong humanitarian rhetoric to legitimize further externalization and an intensification of border controls.

Also in the context of our first Transit Migration research project in the early 2000s, mentioned above, we deduced processes that we called "NGOization" and a "governmentalization of politics," pointing to the fact that the expansion of the border regime not only functioned by means of "security-actors," but particularly operated via the specific appeal to and articulation of humanitarian positions, as in the field of anti-trafficking policies and in the context of asylum (Hess and Karakayali 2007). After the deaths of more than 600 migrants in the Lampedusa shipwreck of 2013, this became a discourse in its own right. This paradigmatic shift seemed to have been possible due to wider hegemonic changes, which in part were also due to incessant migration struggles, transnational solidarity networks, and the professionalized critical knowledge practices of NGOs and legal

6 I follow William Walters's (2011) and Didier Fassin's (2007) conception of humanitarianism, with which both understand more than just "ideas and ideologies" or "simply the activity of certain nongovernmental actors," and rather grasp humanitarianism as a specific form of governance; as a rationality of power, thus situating the debate "in relation to the analytics of government" (Walters 2011, 143). As Paolo Cuttitta (2016) puts it, this results in a specific operational logic, which finds its expression in an "increasingly organised and internationalised attempt to save the lives, enhance the welfare, and reduce the suffering of the world's most vulnerable populations," now also in the field of border control.

interventions, all of which have led to a further juridification of the border regime and human rights based approaches over the last years (see Hess 2016).

Confronted with a vast new quantity of migrant arrival, this architecture of the European border regime collapsed in summer 2015. In the end, the massive movements of migration challenged not only the European Union's border and migration regime, but the EU and the European project as a whole.

3 "Border Work" and the Agency of Migration as an Excluded Category

These new territorially and deterritorialized extended border spaces are also being described using terms such as 'border zones,' 'borderlands' or 'borderscapes.' At the same time, these concepts include ideas of mobile, fluid, selective, and differentiated border situations. Such talk is thus also of "mobile borders" (Kuster and Tsianos 2014, 3) or "networked borders" (Rumford 2006, 153). In this context, Balibar argues in favor of describing borders as "overdetermined, polysemic (that is to say that borders never exist in the same way for individuals belonging to different social groups), and heterogeneous" (cited in Salter 2011, 67). Those who have the relevant economic resources, nationality, and documents have thus over the last years enjoyed the pan-European free travel zone. Others, by contrast, such as those belonging to states in the global South, now face the border in trains and in railway stations, or in airports, schools, and health-care facilities on the level of the municipality. With the *Eurodac* system, the 'border' is now stamped on the body.

This conceptualization also initiated a practice turn in border studies focusing on processes of doing and performing borders (see van Houtum and van Naerssen 2002, 126; Salter 2011). The border is now being conceptualized as an effect of a multiplicity of agents and practices, as becomes clear in the concept of 'border work' (see Rumford 2008; Salter 2011). The concept of border work in particular draws attention to the everyday micropractices of politicians, border guards, journalists, academics, judges, NGO staff, and transport personnel, who need to reinterpret the border again and again in order to enact it. In recent years, the concept has been enlarged also to draw attention to the growing arsenal of technologies such as drones, satellites, heat-detecting cameras, scanners, and databases. Following these perspectives, bordering is to be understood as a performative act. Drawing on Judith Butler's notion of performance, Marc Salter points to the fact that also "sovereignty, like gender, has no essence, and must continually be articulated and rearticulated in terms of 'stylized repetition of acts' of sovereignty" (Salter 2011, 66).

All these recent practice-oriented conceptualizations indeed understand the border as an effect of a multitude of actors and practices, human and non-human alike. However, many of these highly interesting constructivist approaches still completely ignore the constitutive power of migration, or once again conceptualize migrants as structurally powerless and as 'victims.' The dominant focus of border studies, especially these ones following the classical securitization approach on the function of the border as a barrier or filter to exclude people, seem to lead to an epistemological exclusion of the agency of migrants.

By contrast, the recently published volume *Border as Method* by Sandro Mezzadra and Brett Neilson (2013), for example, takes up the standpoint of the autonomy of migration approach. In this respect, the authors define borders as "social institutions, which are marked by tensions between practices of border reinforcement and border crossing" (Mezzadra and Neilson 2013, 3) and emphasize the decisive role that "border struggles" play in constituting a specific border regime and its localized enactments and implementations.

This has many aspects in common with our approach, which we labeled "ethnographic border regime analysis," as a methodology to theorize the border from the perspective of the autonomy of migration (Transit Migration Forschungsgruppe 2007). This approach makes it possible to regard the border regime as a space of constant tension, of conflict and contestation in the face of the power and agency of the migration movements, without minimizing the border regime's militarization and brutality. These conceptualizations represent a methodological and theoretical attempt not only to think about the relationship between migration movements and control regimes in a way that is different from the classical sociological way of object-structure, but also to conceive of migration differently than the previous dominant practice in the cultural and social sciences – namely, not thinking about it in the sense of a 'deviation' from the paradigm of the settled way of life in the modern nation-state, or as a functionalist variable of economic processes and rationalities. Instead, this theoretical and methodological approach represents an attempt to conceptualize migration both historically and also structurally as an act of 'flight' and as 'imperceptible' form of resistance, in the sense of withdrawal and escape from miserable, exploitative conditions of existence such as those described by Dimitris Papadopoulos, Niamh Stephenson and Vassilis Tsianos (2008). Yann Moulier Boutang (2006) described this aspect as the "autonomy of migration." This draws attention to migration as a co-constitutive factor for the border, with the forces of the movements of migration that are challenging and reshaping the border every single day (Hess 2016).

4 The Autonomy of Migration as a Prism

What would change if we were to conceptualize migration in the way expressed by the concept of 'the autonomy of migration'? The concept of the autonomy of migration is often wrongly interpreted to mean the autonomy of migrants. However, this misses the concept's intent, which instead needs to be understood as a structural argument drawn from a historical materialist reading of history. Nor does the concept intend to obscure the suffering and plight of numerous migration projects. Instead, it represents an attempt by Moulier Boutang and other researchers to re-situate migration within the history of labor, capitalism, and modern forms of government that focuses on the previously little-considered ability of living labor to escape from unbearable conditions of (re-)production (Mezzadra and Neilson 2013). Moulier Boutang writes, for example, "If one links it with Foucault's 'desire of the masses not to be governed in this way' and connects it to the concept of flight or exile, it becomes fruitful: because flight is the masses' refusal to let themselves be governed, a response to asymmetric power relations" (Moulier Boutang 2006, 172).

In his theoretical approach, Moulier Boutang draws strongly on the theoretical tradition of *operaismo* (workerism). '*Operaismo*' originated on the one hand as a political movement and on the other emerged from 1960s political theory in Italy in opposition to the Marxist mainstream. Two central insights of this 'workerism' appear to be crucial for the change in perspective emphasized by a migration research approach focusing on the autonomy of migration. First, workerism conceives of the history of capitalism as being driven by workers' struggles. From this viewpoint, for example, both industrialization and the development of the factory appear as a political response to the mass flight from rural regions and workers' resistance. Secondly, 'resistance' is conceived empirically by taking into account silent, unorganized, apparently insignificant forms of subversion and withdrawal such as slow working. Analogously, Moulier Boutang regards capitalist developments not simply as being motivated by the dynamics of the profit rate but as representing reactions to the lived mobility of the labor force, and as a constant attempt to control living labor and its ability and desire to mount resistance and escape from the existing conditions (Moulier Boutang 2006; Papadopoulos, Stephenson, and Tsianos 2008).

This viewpoint of the autonomy of migration does not end by assuming that migration can be understood as an active force and as a form of everyday silent resistance. Instead, this approach asks about the ways in which migration intervenes in the center of knowledge production (Hess 2015). Bernd Kasparek and Maria Schwertl recently summed up the theoretical implications of the autonomy of migration as follows: "The autonomy of migration is less a conclusion to arrive at but a perspective that opens up new ways of interrogation and doing research. Or, to quote Moulier Boutang, autonomy of migration is not a slogan, but a method" (Hess, Kasparek, and Schwertl 2017).

If we follow the concept of the autonomy of migration in the sense of a method or a prism, the question inevitably arises of what such a viewpoint may enable us to see. First, the approach conceives of migration and mobility as a social movement not in the classical sense of an organized, ideological driven movement, but rather in the sense of world-making collective practice, and consequently as a fundamentally political, social, and transformative project. Through migration, social agents escape from their normalized representations and reshape themselves and their own conditions of existence. According to Papadopoulos, Stephenson and Tsianos, migration represents an active transformation of the social space:

> Migration is not the evacuation of a place and the occupation of a different one; it is the making and remaking of one's own life on the scenery of the world. World-making. You cannot measure migration in changes of position or location, but in the increase in inclusiveness and the amplitude of its intensities. Even if migration starts sometimes as a form of dislocation (forced by poverty, patriarchal exploitation, war, famine), its target is not relocation but the active transformation of social space. (Papadopoulos, Stephenson, and Tsianos 2008, 169–170)

Second, when we look at the border and the migration regime, the way in which we conceptualize the border and consequently the way in which we understand the 'state' or 'sovereignty' also changes. The formerly monolithic border apparatus breaks down and dissolves into multiple factors: agents, practices, discourses, technologies, bodies, emotions, processes, and contestations become visible, and migration can be conceived of as one of the driving forces behind this (Heimeshoff, Hess, Kron, and Schwenken 2014, 13–14). This way of conceiving the border discards simplified binary models that locate the structure as a simple opposite of the power to act. Instead, the border is newly conceptualized as a space of challenge, of conflict and negotiation. The ethnographic border regime analysis developed by the Transit Migration Research Group in the 2000s attempts to provide a methodological operationalization of these theoretical implications (Transit Migration Forschungsgruppe 2007; Tsianos and Hess 2010).

The ethnographic border regime analysis draws on a political-science notion of 'regime' as well as a Foucauldian notion in order to take into account the border work of a variety of agents, institutions, and other human and non-human factors, without simplifying the diverse interests and rationalities of these forces into a simple or linear logic or hidden agenda (like that of capital or European racism). The ethnographic border regime analysis is rather based on an empirical and theoretical conceptualization of the border as a location of continual encounters and tensions, so that migration becomes a constitutive component of the border. According to Giuseppe Sciortino, the regime is “rather a mix of implicit conceptual frames, generations of turf wars among bureaucracies and waves after waves of ‘quick fixes’ to emergencies [...] the life of a regime is the result of continuous repair work through practices” (Sciortino 2004, 32–33).

Accordingly to the regime approach, the continual and structurally conflict-ridden reconfiguration of the border must be understood primarily as a reaction to the movements of migration that challenge, cross, and reshape borders. From this point of view, it is the migration movements that produce the socioeconomic phenomenon of the border space: border spaces represent the product of a collectivized excessive desire to overcome borders, of networks of people on the move and of collective knowledge practices of border crossing (Fröhlich 2015). It is this “excess” of autonomy that is the target of control, regulation, and exploitation by state border agencies and policies in order to construct the border as a stable, controllable, and manageable tool for selective and differential inclusion.

References

Andreas, Peter. “Introduction: The Wall after the Wall.” *The Wall around the West: State Borders and Immigration Controls in North America and Europe.* Eds. Peter Andreas and Timothy Snyder. Lanham, MD/New York/Oxford: Rowman & Littlefield, 2000. 1–14.

Andreas, Peter, and Timothy Snyder, eds. *The Wall around the West: State Borders and Immigration Controls in North America and Europe.* Lanham, MD/New York/Oxford: Rowman & Littlefield, 2000.

Balibar, Etienne. *Politics and the Other Scene.* London: Verso, 2002.

Barth, Fredrik. *Ethnic Groups and Boundaries: The Social Organization of Culture Difference* (1969). Long Grove: Waveland Press, 1998.

Bialasiewicz, Luiza. “Off-Shoring and Out-Sourcing the Borders of Europe: Libya and EU Border Work in the Mediterranean.” *Geopolitics* 17.4 (2012): 843–866.

Bigo, Didier, and Elspeth Guild. “Policing in the Name of Freedom.” *Controlling Frontiers: Free Movement Into and Within Europe.* Eds. Didier Bigo and Elspeth Guild. Aldershot: Ashgate, 2005. 1–13.

Borri, Giulia, and Elena Fontanari. “Lampedusa in Berlin. (Im)Mobilität innerhalb des Europäischen Grenzregimes.” *Peripherie* 35.138/139 (2015): 193–211.

Broeders, Dennis. “The New Digital Borders of Europe: EU Databases and the Surveillance of Irregular Migrants.” *International Sociology* 22.1 (2007): 71–92.

Carrera, Sergio, and Leonhard den Hertog. “Whose *Mare*? Rule of Law Challenges in the Field of European Border Surveillance in the Mediterranean.” *Centre for European Policy Studies (CEPS) Paper in Liberty and Security in Europe* 79 (2015): 1–29. <http://www.ceps.eu/system/files/LSE_79.pdf> [accessed: 16 March 2018].

Castells, Manuel. *The Rise of the Network Society.* Cambridge, MA: Blackwell, 2000.

Clifford, James. *Routes: Travel and Translation in the Late Twentieth Century*. Cambridge, MA/ London: Harvard University Press, 1997.
Commission of the European Communities. "Preparing the Next Steps in Border Management in the European Union." *Communication from the Commission to the European Parliament, the Council, the European Economic and Social Committee and the Committee of the Regions* (Brussels, 13 February 2008).
Cuttitta, Paolo. "Zwischen De- und Repolitisierung. Nichtstaatliche Search and Rescue-Akteure an der EU-Mittelmeergrenze." *Grenzregime III. Der lange Sommer der Migration*. Eds. Sabine Hess et al. Berlin: Assoziation A, 2016. 115–125.
Dijstelbloem, Huub, Albert Meijer, and Michiel Besters. "The Migration Machine." *Migration and the New Technological Borders of Europe*. Eds. Huub Dijstelbloem and Albert Meijer, London: Palgrave Macmillan, 2011. 1–22.
Donnan, Hastings, and Thomas M. Wilson. *Borderlands: Ethnographic Approaches to Security, Power, and Identity*. Lanham, MD/New York/Oxford: Rowman & Littlefield, 2010.
Fassin, Didier. "Humanitarianism as a Politics of Life." *Public Culture* 19.3 (2007): 499–520. <https://www.sss.ias.edu/files/pdfs/Fassin/Humanitarianism-as-politics-life.pdf> [accessed: 16 March 2018].
François, Etienne, Jörg Seifarth, and Bernhard Struck, eds. *Die Grenze als Raum, Erfahrung und Konstruktion. Deutschland, Frankreich und Polen vom 17. bis 20. Jahrhundert*. Frankfurt a.M./New York: Campus, 2007.
Fröhlich, Marie. "Routes of Migration – Migrationsprojekte unter Bedingungen europäisierter Regulation." *Movements of Migration. Neue Perspektiven im Feld von Stadt, Migration und Repräsentation* [exhib. cat. "Movements of Migration," 3–30 March 2013, Kunstverein Göttingen]. Eds. Sabine Hess and Torsten Näser. Berlin: Panama-Verlag, 2015. 150–162.
Glick Schiller, Nina, Linda Basch, and Cristina Szanton Blanc. "From Immigrant to Transmigrant: Theorizing Transnational Migration." *Anthropological Quarterly* 68.1 (1995): 48–63.
Glick Schiller, Nina, and Noel B. Salazar. "Regimes of Mobility Across the Globe." *Journal of Ethnic and Migration Studies* 39 (2013): 183–200.
Hall, Stuart, and Bram Gieben. *Formations of Modernity*. Cambridge: Polity Press in association with the Open University, 1992.
Heimeshoff, Lisa-Marie, Sabine Hess, Stefanie Kron, and Helene Schwenken. "Einleitung." *Grenzregime II: Migration – Kontrolle – Wissen. Transnationale Perspektiven*. Eds. Lisa-Marie Heimeshoff, Sabine Hess, Stefanie Kron, and Helene Schwenken, Berlin/ Hamburg: Assoziation A, 2014. 5–26.
Hess, Sabine. "De-naturalising Transit Migration: Theory and Methods of an Ethnographic Regime Analysis." *Population, Space and Place* 18.4 (2012): 428–440.
Hess, Sabine. "'Citizens on the Road': Migration, Borders, and the Reconstruction of Citizenship in Europe." *JEECA: Journal of European Ethnology and Cultural Analysis* 1 (2016): 7–22.
Hess, Sabine, and Vassilis Tsianos. "Europeanizing Transnationalism! Provincializing Europe! – Konturen eines Neuen Grenzregimes." *Turbulente Ränder. Neue Perspektiven auf Migration an den Grenzen Europas*. Ed. Transit Migration Forschungsgruppe. Bielefeld: transcript, 2007. 23–38.
Hess, Sabine, and Serhat Karakayali. "New Governance oder die imperiale Kunst des Regierens." *Turbulente Ränder. Neue Perspektiven auf Migration an den Grenzen Europas*. Ed. Transit Migration Forschungsgruppe. Bielefeld: transcript, 2007. 39–56.

Hess, Sabine, Bernd Kasparek, and Maria Schwertl. “Regime ist nicht Regime ist nicht Regime. Zum theoriepolitischen Einsatz der ethnografischen (Grenz-)Regimeanalyse.” *Was ist ein Migrationsregime? What is a Migration Regime?* Eds. Andreas Pott, Christoph Rass, and Frank Wolff. Wiesbaden: Springer VS, 2018. 257–283.

Hess, Sabine, and Bernd Kasparek. “De- and Restabilising Schengen: The European Border Regime after the Summer of Migration.” *Cuadernos Europeos de Deusto* 56 (2017) (special issue *Governing Mobility in Europe: Interdisciplinary Perspectives*): 47–78. <http://ced.revistas.deusto.es/issue/view/196> [accessed: 16 March 2018].

Huysmans, Jef. “The European Union and the Securitization of Migration.” *Journal of Common Market Studies* 38.5 (2000): 751–777.

Karakayali, Serhat, and Vassilis Tsianos. “Movements that Matter. Eine Einleitung.” *Turbulente Ränder*. Ed. Transit Migration Forschungsgruppe. Bielefeld: transcript, 2007. 7–22.

Kasparek, Bernd. “Complementing Schengen: The Dublin System and the European Border and Migration Regime.” *Migration Policy and Practice*. Eds. Harald Bauder and Christian Matheis. London: Palgrave Macmillan, 2016. 59–78.

Kearney, Michael. “Borders and Boundaries of State and Self at the End of Empire.” *Journal of Historical Sociology* 4.1 (1991): 52–74.

Koslowski, Rey. *Smart Borders, Virtual Borders or No Borders: Homeland Security Choices for the United States and Canada*. 2005. <http://www.albany.edu/~rk289758/documents/Koslowski_Smart_Borders_SMU_law_Review05.pdf> [accessed: 16 March 2018].

Kuster, Brigitta, and Vassilis Tsianos. “Zur Digitalisierung der europäischen Grenze. Der Fall Eurodac.” *Cilip. Bürgerrechte und Polizei* 105 (2014): 61–68.

Lahav, Gallya, and Virginie Guiraudon. “Comparative Perspectives on Border Control: Away from the Border and Outside the State.” *The Wall around the West: State Borders and Immigration Controls in North America and Europe*. Eds. Peter Andreas and Timothy Snyder. Lanham, MD/New York/Oxford: Rowman & Littlefield, 2000. 55–77.

Lavenex, Sandra. “EU External Governance in ‘Wider Europe.’” *Journal of European Public Policy* 11.4 (2004): 680–700.

Mezzadra, Sandro, and Brett Neilson. *Border as Method, or, the Multiplication of Labor*. Durham, NC/London: Duke University Press, 2013.

Morris, Lydia. *Managing Migration: Civic Stratification and Migrant Rights*. London/New York: Routledge, 2002.

Moulier Boutang, Yann. “Europa, Autonomie der Migration, Biopolitik.” *Empire und die biopolitische Wende. Die internationale Diskussion im Anschluss an Hardt und Negri*. Eds. Marianne Pieper, Thomas Atzert, Serhat Karakayali, and Vassilis Tsianos, Frankfurt a.M.: Campus, 2006. 169–180.

Newman, David. “The Lines that Continue to Separate Us: Borders in a Borderless World.” *Progress in Human Geography* 30.2 (2006): 1–19.

Ong, Aihwa. *Flexible Citizenship: The Cultural Logics of Transnationality*. Durham, NC/London: Duke University Press, 1999.

Papadopoulos, Dimitris, Niamh Stephenson, and Vassilis Tsianos. *Escape Routes: Control and Subversion in the Twenty-First Century*. London: Pluto, 2008.

Parker, Noel, and Nick Vaughan-Williams. “Lines in the Sand? Towards an Agenda for Critical Border Studies.” *Geopolitics* 14.3 (2009): 582–587.

Robertson, Roland. “Glokalisierung. Homogenität und Heterogenität in Raum und Zeit.” *Perspektiven der Weltgesellschaft*. Ed. Ulrich Beck. Frankfurt a.M.: Suhrkamp, 1998. 192–220.

Rumford, Chris. "Theorizing Borders." *European Journal of Social Theory* 9.2 (2006): 155–169.
Rumford, Chris. "Introduction: Citizens and Borderwork in Europe." *Space and Polity* 12.1 (2008): 1–12.
Salter, Mark B. "Places Everyone! Studying the Performativity of the Border." *Political Geography* 30 (2011): 66–67.
Sassen, Saskia. *Global Networks, Linked Cities*. New York/London: Routledge, 2002.
Schuster, Liza. "Dublin II and Eurodac: Examining the (Un)intendend(?) Consequences." *Gender, Place & Culture: A Journal of Feminist Geography* 18.3 (2011): 401–416.
Sciortino, Guiseppe. "Between Phantoms and Necessary Evils: Some Critical Points in the Study of Irregular Migrations to Western Europe." *IMIS-Beiträge* 24 (2004) (special issue *Migration and the Regulation of Social Integration*. Eds. Anita Böcker, Betty de Hart, and Ines Michalowski): 17–43.
Smith, Michael Peter, and Luis Eduardo Guarnizo. *Transnationalism from Below*. New Brunswick/London: Transaction Publishers, 1998.
Transit Migration Forschungsgruppe. *Turbulente Ränder. Neue Perspektiven auf Migration an den Grenzen Europas*. Bielefeld: transcript, 2007.
Tsianos, Vassilis, and Sabine Hess. "Ethnographische Grenzregimeanalyse." *Grenzregime: Diskurse, Praktiken, Institutionen in Europa*. Eds. Sabine Hess and Bernd Kasparek. Berlin/Hamburg: Assoziation A, 2010. 243–264.
Urry, John. *Sociology Beyond Societies: Mobilities for the Twenty-First Century*. New York: Routledge, 2000.
Van Houtum, Henk, and Ton van Naerssen. "Bordering, Ordering and Othering." *Tijdschrift voor economische en sociale geografie* 93.2 (2002): 125–136.
Walters, William. "Mapping Schengenland: Denaturalizing the Border." *Environment and Planning D: Society and Space* 20.5 (2002): 561–580.
Walters, William. "Border/Control." *European Journal of Social Theory* 9.2 (2006): 187–203.
Walters, William. "Foucault and Frontiers: Notes on the Birth of the Humanitarian Border." *Governmentality: Current Issues and Future Challenges*. Eds. Ulrich Krasmann and Thomas Lemke. New York: Routledge, 2011. 138–164.
Walters, William, and Jens Henrik Haahr. *Governing Europe: Discourse, Governmentality and European Integration*. New York: Routledge, 2004.
Zolberg, Aristide R. "Managing a World on the Move." *Population and Development Review* 32.1 (2006): 222–253.

Charlton Payne

Displaced Papers

Keeping Records of Persons on the Move

How do we document the identity of persons whose lives do not follow readily intelligible paths? Our world is populated by many persons who have been displaced by (civil) war, religious conflict, extreme weather events or environmental conditions, or, despite an unwillingness to acknowledge it in international law, by economic necessity. We can describe this vast and diffuse multitude of persons, whose lives have been shaped by the many possible forms of displacement we have witnessed in the last century, as displaced persons. The term, which entered official as well as common parlance as the acronym 'DP,' was coined during WWII by the US military as it prepared for the challenges of a rearranged postwar Europe. At that time, the term 'DPs' referred primarily to slave laborers, prisoners of war, and liberated concentration camp inmates. Included among these persons were Jewish and non-Jewish asylum seekers who did not wish to be repatriated after the war. In the immediate postwar years, then, the acronym became shorthand for certain victims of Hitler and Stalin (Cohen 2012, 4–7). With this historical coinage in mind, I wish to reclaim the greater potential resonance of the term 'displaced persons' to describe the countless persons affected in so many drastic ways over the last century by widespread, and even systemic, forms of displacement.

One of the major effects of these displacements is to unsettle seemingly stable categories and previously recognizable pathways of identity. For one, the legal status of many displaced persons is uncertain. Many are unwilling or unable to lay claim to the protections of a native country; many have fled, have decided to or are forced to travel 'irregularly' – often placing their fates in the hands of human traffickers – and thus they enter into domains of illegality and lawlessness, not to mention the very threshold of survival. Some displaced persons are even unable to cross national borders, or once they have, they find themselves sequestered in camps and detainment centers for an indefinite period of time and without recourse to established legal protections. Families are torn apart, new affiliations are formed. Such stories of displacement are becoming increasingly familiar to us through media images, news reports, political debates, personal testimonies, as well as an array of depictions in film and literature.

On the one hand, our familiarity with displaced persons in their different manifestations – as refugees, stateless persons, expellees, irregular migrants – is indeed growing, whether due to the sheer numbers, a heightened attention gen-

https://doi.org/9783110600483-007

erated and guided by media, or the sobering awareness of the inability of nations and the international community to imagine political solutions to this unrelenting fact of a globalized world. But on the other hand, our knowledge of displaced persons arrives in the form of an interruption of social perception: that is to say, only to the extent that they do not easily fit within the standardized and officially codified categories used to establish what counts as an intelligible social and political identity, namely categories of age, ethnicity, nationality citizenship, marital status, and profession. Our perceptions of displaced persons are, in other words, shaped by the ways in which their identities elude the patterns by which these guiding ciphers of identity are plotted as pathways for the lives of persons.

In what follows, I wish to explore this fundamental predicament of the social legibility of displaced persons by focusing on the underlying medial presuppositions and the reasons for their breakdown. I trace this predicament back to the effort to document the identities of persons. In its modern forms, governance has been constituted by techniques developed to record the existence and affiliations of persons. These techniques of registration and tracking have given rise to one of the most formative institutions of modernity: the identity document. The passport, in particular, encapsulates in object-form the creation and organization of a system of reference upon which modern institutions rely in order to vouch for the identities of persons. The systematization of those techniques goes hand in hand with the rise of the modern nation-state, the emergence of a biopolitically-driven administration of society, and the pursuit of global commerce in labor and goods. Accordingly, the migration of persons is always informed in some way or another by systems of documentation. At the same time, however, migration presents challenges to the classifications used to identify persons as both citizens of a particular nation-state and beings on the move. The history and theory of migration is thus inseparable from practices of documentation.

My discussion begins with a concept that I argue is crucial for the critical study of migration: documentality. Drawing on the recent work of the Italian philosopher Maurizio Ferraris, from whom I borrow the term documentality, I take up the idea that the ability to leave and record traces supplies the performative basis of social institutions. I then point to limitations in Ferraris's model as it applies to displaced persons, for whom a document of identity may or may not be on record, and recast the question of traces with a view to their political implications. How do the techniques of documentation make certain traces legible and what gets overlooked in the process? I turn to writings by Kurt Tucholsky and Joseph Roth for critical historical perspectives on these questions. These two German-language authors observed the formation of a bureaucratic apparatus for registering identities in the aftermath of the First World War. Linking the identity document to reductive identity ascriptions and a militarization of society, they call attention to the mechanisms of exclusion, structures of complicity, and inherent instabilities this system of reference engenders.

1 Documentality

The issue of social legibility brings us to the theory of documentality. I use the term here to designate the very effort to leave, secure, and follow traces as an epistemic practice of establishing reference. According to the philosopher Maurizio Ferraris, one of the few thinkers to attempt a systematic philosophy of the document, the documental function is a fundamental condition of social reality. On the one hand, the registration of traces allows us to give permanence to thoughts and memories which are otherwise ephemeral, unpredictable, and unreliable. On the other hand, inscriptions also enable us to endow those traces with social significance. The legibility of the trace is what makes individual thoughts and actions accessible to other persons and opens up the potential for them to enter a public domain. Traces understood as inscriptions, in this sense, supply the material conditions for creating social objects – and hence documents – since they are "open to access by more than one person" and can be "codified in objects that are essentially designed to be exhibited" (Ferraris 2013, 178).

Because such inscriptions enable us to fix performatives, they also supply the basis for building social institutions and conferring institutional value upon certain acts and objects.

> The fact of being written, and hence shared or shareable allows the registration to fix values, to integrate different values in a single system, to mobilize resources, to put people into relations with each other, to protect transactions as well as to certify identity, and to confer or safeguard determinate statuses. (Ferraris 2013, 236)

Ferraris's theory helps us to think the transitions from mental and material traces, via inscriptions and registrations, to social objects, documents, and institutions. Documentality is his term for the social activity of record-keeping as it emerges from inscriptions. Within this framework, documents themselves are inscriptions endowed with an institutional value and hence the trace of social act. A social collectivity, a physical medium, and usually an idiom, such as a signature – defined by Ferraris as an "iterable singularity" (Ferraris 2013, 254) – are necessary for traces to transform into documents. Once these conditions are met, a document, derived from the Latin *docēre*, meaning to teach or show, is a material object that has been processed and framed as a piece of evidence with the epistemic function of enabling us to know something by showing us something (Gitelman 2014, 1–3).

Yet the problem with Ferraris's account, as I see it, is that it does not account for the validity or the authority of the documents upon which our institutions are built. His effort to unseat the metaphysical commitment to a fiction of collective intentionality by emphasizing the material conditions of social interaction is certainly laudable. But by privileging the materiality of the trace, the argument too readily sidesteps the question of the source of authority that endows certain inscriptions with a documental status not granted to others. The argument is in fact circular. Ferraris wants to position the exchange of documents as the enabling condition of a collective. Yet the existence of such a social collectivity is one of the necessary conditions for traces understood as inscriptions to be granted the status of documents in the first place. The validity of not only what is written in a document but even of the document's legibility as a document in contrast to various inscriptions of whatever kind depends on a system of practices embedded in an institutional framework with the authority to make certain inscriptions normative. In what follows, I will examine how identity documents exemplify this fundamentally instable slippage between the indexicality and validity of the trace within documentality.

Identity documents provide knowledge in the form of a written record about a person's date and place of birth, parentage, citizenship, and so on, that can be shared with others, and they confer, in a performative sense, such social markers on persons by being produced as written records by state authorities in the first place. The existence of official documents, such as a birth certificate, is coeval with the establishment of a person's social and political status (for instance, being born as a citizen of a certain nation). Conversely, in the absence of official documentation, the validity of a person's claims about their social status, even that person's credibility, can be cast into doubt by others. Since social identities understood in this way both generate and require a paper trail, we cannot, in the social realm, maintain a distinction between being born and being made, for either documents or persons.

Let us return briefly to Ferraris in order to tease out some sociopolitical implications of his account of documentality as they apply to the identity document. Ferraris wants to provide an account of the social that replaces the notion of collective intention with a materialist account of the trace. In Ferraris's Baltheory, writing and record-keeping are more fundamental features of social institutions than communication. Registrations are a condition of possibility for communication, but they are not reducible to it. According to him, restaurant receipts and train tickets, to cite some of his examples, involve leaving traces in the form of written records rather than acts of communication. They represent instances of documenting a transaction or storing information to later be transmitted. Such operations are as crucial to society today as ever before. This seems uncontroversial. However, to raise the stakes, I want to take a closer look at one of Ferraris's passing remarks in this discussion of the primacy of registration over communication. Ferraris refers to having little money and "the social exclusion of clandestine immigrants" as "lacks and difficulties [that] derive from a defect in registration and not in communication" (Ferraris 2013, 186–187). Social inclusion, Ferraris suggests, depends on the efficacy of our documental practices. Accordingly, we seem to require a more comprehensive system of registering identities for our institutions to be able to embrace all persons within their fold. But to what extent can we attribute the exclusion of undocumented persons from participation in the social world to a 'defect in registration'? Wherein lies the defect? Is the point that once documented on paper, everyone has the ability to participate in social activities; or is it rather that everyone's activities – especially in their interactions with one another – leave myriad traces that require a more accurate or all-encompassing system of documental registration that can account for the varieties of social activity? To even begin to tackle this conundrum, which in my view Ferraris does not resolve, we need to take a closer look at the points where practices of registration break down and certain traces withhold their documentality.

One place to start is with the passport, perhaps the most significant identity document in the modern age of migration. The passport encapsulates in object-form the creation and organization of a system of reference which modern institutions rely upon in order to establish and vouch for the identities of persons. Though the passport as we know it today has a long history filled with various precursors, with the emergence of modern citizenship as a result of the French Revolution it gained in institutional importance as a tool for implementing uniform regulations on movement and for distinguishing between foreigners and nationals. The passport subsequently transformed into the unique document of an international regime with standardized formats and rules in the aftermath of the First World War. As an assemblage of a picture, stamps, watermarks, seals, and a signature, the passport combines iconic, indexical, and symbolic signifiers in one portable documental object. The efficacy of this collage of signifiers depends, of course, upon the existence of state-sponsored bureaucratic apparatuses and established procedures for guaranteeing the passport's authority and authenticity. Further presupposed by this regime of signification are a standardized orthography and the creation of legal criteria for proper names (Caplan 2001, 54–55). States harnessed these signifying practices for the administration of resources human, natural, and monetary along national lines. The distinguishability of individual persons therefore went along with the biopolitical desire to implement a distinction between foreigners and nationals. Documents enabled governments to create knowledge about the identity of persons and, in the condensed form of the passport, made this knowledge portable and showable in the hands of its citizens as they moved within territories and crossed national borders (Torpey 2003, 83). Whether in its early forms of paper knowledge or in its increasingly prevalent mediations as digital data, exemplified today by the EU's fingerprint database Eurodac,[1] the legibility of distinct bodies of persons and citizens has been a function of documentality.

2 Displaced Persons

For many displaced persons, there is no legible paper trail documenting their identities and movements.[2] The existence of refugees, for instance, reveals the cracks in the system of documentality; they embody moments in which social relations and institutions break down, identities can no longer be certified, and legal statuses are no longer safeguarded or become indeterminate. This is why the invention of the modern passport system coincided with its breakdown. The passport system aims not only to identify and manage persons on the move but also to identify the specific nation-state ultimately responsible for persons defined as citizens. Shortly after its inception, however, the limitations of the passport system were exposed when individuals and populations appeared who were without a verifiable nationality or citizenship, were on the move without a nation-state that was willing to claim responsibility for them, or were unwilling or unable to return to the country of their former citizenship due to fear of persecution. The massive refugee crisis of the 1920s and 1930s brought to the forefront the paradoxes of the passport system at the same time that it contributed to its standardization and internationalization. Nation-states still struggle to administer and manage such persons on the move. Their authorized agencies seek to establish, track, and document the identities of displaced persons for purposes of travel, immigration, asylum, or humanitarian aid. Yet at the same time, it is this very effort to classify persons according to categories of ethnicity, nationality, marital status, and age that can be used to discriminate against them, drive them out, or threaten their lives in the first place.

1 Launched online in January 2003, Eurodac stores data about persons and fingerprints organized into three categories: for individuals over the age of 14 who apply for asylum in an EU member state for use by the Dublin system of regulation; for irregular migrants arrested crossing the border illegally; for irregular migrants arrested within a member state (Broeders 2011, 52–55).
2 To this extent, their status overlaps with that of many undocumented migrants.

3 Displaced Papers

The refugee, displaced or stateless person embodies, as it were, a reminder of the gaps within the documental regimes of nation-states. For one, they remind us that personal identity only tenuously rests upon a presumed identity between nativity and nationality; in other words, they challenge one of the referential bedrocks of modern politics, namely, that a human being is born as a citizen of rights into a nation-state. For Giorgio Agamben, they therefore expose the biopolitical fiction underpinning the logic of sovereignty at the same time that they point to how the human being can only exist as an exception within a political order divided up among sovereign nation-states (Agamben 2000, 20–21). It is not my aim here to re-launch a complex discussion of the place of human rights within the discourse of law and sovereignty. Rather, I wish to highlight the precarious documentality of persons revealed by the phenomenon of displacement. I want to emphasize the constitutive role of documental practice within the administrations of nation-states, in the hope that by calling attention to the heterogeneity of officially ratified signifiers of identity and the traces of persons on the move, we can get a better glimpse of both the shortcomings and political stakes of our techniques for establishing reference. In short, I propose that we regard the trace as a highly politicized phenomenon at the intersection of migration and modern bureaucracy.

The desire to create paper trails vouching for the identities of persons and to make them evident and portable in passports has generated a parallel set of phenomena which challenge and potentially undermine the sources of referential authority within documentality. Identity documents only wield signifying authority if a regime of bureaucratic reference is already in place. This presupposes an established state with considerable resources and an administrative apparatus vested with the authority to wield state power. Yet there is more to the story. The rise of a documental regime making officially authorized identities legible has been accompanied by a history of counterfeiting, forgery, and impersonation, not to mention exclusion, exploitation, and illegibility. The various efforts to establish the likeness of a person and to classify sameness, as the historian Valentin Groebner argues, reveals the "history of identification as fiction, deception, pretense and ambivalence" (Groebner 2001, 27). The documental regime not only enables nation-states to keep records about citizens and track the movements of migrants both 'regular' and 'irregular,' but it also generates clandestine persons and uncanny double identities. In addition to recording, transmitting, and tracking the traces of persons on the move, the documental regime inadvertently bears traces, as it were, of its own instability.

4 Bureaucracy, Credibility, and Compliance: Kurt Tucholsky's Critique

The traces of that uncertainty and instability – manifest in instances of deception and ambivalence – derive from the fact that signifiers of identity are themselves conventional and hence rely upon leaps of faith. Their credibility as evidence of personal identity rests upon our general acceptance of their signifying efficacy. Indeed, it is because nationality is an ascribed status that cannot be determined on the basis of appearance and language alone that state administrations have developed systems of documentation to supply evidence of national identity (Torpey 2001, 269). Citizenship and nationality are social conventions whose signifiers are accordingly unmotivated. This is why they have to be coupled with those signifiers of personal identity regarded as motivated – pictures, signatures, fingerprints – in order to index a person with an ascribed nationality. Proper names exemplify the intersection of these semiotic registers: are we attached to our names or are they attached to us? We also know that appearances change, pictures can deceive, and signatures can be forged. The interplay of motivated and unmotivated signifiers can therefore both bolster belief in the evidentiary quality of identity documents and foster disbelief in their authenticity. The efficacy of identity documents thus hinges to a great extent upon our acceptance of their ability to show us knowledge about persons.[3]

In this regard, the signifying power of identity documents is not only a function of a documentality that supplies the medial conditions of our social institutions, but also of what the anthropologist David Graeber calls the "total bureaucratization" of social life. Traces are not merely recorded on various surfaces or registered on papers recognized as documents without a framework that confers institutional significance upon them. We also have to be willing to accept their inscriptions as valid. According to Graeber, bureaucratic systems "always create a culture of complicity," in which we act as if we believe in their abstractions (Graeber 2015, 26–27). Our personal investment in identity documents, our acceptance of their ideological and semiotic presuppositions, is yet another example of how the logic of bureaucracy infiltrates our lives.

Graeber is not the first to call attention to this 'as-if' structure of belief within bureaucracy. We find it articulated in the immediate aftermath of the First World

3 In their portable and material immediacy, identity documents thus aim to be more than encoded and stored information. They have already processed such information into a structure of intelligibility, that is, into knowledge. For a useful discussion of the difference between information and knowledge, as it pertains to documents, see Guillory 2004.

War, when the period of xenophobic border controls and mass displacement gave rise to the standardized international passport system and modern refugee regime, leading writers to reflect intensely on the facets and implications of the new prevalence of identity documents. For example, the German writer Kurt Tucholsky (1890–1935), one of the leading critical journalists of the Weimar Republic, penned a satirical piece on the identity document for the left-liberal daily *Berliner Volkszeitung* in June of 1920. His short piece, laconically titled "Identification" ("Der Ausweis"), describes the humiliating and, as it were, grotesque quality of the attempt to acquire an identity document – "sometimes called a passport or a registration card or an entry card or personal documents" – at any of the bureaus issuing such papers (Tucholsky 2017a, 1). In Tucholsky's account, the unwitting traveller, seeking for instance to drive to Upper Silesia with an aunt, can expect to encounter an indifferent man-child at the office as well as a miserable back and forth of paperwork before being able to obtain the necessary identity document.

But in addition to remarking on its ubiquity, Tucholsky diagnoses the causes of this "identification epidemic" which had broken out in Germany around the time of the First World War. He attributes this social malady to a widespread bureaucratic mindset perpetuated by structures of complicity and enforced by the militarization of society. Writing at a time when these administrative measures had only recently been adopted as a feature of civilian life in Germany, Tucholsky characterizes this bureaucratic mentality in terms that have long since become familiar to many of us well beyond the confines of the stereotypical Prussian discipline he invokes: "the Prussian mind is so construed that it simply cannot proceed without lists, registry apparatuses, and – identification" (Tucholsky 2017a, 2). What Tucholsky imputed to a peculiarity of the Prussian administrative culture of Germany at the time, applies to a pervasive bureaucratic mentality upon which the documental system of governance rests in general. The same can be said for the ways in which the administrative apparatus of identity maintains its existence by establishing "its own immoderate importance" (Tucholsky 2017a, 2). Again, Tucholsky's remarks about police registration in the crisis-ridden Weimar Republic ring just as true today for the activities of such agencies elsewhere:

> We recognize identification in a thousand different forms. In the hundred unnecessary registration and deregistration certificates, in that restlessly absurd police registration system which only blooms in this blessed country (and which inhibits not a single criminal, yet burdens thousands of respectable citizens) – we find it in the senseless communications and reports which no one reads and which only become important when they go unfinished – we find it in the lists and minutes which everyone finds necessary to complete – in the ragged forms and permission slips which must be produced, filled out, signed, stamped, and submitted. And meanwhile the country drowns in disarray. (Tucholsky 2017a, 3)

However, it is Tucholsky's emphasis on the twin structures of complicity and militarization that resonate with Graeber's more recent analysis of the total bureaucratization of society. Having served in WWI, Tucholsky was well aware of the passport's function as a military regulation. However, what was introduced for a society at war, in which neither soldiers nor civilians could move freely without first applying for and receiving the official paperwork, had after the war become the standard model of civilian life:

> This identification affair is part of the wartime and controlled market system, and the funniest (and saddest) aspect of this is that 'bureaus' know precisely that they will always lose their paper war against life, the practical challenges of the day will not be just, and that in the end all these racketeers will slip right through the cracks. (Tucholsky 2017a, 2)

Tucholsky's reference to the military context of these regulations underscores how the registering and policing of identities is rooted in the monopolization of violence by the state. Justified in the name of securing the nation's borders in order to protect the nation from foreign enemies, this bureaucratic practice rests on the threat of violence against anyone who does not comply with its rules. At the same time, his example of racketeers slipping through the cracks acknowledges that bureaucracy's standards do not apply equally to everyone. More by virtue of than despite its inefficiency, then, the administration of identity documents is another instance of "the link between coercion and absurdity" upon which bureaucracy feeds (Graeber 2015, 66). In Graeber's view, bureaucracy requires onerous expenditures of interpretive labor on the part of those trying to navigate bureaucracy's absurdities and thereby taxes their creative imagination. Coupled with the threat of meaningless police violence impervious to any alternative arguments or perspectives, bureaucracies "are ways of organizing stupidity – of managing relationships that are already characterized by extremely unequal structures of imagination, which exist because of the existence of structural violence" (Graeber 2015, 81).

Tucholsky, as these passages show, anticipates Graeber's analysis of bureaucratic compliance. Toward the end of his piece on the identity document, Tucholsky even alludes to how the, as I call it, 'as-if' structure of belief in documental identification encourages otherwise intelligent people to conform to its rules and, as a result, the perpetuation of the bureaucratic administration of identity documents: "All who read this will quite possibly nod their heads, smile, and say 'Yes: He's right!' But does this rid us of even one single identification? Everyone considers all these identifications unnecessary – every identification but one's own" (Tucholsky 2017a, 3). Tucholsky challenges his readers to examine their own complicity with the structural inequalities (and violence) instituted and perpetuated by the administration of identity documents. In another short satire published with the title "The Border" ("Die Grenze") in the same newspaper only a few days after "Identification," Tucholsky caricatures the xenophobia entailed in the arbitrary drawing and policing of borders by nations and empires, where hospitality and neighborliness have been replaced by violent division and discrimination.

> Bah, foreigner! You are the most miserable being under the European sun. Foreigner –! The ancient Greeks called foreigners "barbarians" – yet treated them hospitably nonetheless. You, however, are hunted from place to place, you – foreigner of our time – you will receive no right of entry here, there no residency permit; you are forbidden to eat bacon in one place, nor may you bring it with you on your way from one spot to the next – foreigner! (2017b, 1)

Here Tucholsky connects the deadening militarist administration of identification documents he described in the first piece to the logic of border control with its jingoistic divisions of persons into residents and foreigners. Furthermore, he links the xenophobic magnification of conflict to the massive displacements and regnant tribalism unleashed by the First World War:

> After this war, after such displacements – compared to which the tiny day trips of the Great Migrations were naught but child's play – after the bloody marches of peoples across half of Europe, the small disputes around the church steeples of every parish have attained hellish significance. Lineage more ancient than the Greiz-Schleiz-Reuss and the People's State of Bavaria and an autonomous Upper Silesia and France and Congress Poland – it's always the same. (2017b, 2)

But in both pieces he also exhorts his readers not to blindly accept these bureaucratic fictions just because it is convenient for those of us who hold an identity document to do so. For the extent to which the absurd administration of identification could translate into the sociopolitical exclusion of persons was already apparent in Tucholsky's time and it continues to haunt political life to this day.

5 Repetitions and Disappearances of the Person: Joseph Roth's Critique

Contemporaries of Tucholsky, as well as many writers since, have explored the asymmetrical relationships, subaltern identities, and uncanny doublings brought about by the reliance on identity documents to govern the registration and movement of persons. I want to now consider how an unfinished novel by one of Tucholsky's contemporaries, the Austrian-Galician writer Joseph Roth, represents these uncontrollable byproducts of identity documentation. Roth is one among a number of other authors writing in the 1920s–1940s who take up both the dehumanizing absurdities and the potential for forged identities produced by documental identification. These themes come to light in the fictional stories about stateless protagonists from this period by such authors as B. Traven, Erich Maria Remarque, and Anna Seghers (see, respectively, Gulddal 2013; Payne 2013; Hofmann 2017). Written between 1927 and 1929, Roth's abandoned novel *The Silent Prophet* (*Der stumme Prophet*) was rediscovered and pieced together in the archive and published in 1966. At first glance, it gives a strange and morose account of a revolutionary modeled on Trotsky, presumably informed by Roth's observations as a reporter travelling through the Soviet Union in 1926. For the issues I am raising here, however, the novel is important for the ways in which it links the protagonist's motivation, disillusionment, and banishment to the phenomena of statelessness, displacement, and marginalization. Moreover, Roth's novel sets this tale of a revolutionary outlaw among the movements of displaced and clandestine persons as well as profiteering smugglers and forgers, as these figures travel across the borders of the collapsing Austro-Hungarian and Russian Empires.

As much more than a loosely biographical examination of a revolutionary whose ideals were betrayed by party functionaries, then, *The Silent Prophet* explores what types of traces persons might leave behind in a society obsessed with administering identity. The novel directs the reader's imagination towards the marginalized and excluded persons inhabiting the European landscape in the time around WWI and the Russian Revolution. The protagonist, Friedrich Kargan, was born in Odessa as the illegitimate son of an Austrian piano teacher and the daughter of a wealthy Russian tea merchant. Rejecting the piano teacher as an unsuitable husband for his daughter, the tea merchant sends her and the child to live with her wealthy businessman uncle in Trieste. After Friedrich's mother dies at an early age, he moves into the servants' quarters at the uncle's house and spends his childhood separated from the rest of the family. Denied the opportunity to study at university, Friedrich enters instead the world of bureaucracy as an apprentice in a shipping bureau. He then transfers, in 1908, to a border affiliate of the shipping agency, where he comes into contact with migrants and smugglers of varying shades. After finding a way to study in Vienna, he meets his love interest and joins the revolutionary cohort. We follow him as he travels around clandestinely on behalf of the revolution, is disappointed by the woman he loves, and is arrested and banished to Siberia, first by imperial authorities and later by his former revolutionary party after they have taken over the government.

Through the narration of Friedrich's biography, Roth's novel provides scenarios in which the bureaucratic administration of identity documents produces a doubling of identities. Examples of such ambivalent identities in the novel include displaced and migrant persons either without a recorded identity or seeking a new one, the creation of fraudulent identities, and the bifurcation of an individual identity into official and unofficial, public and private, selves. Dissimulation is represented as a lucrative business and a defining feature of Friedrich's character made possible by documentality. Without any familial or national affiliations to tie him down, Friedrich decides to transfer to the company's shipping office on the eastern border of the Austrian monarchy as a chance to alter the course of his life. In this unnamed border town he meets the Parthagener family who runs the affiliate of the shipping agency as well as an inn called "The Ball and Chain" ("Zur Kugel am Bein"). Here he comes into contact with "deserters, emigrants and refugees from the pogroms" making their way across the Austrian border from the Russian empire (Roth 1979, 13). Within this constellation, he becomes acquainted with the lucrative trade of human trafficking in an era of mass migration. This form of entrepreneurship operates not only at the borders of nation-states but also at the crossroads of legality and illegality. Such border agencies, the narrator matter-of-factly informs us, "enjoyed the goodwill of the authorities":

> It was the government's unconcealed desire to remove these poor, unemployed and not altogether innocuous refugees from Austria as quickly as possible; but also to convey the impression that Russian deserters would be supplied with sailing tickets and recommendations to countries overseas – to such an extent that the desire to quit the army should affect an increasing number of Russian malcontents. The authorities were probably tipped off not to keep too close an eye on the shipping agents. (Roth 1979, 14)

In this account, it was the unofficial job of the shipping agencies to intercept these migrants and redirect them to North and South America (Roth 1979, 14). Since European nation-states had already begun to control the flow of Eastern European migrants across their borders in the decades leading up to the First World War, these migrants, refugees, and fugitives required the necessary papers to travel (Gerhard 2006, 21–22). However, because of their dubious status vis-à-vis state authorities, many lacked such official documentation. Therefore, "for those who could pay, papers were manufactured at the frontier" (Roth 1979, 16). A subterranean economy of false documents was born.

By naming the inn owned by the Parthageners "The Ball and Chain," Roth aligns this scene of border-crossings and forged documents with his critique of the modern identity document as a stultifying practice of identification. On 28 September 1919, Roth published an article with the title "The Ball and Chain" ("Die Kugel am Bein") in a short-lived daily newspaper in Vienna called *Der Neue*

Tag. In this article, in which he formulates perhaps his most explicit critique of the dehumanizing effects of the new passport requirements, the ball and chain serves as a metaphor for the passport. Through its use of classifications to pattern fixed knowledge about a person's origin and physiognomy, the passport territorializes individuals and renders bundles of physiological data into representations of abstract citizens, *Staatsbürger*, subtracted of the dynamism proper to a biography and hindered in their ability to move through physical space (Roth 1989a; Gerhard 1998, 210–211; Rahn 2008, 116–122).

In light of Roth's deployment of the ball and chain metaphor to describe the (re-)territorializing functions of the passport, the inn takes on antithetical connotations, which I would argue reflect the ambivalences within documentality. That the inn is named for the ball and chain fastened to a prisoner's leg stands in stark contrast to its function as a non-place of transience and the transnational movement of goods and persons in *The Silent Prophet*. For Friedrich the antithetical inn is both a crossroads and a site of personal repetitions and returns: it is where he "learned how to lie, to forge papers, to exploit the impotence, the stupidity, and even sometimes the brutality of the officials" (Roth 1979, 23); where he first meets Savelli (the revolutionary modeled on Stalin); where he acquires false identity papers that allow him to go study in Vienna; where he is arrested by border police who send him to Siberia; and where he would return time and again during his clandestine travels. There is even something uncanny about these repetitions and returns. Because his "birth had not been registered anywhere," he has the local forger give him the surname and nationality of his Austrian father, Zimmer, with which he travels to Vienna and studies at the university (26). In this way, he illegally assumes the name and nationality that would have been his birthright. Through this iterative repetition with a difference, he joins the ranks of all the other assumed and doubled identities that have emerged from "The Ball and Chain." In light of Roth's critique of state-administered identity papers, we can read the inn's name as suggesting however that even the adoption of false personas is nonetheless a form of subjugation to the dictates of documental identity. The narrator echoes Roth's critique when, after the outbreak of the First World War and the introduction of passport requirements in the novel, he remarks: "People – as one knows – had all become the shadows of their documents" (Roth 1979, 100). Here, the image of the person as a shadow is a metaphor for the reversal of indexicality under conditions of documentality. The document does not consist of assembled traces of a person, it is not the representation of a person, but rather the person, as a shadow, has his or herself become a trace of the document.

We can thus read *The Silent Prophet* as exploring the ambivalent kinds of traces left behind by persons in a society in which efforts by state bureaucracies to keep records of their citizenry and to police national borders give rise to displaced persons, false papers, and doubled identities – in short, shadow existences. At the same time, the novel is concerned with the threat of certain persons disappearing from the record altogether. It thus also presents alternative techniques for preserving and transmitting the traces of displaced persons in opposition to those dictated by state bureaucracies. Given that signifiers of identity only function within an institutional framework that endows them with their signifying power, it is not surprising that Roth's novel, like many other tales of displaced persons, is structured by a frame narrative.[4]

At the outset of the frame narrative, the first-person narrator explains how the embedded narrative reconstruction of Friedrich Kargan's story came about. The framework in question is provided by the official Soviet political apparatus: a regime consisting of party functionaries, secret police, and their numerous reports. The once revolutionary movement has become a modern state bureaucracy. On New Year's Eve of 1926–1927, the narrator tells us, some friends and acquaintances have retreated from this official Soviet world to celebrate privately in a hotel in Moscow. When Friedrich Kargan's name comes up in conversation, the narrator comes to his defense against the charges of anarchism and intellectual individualism uttered by others in the room. The narrator then undertakes, with the help of a secret service agent whose business was to know details about Kargan, what he calls "an attempt at a biography" (Roth 1979, 8). The ensuing narration took three nights, he tells us, and by the end the audience had dwindled to only two listeners. Apparently concerned that Friedrich Kargan's story might not be heard by a wider audience, the narrator decided to transcribe his account. Yet he puts an ironic twist on the familiar convention of the framework narrative to insist on both the fictional status of this biography and a greater truth about the fate of individuals in society:

> Omitted [...], even deliberately suppressed, are certain indications that might lead to Kargan's identification and might further the reader's natural impulse to recognize in the individual portrayed a definite, historically existing personality. Kargan's life story is as little related to actual events as any other. It is not intended to exemplify a political point of view – at most, it demonstrates the old and eternal truth that the individual is always defeated in the end. (Roth 1979, 9)

Roth's narrator self-consciously marks the embedded narrative as fiction. The truth it claims to present us with is a matter of significance rather than refer-

4 Other examples include Erich Maria Remarque's *The Night in Lisbon* (see Payne 2013), Anna Seghers's *Transit*, and more recently Abbas Khidder's *The Village Indian* (see Hofmann 2017).

ence. Its significance consists, to a large extent, in resisting society's obsession with identification. When the narrator asks, "Is Friedrich Kargan destined finally to sink into oblivion?" he really is pondering whether society can accommodate individuals without a citizenship status, or whether persons who are not defined by a nationality, like the stateless Friedrich, "will be engulfed in empty solitude, unnoticed and without a trace, like a falling star in a silent obscure night" (Roth 1979, 9).

By transcribing his biography of Friedrich, however, the narrator does not merely add more paper to what Roth described in another article from October 1919, bluntly titled "Paper," as a "vicious circle" in which persons have been rendered "a scrap of paper, no longer flesh and blood but a legitimation, a passport, a registration form."[5] The narrator does not wish to salvage the traces of Friedrich's origins and movements from oblivion by documenting them on paper as if he were attesting to an unofficial record of this private history to counteract Friedrich's absence from the public record. Rather, by bracketing the question of reference in this ironic framing of the trace, the novel presents us not with traces understood as indices of a really existing person but only with mere traces. These fictional traces mark, if anything, an absence. What emerges out of this constellation is an insistence on the ways in which a trace marks just as much the disappearance of an origin as it does the possibility of repetitions that can give rise to both records and forgeries. It is in this sense, "namely, that there is no presence without trace and no trace without a possible disappearance of the origin of the said trace, thus no trace without a death" (Derrida 2005, 158), that we should interpret the narrator's statement about the demise of the individual. The individual in Roth's sense is only present as a trace – has disappeared in a world populated by identities recorded on paper and the displaced persons to which that documental regime gives rise. This ironic framing of the trace as fiction is then mirrored by the novel's anti-fictional devices: it reproduces letters by Friedrich as evidence about his person. The novel deploys an ironic discourse of the trace in order to critique the state's bureaucratized documentation of traces for the purpose of identification.[6] In Roth's critique, forgery's doublings and the dis-

---

5 "Ewig, ewig ist der Kreislauf des Papiers. Ein papierener *circulus vitiosus,* in dem du dich mitdrehst, ratlos, ohnmächtig, selbst ein Fetzen Papier, nicht Fleisch und Blut mehr, sondern eine Legitimation, ein Paß, ein Meldezettel ..." (Roth 1989b, 160).

6 I want to thank the colleagues from the Department of German and Russian Studies at the University of Missouri, Columbia, for calling my attention to the ironic framing of the trace in literary texts as a critical tool for reflecting on documentation. The point about irony was raised in response to my presentation of some works by Arno Schmidt, but the insights apply to many other examples.

appearance of individuals are the front and back sides of the same government document.

6 Conclusion

In a world that defines persons as national citizens, displaced persons bring to light myriad ways for conceptualizing the traces of persons. The dynamics of the trace sketched above with the help of Tucholsky and Roth have concrete political implications for the theory of documentality upon which state practices of record-keeping are based. They show us that documental reference seeks to suppress but cannot keep up with the heterogeneous traces of persons and unpredictable iterations of signifiers of identity. My discussion of the texts by Tucholsky and Roth sought to situate Ferraris's theory of documentality within historical practices of state bureaucracy and border control. These two authors belong to a transnational cohort of writers who have responded critically to the reductive version of making identity legible that proceeds, as patterned by the passport, from a set of ascribed origins and "marks out the path taken by the individual up to the moment of inspection" with reference to visa stamps and categories of identity such as marriage, occupation, and kinship (Chalk 2014, 28). They expose the ways in which these procedures generate signifying gaps and onerous absurdities as well as political discrimination and social exclusion.

These criticisms still hold today. Similar logics inform the newly developing technologies and purposes of identification. One of the passport system's jobs is to manage mobility across borders, to identify who is within but also to keep unwanted persons from entering a national territory. The distinction between citizens and aliens, foreigners and nationals, has been primary to the self-definition of the modern nation-state and the policing of its borders. But it has become even more vehement with the heightened securitization of the nation-state in the effort to combat terrorism and the desire to curtail mass migration from the global South to the North. Modern migration technologies are now used for 'profiling' persons on the move in accordance with three broad categories: privileged (trusted) travellers, the unwanted, and those who are suspect and hence requiring further scrutiny. The construction of such profiles increasingly relies on biometric signifiers of personal identity and digital storage and communication technologies to assist in the processing of this information as knowledge about persons. One result is a new permutation of the splitting and doubling of identities already criticized by Roth: by transforming bodies into information that can be stored and circulated in order to regulate people's access to territories, these technologies create "data doubles" of persons on the move (Dijstelbloem and Broeders 2015,

28). Such scrutiny of travellers and migrants overlaps with tighter restrictions on the ability of displaced persons to move and find asylum. The technology might be changing, but government bureaucracies and their police forces are still in pursuit of the elusive dream of a seamless paper trail.

References

Agamben, Giorgio. *Means Without Ends*. Trans. Vincenzo Binetti and Cesare Casarino. Minneapolis/London: University of Minnesota Press, 2000.

Broeders, Dennis. "A European 'Border' Surveillance System under Construction." *Migration and the New Technological Borders of Europe*. Eds. Huub Dijstelbloem and Albert Meijer. Basingstoke/New York: Palgrave Macmillan, 2011. 40–67.

Caplan, Jane. "'This or That Particular Person': Protocols of Identification in 19th-Century Europe." *Documenting Individual Identity: The Development of State Practices in the Modern World*. Eds. Jane Caplan and John Torpey. Princeton/Oxford: Princeton University Press, 2001. 49–66.

Chalk, Bridget T. *Modernism and Mobility: The Passport and Cosmopolitan Experience*. New York: Palgrave Macmillan, 2014.

Cohen, Gerard Daniel. *In War's Wake: Europe's Displaced Persons in the Postwar Order*. New York: Oxford University Press, 2012.

Derrida, Jacques. *Paper Machine*. Trans. Rachel Bowlby. Stanford: Stanford University Press, 2005.

Dijstelbloem, Huub, and Dennis Broeders. "Border Surveillance, Mobility Management and the Shaping of Non-Publics in Europe." *European Journal of Social Theory* 18.1 (2015): 21–38.

Ferraris, Maurizio. *Documentality: Why It Is Necessary to Leave Traces*. Trans. Richard Davies. New York: Fordham University Press, 2013.

Gerhard, Ute. *Nomadische Bewegungen und die Symbolik der Krise. Flucht und Wanderung in der Weimarer Republik*. Opladen/Wiesbaden: Westdeutscher Verlag, 1998.

Gerhard, Ute. "Neue Grenzen – andere Erzählungen? Migration und deutschsprachige Literatur zu Beginn des 20. Jahrhunderts." *Literatur und Migration*. Ed. Heinz Ludwig Arnold. München: Ed. Text + Kritik, 2006. 19–29.

Gitelman, Lisa. *Paper Knowledge: Toward a Media History of Documents*. Durham, NC/London: Duke University Press, 2014.

Graeber, David. *The Utopia of Rules: On Technology, Stupidity, and the Secret Joys of Bureaucracy*. Brooklyn/London: Melville House, 2015.

Groebner, Valentin. "Describing the Person, Reading the Signs in Late Medieval and Renaissance Europe: Identity Papers, Vested Figures, and the Limits of Identification, 1400–1600." *Documenting Individual Identity: The Development of State Practices in the Modern World*. Eds. Jane Caplan and John Torpey. Princeton/Oxford: Princeton University Press, 2001. 15–27.

Guillory, John. "The Memo and Modernity." *Critical Inquiry* 31.1 (2004): 108–132.

Gulddal, Jesper. "Passport Plots: B. Traven's *Das Totenschiff* and the Chronotope of Movement Control." *German Life and Letters* 66.3 (2013): 292–307.

Hofmann, Hanna Maria. "Erzählungen der Flucht aus raumtheoretischer Sicht. Abbas Khiders *Der falsche Inder* und Anna Seghers' *Transit*." *Niemandsbuchten und Schutzbefohlene. Flucht-Räume und Flüchtlingsfiguren in der deutschsprachigen Gegenwartsliteratur*. Eds. Thomas Hardtke, Johannes Kleine, and Charlton Payne. Göttingen: V&R unipress, 2017. 97–121.
Payne, Charlton. "Der Pass zwischen Dingwanderung und Identitätsübertragung in Remarques *Die Nacht von Lissabon*." *Exilforschung. Ein internationales Jahrbuch* 31 (2013): 335–346.
Rahn, Thomas. "Aufhalter des Vagabunden. Der Verkehr und die Papiere bei Joseph Roth." *Unterwegs. Zur Poetik des Vagabundentums im 20. Jahrhundert*. Eds. Hans Richard Brittnacher and Magnus Klaue. Cologne/Weimar/Vienna: Böhlau, 2008. 109–125.
Roth, Joseph. *The Silent Prophet*. Trans. David Le Vay. London: Peter Owen, 1979.
Roth, Joseph. "Die Kugel am Bein." *Werke I: Das journalistische Werk 1915–1923*. Ed. Klaus Westermann. Cologne: Kiepenheuer & Witsch, 1989a. 145–148.
Roth, Joseph. "Papiere." *Werke I: Das journalistische Werk 1915–1923*. Ed. Klaus Westermann. Cologne: Kiepenheuer & Witsch, 1989b. 159–160.
Torpey, John. "The Great War and the Birth of the Modern Passport System." *Documenting Individual Identity: The Development of State Practices in the Modern World*. Eds. Jane Caplan and John Torpey. Princeton/Oxford: Princeton University Press, 2001. 256–270.
Torpey, John. "Passports and the Development of Immigration Controls in the North Atlantic World during the Long Nineteenth Century." *Migration Control in the North Atlantic World: The Evolution of State Practices in Europe and the United States from the French Revolution to the Inter-War Period*. Eds. Andreas Fahrmeir, Olivier Faron, and Patrick Weil. New York/Oxford: Berghahn, 2003. 73–91.
Tucholsky, Kurt [Peter Panter]. "Identification." Trans. Jon Cho-Polizzi. *Transit* 11.1 (2017a): 1–3.
Tucholsky, Kurt [Peter Panter]. "The Border." Trans. Jon Cho-Polizzi. *Transit* 11.1 (2017b): 1–2.

Paul Mecheril

Orders of Belonging and Education

Migration Pedagogy as Criticism

1 Introduction

In this paper, migration is conceived not only as a phenomenon involving movements of people across borders, but also as a phenomenon of discourses, and to that extent as a phenomenon of hegemonic power relations.[1] The discursive construction of migration is of particular interest to the field of pedagogy, as discourses can be understood to frame educational processes. Furthermore, education is itself part of these discursive relationships. This paper will explore what this means and what terminology can be used to describe this, paying special attention to critical references to discourses and discursive orders that transform people into subjects. This will be done in four sections. The first section will start with a general outline of the existing understanding of migration, and then focus on two examples that suggest that migration also generates reactions that involve the hegemonic preservation of natio-racial-culturally coded orders of belonging. Migration can be understood as a form of disruption of social orders. Here, education can potentially function both as a site of the reproduction of natio-racial-culturally coded orders of belonging, and as a mechanism to transform such orders. This is explained in the fourth section of the paper, which ends with a brief idea of migration-pedagogical research as criticism (of hegemony). In general, this paper aims to characterize migration pedagogy as a programmatically distinct approach.

2 Movements and Discourses

Movements of people across borders have taken place almost everywhere and in every historical era. Migration is a universal human activity. In this sense, it exhibits a dimension relating to time and space: "Migration means crossing

1 This paper builds upon parts of previously published texts that have been reworked to fit the scope of this publication: Mecheril, Castro Varela, Dirim, Kalpaka, and Melter (2010); Mecheril (2014); Mecheril (2015).

https://doi.org/9783110600483-008

the boundary of a political or administrative unit for a certain minimum period" (Castles 2000, 269). The social significance of the crossing of borders is not simply given, but rather generated in complex processes in which social reality is affirmed, negotiated, and changed. Phenomena of crossing borders have long been and are currently significant drivers of societal change and modernization. The consequences of movements that cross, constitute, and weaken borders can be studied and understood as phenomena in which new knowledge, experiences, languages, and perspectives have been introduced into different social contexts, which have in turn been rearranged, modernized, and renovated.

Even if migration is not an exclusively modern phenomenon, it is nonetheless characterized by specifically modern conditions. At no point in history were so many people worldwide *prepared* to migrate; *compelled* to migrate due to environmental disasters, wars (civil or otherwise), and other threats; and *able* to shift their location of work and daily life across great distances thanks to technological changes that mitigate the limits of space and time: we are living in an age of migration (Castles and Miller 2009). In recent times, cross-border movements of people have gained a particular significance for individuals and societies worldwide. This is connected to at least three main factors:

a) Migration increases with the proliferation of modern thought and vice versa: this characteristically 'modern' idea is increasing in importance due to migration phenomena. Migration can be understood as the attempt in a very basic sense to take charge of one's own life with regard to geographical, ecological, political, and cultural location, and it thus serves as a model of and also for a modern lifestyle – with all of its ambivalences, illusions, and questionable incidental consequences.
b) Migration increases with the consciousness of injustice and vice versa: due primarily to the brutality of modern warfare's weapons technologies, the uneven distribution of poverty and wealth in the world, and varying degrees of ecological change and its associated destruction of natural resources, the intensity of global inequality grows. Given this manifestation of inequality, the total number of people living in this world, and the spread of global knowledge (which increases the representation of the world in people's minds, through information technologies such as television and computers), global maladjustment and inequality have never been so intensive as in the present.
c) Migration increases with the modification of time and space and vice versa: the 'shrinking' of the world with regard to space and time due to technological developments in transport and communications is characteristic of the present, particularly regarding economic resources. This is quite significant for people's understandings and perceptions of themselves and their oppor-

tunities in this world of transformed time and space relations. Furthermore, this facilitates movements across borders, or at least encourages attempts to cross borders.

Migration as the act of crossing borders goes along with both the transformation and the confirmation of existing conditions. As they are crossed, borders (such as those of nation-states) become visible in a particular way: they become weak and modified while at the same time their power and claim to validity is reinforced. Migration leads to the questioning and strengthening of borders and their validity in much the same way. As such, migration must be understood as a source of disquiet and perturbation; it is the object of discourses as well as the subject of conflicts of both political and everyday nature. The term 'discourse' is particularly meaningful here, as migration is not simply the physical movement of bodies. Rather, phenomena arising from the crossing of borders are generated by discourses, understood here as socially constructed systems of knowledge and understanding that label these phenomena: politically, aesthetically, educationally, and in everyday life, for example, as 'forced migration,' as 'mobility,' or as labor 'migration.'

In general, the term 'discourse' refers to the dynamic complex of knowledge of a given topic. Discourses can be conceived as flows of knowledge: there are discourses on irregular migration, discourses on European values, discourses on economic migration and social inequality, discourses regarding the question of which migrants are welcome and which ones are regarded as dangerous. The subject of a discourse first comes into being within and from the discourse itself. Discursive knowledge generates social reality and gives rise to contexts, circumstances, and surroundings that enable or hinder the activity of real people. Discourses create topics and objects, and at the same time they create knowing subjects, people who by virtue of their knowledge and the use of such knowledge become what they are. In this respect discourses are doubly productive.

To the degree that knowledge and power, according to Foucault, are two sides of one coin, discourses are always powerful. They come about in specific power relationships and also produce power relationships. Power relationships are articulated in interactions by individuals as well as institutions that impact the addressed other and her or his capacity of activity in constitutive, restrictive, negating, or encouraging ways. For Foucault, power is a 'total' phenomenon; not only does it appear where repression is visible, but rather it is a constitutive dimension of the social and the symbolic. As Hannelore Bublitz states, "Power does not operate [...] primarily in an oppressive manner, but rather in a generative one. It is not simply that which individuals struggle against, but rather, strictly speaking, what makes them what they are" (Bublitz 2003, 69; my translation).

Power has the effect of creating subjects, turning individuals into subjects. In Stuart Hall's usage, discourse as "one of the 'systems' through which power circulates" (Hall 1996, 204) produces differing opportunities for action. In his reflections on discourses, Hall (1996) writes about "the West and the Rest," referring to the special way in which "the West," "the Rest," and their relationships are represented, and by extension the way in which this knowledge produces a discourse that "constitutes a kind of power, exercised over those who are 'known.'" When that knowledge is exercised in practice, those who are "known" in a particular way will be subject (i.e., subjected) to it. Those who produce the discourse also have the power to *make it true* – i.e., to enforce its validity, its scientific status" (Hall 1996, 205). Discourses about those who count as 'Others' make the 'Others' what they are, and likewise produce categories of 'non-others.' Discourses on migration are definitely not all similar or uniform in meaning; they compete with one another, and this competition can be described as a struggle for symbolic dominance or hegemony. Within these struggles, the question whether social, institutional, or identity-based preservation or transformation should be the political aim is discussed with much controversy. Outcomes of these struggles in turn have innovative and restorative effects with regard to social orders.

This will be briefly illustrated using two examples. Both examples operate on the basis of the concept of 'Othering,' a tool from postcolonial studies. Based upon the psychoanalytical theoretical concepts of Jacques Lacan, the term was coined in the theoretical post-colonial context by Gayatri Chakravorty Spivak, and has been widely adopted since the 1970s, particularly in anthropological studies. Lacan's ideas establish the theoretical framework in which subjectivization and identity formation can be understood not merely in the solipsistic process of the self, but rather as a constant 'mirror dynamic.' According to Spivak, colonized subjects can only be recognized as such through the dominant discursive practices of the colonial *center*, and indeed in a dependent relationship with it (Spivak 1985). Colonizing practices create subjects. Another perspective focuses on those discursive practices that designate some as 'Others,' and in so doing, create a collective identity. This perspective has become widely known and influential largely on account of Edward Said's works on the construction of the 'Orient' as an antagonistic foil to the 'Occident.' Said did not use the term 'Othering' himself. However, in the post-colonial context, his thesis was understood as an analysis of a paradigmatic 'Othering' practice and further developed theoretically. In his work *Orientalism* (Said 2003, 1978), Said analyses the discursive practices that first created 'the Orient' and 'the Orientals' and places them in a constitutive relationship with the self-perception of the 'West.' The mechanisms and the efficacy of these practices can only be understood, according to Said, in the context of European imperialism and thereby as the legitimation and stabilization practices of claims to dominance in relation to the constructed 'Other.' In this context, 'Othering' can be understood as a double process; the 'Others' are constructed by means of specific knowledge-production practices that legitimate the establishment of colonial dominance, and likewise it is this hegemonic (political, economic, cultural) intention that makes these epistemic practices appear 'plausible' and 'useful.'

2.1 Example 1: Sexual Othering

Racist representations and speech have become socially acceptable in 21st-century post-National Socialist Germany.[2] One example may be found in *Focus* magazine's thoroughly biased and highly sexualized depiction of the widespread sexual assaults that took place in Cologne on New Year's Eve in 2016. On the title page of the January 8, 2016 issue, we encounter the naked body of a young, white, blonde woman: a red beam runs diagonally across her body; her breasts are covered by her right arm, and her left hand coyly shields her pubic area from view. Her eyes are cropped out of the frame but her mouth is visible; her lips are slightly parted. Five black paw-like imprints of male hands mark her body: appearing at once both oily and dirty, they seem to declare her body a male possession. The headline asks: "After the sex-attacks of migrants: Are we still tolerant or already blind?"

Focus magazine's portrayal of this issue is racist because of the lurid, obtrusive, and emotionally manipulative way in which, with the help of such sexualized representations, migrants are demonized and a white 'us' ("Are we still tolerant or already blind?") is constructed. The title page plays the black versus white game: the 'Others' are black, violent, faceless, brazen, dangerous, and dirty. 'We,' on the other hand, are white, pure, vulnerable, civilized, chaste, and exalted, even sublime, superior. The 'us' that asks itself if it is tolerant or already blind – the 'us' that *Focus* is addressing – consists of white women who are groped by black migrant hands, and the white men tasked with protecting these women. The protection of 'our' women from the sexuality of the other race has always been a constituent element of simultaneously racist and racist-patriarchal traditions. The fact that such blatantly racist media representations can and have become socially acceptable points to an entrenched historical amnesia. The powerfully subjectifying and disciplining potential of the discursive interplay of sexuality, race, nationality, and religion in producing, reproducing, and inscribing

2 See Mecheril and van der Haagen-Wulff 2016.

racist patriarchal white European positions of hegemonic dominance in Germany has a clear history, which says something about the function of these practices.

During the French occupation of the Rhineland following World War I, 30,000–40,000 of the roughly 85,000 French soldiers present were from French colonies of North and West Africa (Algeria, Madagascar, Morocco, Senegal, Tunisia, and several other countries). Here, much like in Cologne in 2016, a single event tipped the tide of social tolerance, when a French Moroccan soldier fired his machine gun into a group of civilians, and in the process killed several German nationals. The repercussions were an enduring, statewide coordinated protest campaign that had widespread international support against the presence of 'colored' (*farbige*) soldiers in the Rhineland. In the national parliament, all parties except the USPD (the Independent Social Democratic Party of Germany) and the KPD (Communist Party of Germany) declared the presence of colored soldiers as an ineradicable humiliation (Wigger 2007). Many politicians, including Friedrich Ebert and Adolf Köster, tried to rally support from the "white world" in the fight to eliminate this "black humiliation" from the Rhineland (Wigger 2007, 11). Their justification was that "the use of colored troops of the most inferior culture to guard a population of high mental capabilities and economic significance such the Rhinelanders" would seriously undermine and harm "the laws of European civilization" (Wigger 2007, 12). In the German public propaganda, it was the assumed inferior cultural position of the black French soldiers that was used as an official justification for public outrage. It was in the informal propaganda representations, however – in popular magazines, newspapers, flyers, posters, films, postcards, postal stamps, and even elaborately hand-crafted and mass produced copper coins – that derogatory stereotypes and portrayals of these men as sex-offending, sexually libidinous and uncontrolled, race-defiling rapists of white German women were sold and consumed en masse. In these representations, colonized French soldiers are violently stripped of all humanity, agency, and human dignity in a colonially inspired project of justifying and maintaining white European supremacy. When we look at these historic images side by side with the media representations of the recent Cologne New Year's Eve events, the similarity of patterns, justification mechanisms, symbolic orders, and representations, and their hidden agendas in securing and maintaining positions of power are impossible not to recognize. The fact that this history is ignored in association with the recent Cologne events is evidence of an institutionalized, pathological repression: a kind of historical amnesia (see Stoler 2011).

Against this background it becomes clear that the current threat scenarios are linked to historical forerunners that continue to be productive as they are modified and re-inscribed in the present. The racist speech and affect among the German public today can be understood as practices of Othering in the form of threat scenarios. The affect that is currently intensely articulated against the discursively constructed others, and the intensity with which an imagined group can be condemned, can be explained if we are clear that we are dealing with a struggle for societal order in which privileges are distributed differentially.

2.2 Example 2: Religious Othering

In her analysis of current hegemonic discourse concerning 'Islam' and 'the Muslims,' Iman Attia (2009) demonstrates that generalizing speech about 'the Muslims' is something different than the effect of 'anti-Muslim stereotypes.'[3] The interconnections between discourses on Muslims that have been passed down in cultural and academic spheres (Orientalist, Islamic, and religious studies) and powerful self-perpetuating negative narratives (the 'oppressed Muslim woman,' the 'backward Muslim migrant,' etc.), political dispositions, and everyday discourses point to the efficacy of a particular hegemonic discursive practice that defines a group through essentialization, attribution, and representation. This discursive creation of the 'religious identity' of Muslims is not a random phenomenon. It is constitutively bound to the creation of a non-Muslim 'we' narrative. The fixation of 'religious identity' makes Muslim subjects possible by constructing the self-perception of a complementary 'we.'

Thus, the discursive practice of Othering includes the category of religion; this practice marks and reifies an epistemic, political, anthropological, and thereby quasi-ontological distinction between a more or less explicit 'we-group' and the 'Other.' This takes place in a context that is paradoxically not spatially (geographically, territorially) bound, but rather marked by the de-territorialization of identity-political relations. The defining characteristic of religious Othering is that it does not generally concern sociological issues on religion or the historical development of religious groups, but rather the delineation of 'another religion.' The 'Otherness' of the religion in this case is not defined in terms of categories one would find in religious studies, but rather on a more fundamental level. Thus the 'other religion' in these discourses is always the 'religion of the others,' with the Otherness of the others taking on a 'quasi-religious' quality. Attia (2009) demonstrates that diverse and contradictory tendencies in the discourse on Islam are all built around the essentialization of 'the Muslims,' even when their linguistic, pedagogical, and political intentions diverge. In order to appropriate, understand, discriminate, tolerate, integrate, oppose, or identify

3 See Mecheril and Thomas-Olalde 2011.

'distinctions and differentiation' among them, it is necessary to have understood and formulated their 'essence' (see Attia 2009, 7).

'Othering' represents a practice of objectification by and a means of dominance. Objectification is an action and practice that transforms 'Others' into subjects under the conditions of "positional superiority" (Said 2003, 5). Practices of objectification are forms of symbolic violence that make 'examples' of subjects. Said pointed out that these practices primarily use description to produce facts. The use of the so called "ethnographic present" (Fabian, 1983, 81), which 'freezes' subjects (and collectives) into specific temporal and political contexts, can be conceived as a method of objectification. Objectifying practices not only elevate momentary and subjective experiences or observations to the level of objective knowledge, they both abstract and solidify the (power-political) differences that such observations and their articulation facilitate. According to Laclau (1991), hegemonic discourses require a radical other, a corrupt non-identity, a constitutive 'outside.' This 'outside' first makes it possible to demarcate 'the inside of the society' and make a vacuous and unfathomable yet apparently unquestionable 'us' appear to be a given.

With this in mind, it seems plausible to assume that discussions on the 'religion of the others' represent a hegemonic discourse. Indications of this are evident in the currently dominant discourse and representation practices (in academic discourse, mass media, literature, public debate, etc.) that portray 'the Muslims' as a homogeneous and closed group. This knowledge of Muslims in European migration societies is the knowledge of the 'Otherness of the others.' On the basis of this binary thinking, it is possible for very different and even contradictory practices to develop; these can range from a caring epistemic position regarding Muslims ('one only has to understand them') to constant calls for differentiation, and ultimately to manifest cultural racism. What these practices have in common is that they create specific subjects (Muslims) through epistemic dominance and use specific representation practices to transform them into a subjectified "constitutive other" (Laclau 1991), into 'radically other subjects.'

Against the backdrop of the understanding of migration outlined here as both bodily and discursive, it becomes clear that scholarship on migration in general and migration pedagogy specifically are not only concerned with the conditions, forms, and consequences of movements of people across borders. They also direct critical attention to the discourses on migration marked discursively as legitimate and less-legitimate forms of belonging. In this context, migration pedagogy sheds light on what happens when migration becomes an issue in educational fields, and to what extent pedagogical protagonists and institutions weaken or strengthen natio-racial-culturally coded concepts of 'us' and 'them.' Education and pedagogy can be regarded in this context as both a mirror and a playground for those symbolic practices that differentiate between a natio-racial-culturally coded 'us' and 'them.'

3 Education as a Site of the Reproduction and Transformation of Natio-Racial-Culturally Coded Orders of Belonging

Migration phenomena challenge, unsettle, and disrupt the legitimacy and functionality of the natio-racial-culturally coded 'us.' In addition, the legitimacy and functionality of institutional routines such as linguistic practices or practices of collective memory (see Lücke 2016) are called into question. Finally, migration as the movement of people across borders, as well as a discourse that shapes new knowledge about belonging and citizenship, troubles culturally dominant views and opinions about the legitimacy and functionality of individual privileges, for example the privilege to not only expect but to claim that one's own language is also the language of the other.

In this context, studying migration as a societal phenomenon cannot be reduced to the study of migrants. The basic analytical approach of migration research is the focus on individuals and groups in their relationship to natio-racial-culturally coded orders of belonging as well as the transformation of this relationship (Mecheril et al. 2010). The expression 'natio-racial-cultural' (*natio-ethno-kulturell*, see Mecheril 2003, 118–251) refers on the one hand to the fact that the concepts of nation, ethnicity, race, and culture are often used in a diffuse and undifferentiated way, both in research and everyday communication. On the other hand, this term indicates the fact that concepts of nation, ethnicity, race, and culture are manifested formally in laws and regulations, materially through border controls and identity documents, and also socially through symbolic practices that generate rather blurred meanings and outcomes that are subsequently used politically.

In natio-racial-culturally coded conceptions of affiliation, constructions of 'race' can have a significant impact, as can forms of religious references. Studies and analyses indicate an association between conceptions of national or cultural belonging and nationalist or racist images (Hormel and Jording 2016). Anti-Muslim racism in connection to the idea of the 'West' has been and is increasingly evoked in public arenas to negotiate issues of national belonging (Attia 2016). Migration movements take place in the context of natio-racial-culturally coded orders of belonging; they activate these orders and change them ('mobility' is a form of movement which, unlike 'migration,' does not necessarily suggest this natio-racial-culturally coding). The ideal type of natio-racial-culturally coded orders makes it possible to describe and study migration and its consequences for the subjects, spaces, and places of migration. These orders are created by "glocal" (Robertson 1995) processes – complex and dynamic, yet nonetheless displaying a degree of inertia – concerning the destabilization of identity and affiliation, as well as conceptions of location and space.

The concepts and practices in which a distinction is made between the natio-racial-culturally coded 'us' and 'them' mediate and influence the experiences, self-understanding, and activities of everyone. Migration pedagogy therefore focuses on the analysis and description of these concepts, experiences, and practices, as well as on the analysis of the empirical and possible conditions under which these concepts, experiences, and practices become more fluid. Migration pedagogy thus does not focus on a particular target group and is not a program of education and treatment of migrants or integration with the intention of changing (assimilating) migrants. Migration pedagogy may be considered in contrast to educational approaches, which focus primarily on improving the skills of migrants (for example, language competences in the hegemonic language or in the standard register) and the question of how the integration of migrants can be optimized. Instead, the migration pedagogy approach analyzes the power of institutional and discursive orders of belonging, and explores the question of how the capacity to act with dignity might be cultivated under given conditions without unreservedly affirming and accepting these conditions.

The decision to examine the relationship between migration and education from the perspective of migration pedagogy goes along with the interest in examining orders of belonging in the migration society, and the power relations that arise from them, as well as the investigation of enabled or hindered learning processes of *all* people – no matter what position or status they have in the respective migration society. The orders of belonging are historically developed structures that create subjective experiences of symbolic distinction and classification, experiences of empowerment and efficacy, and biographical experiences of contextual location. Membership, efficacy, and connectedness are the constitutional analytic elements of belonging (Mecheril 2003, 118–251).

Modern states and societies distinguish between those who belong to the particular societal context and those who do not. They accomplish this using natio-racial-culturally coded patterns in a highly complex, often contradictory, and ever-changing context. The education system and educational approaches play an important role in the affirmation and reproduction of these patterns of distinction in part due to the institutionalization of a special form of social work involving 'work with migrants,' or because schools may employ the mechanisms of institutional discrimination. Pedagogy also has the potential and possibilities to reflect upon these patterns and the practices that confirm them, as well as to consider and promote alternatives. Migration pedagogy focuses on the effects of natio-racial-culturally coded orders on people and their learning processes, on processes of becoming a subject as well as on educational practices that reaffirm, yet also shift and sometimes transform these orders. The societal circumstances and realities that are connected to postcolonial (see Dhawan and Castro Varela 2015) and transnational migration concern all areas of education, including elementary education, art education, adult education, and all levels of educational activity, including organizational forms, methods, contents, and the skills of professionals in the educational field. Educational institutions play a central role in the processes of affirming, iteratively generating, and, often, 'naturalizing' natio-racial-culturally coded orders of belonging. Not least, the most important pedagogical institution, the school, is a sphere in which individuals are introduced to preconceived understandings and practices structured by natio-racial-culturally coded orders that are intersectionally connected to other classifications such as gender and class. Educational institutions are productive with regard to the symbolic positioning of pupils. These positions – for example, as 'migrant' or 'non-migrant,' or 'verbally limited' or 'able to speak' – must be understood as the effects of practices of societal distinction, which, both existing within and extending beyond educational fields such as school-teaching, are taken up and affirmed by pedagogy.

This can be exemplified by taking a brief look at the portrayal of contents relating to migration society in German textbooks. Franz Pöggeler asserted that the existence of "migrants and foreigners was first dealt with in textbooks in the late 70s" (Pöggeler 1985, 35). Today migration is a well-established topic in German textbooks. For example, in the context of the project "Migration in Textbooks at the Georg Eckert Institute," Hanna Schissler (2003) points out that it would be "unfounded [...] to assume that German textbooks do not address the topic of migration," adding that textbooks are better than their reputation suggests in that they have adopted the desired didactic shift in perspective from the majority society to 'migrants' (Schissler 2003, 43–44). However, Dirk Lange and Sven Rößler (2012) assert that education authorities and textbook publishers have shown increased sensitivity to the significance of migration, and migration appears on its own and in connection to other topics in the textbooks examined, though in quantitative terms (that is, the number of dedicated pages), its treatment was brief in comparison to other subjects. There are numerous findings available in the existing literature that describe the way people are portrayed and represented as migrants or as non-migrants, in a particular natio-racial-culturally coded symbolical status. One pattern of representation that has remained unchanged for decades is the construction of 'migrants' as 'foreigners' or 'Others.' Various studies point to a particular dichotomy, wherein 'migrant' is coded as its own natio-racial-cultural group and is contrasted against the 'German national community' (e.g., in Höhne, Kunz, and Radtke 2005, 592; Lange and Rößler 2012, 148). The discourse-analytical study on "images of strangers" in German textbooks conducted by Höhne, Kunz, and Radtke (2005) showed that over time, the attributed and ascribed 'foreignness' of this group increased steadily; first 'guest worker,' then 'foreigner,' and finally 'asylum seeker' (Höhne, Kunz, and Radtke 2005, 598). Additionally, the authors discovered that the figure of the 'migrant as a foreigner' previously found in books was also present in other media, reflecting a sort of national consensus.

The differentiation between 'aliens,' 'migrants,' and/or 'foreigners' and the society of 'Germans' results in an additional dichotomy: the contrast between 'us' and 'them.' The construction of an 'us' is grounded in the conception of an ethnic-racially and culturally coded homogeneous nation. The indirect construction of this dichotomy is evident in the treatment of issues such as homeland, foreignness, immigration, integration, the acquisition of German citizenship, xenophobia, and racism (see Pokos 2011). Such conceptions of homogeneity are problematic first because migration is dealt with as a constant exception (Stöber 2006, 77). Second, such conceptions lead to self-perpetuating stereotypes and reductive descriptions of the 'Others' as the only bearers or members of another 'culture' (Stöber 2006, 78). Even the supposedly positive function of cultural enrichment reinforces a thinking that distinguishes between 'us' and 'them' and leads to a reification of difference.

The cultivation of difference and cultural normalization of dichotomy in textbooks must, according to research, have significant effects on pupils whose 'foreignness' is addressed in the book. Taking difference into consideration can have exclusionary effects; even recognition is not a practice that is free from power. Textbooks are a specific medium that is meant to initiate processes of learning, and thus in part to 'speak' directly to the reader in the form of questions and tasks: "In this arrangement, the migrant children are given the task of giving information about 'their culture'" (Höhne, Kunz, and Radtke 2005, 602). The reading book *Zebra* asks second graders "What countries are the children in this class from?" (Zebra 2010, 87). Asking children to see and understand themselves as 'foreign' or as 'migrant' children in a school environment creates a powerful direct and indirect appeal to the pupil to position himself or herself in a natio-racial-culturally coded symbolic space. However, educational relationships and pedagogical institutions were not adequately understood, if they would not be conceived with regard to their contribution to the transformation of these existing patterns of distinction and positioning. A critical approach to education in a migration society seeks to strengthen this understanding.

4 Migration-Pedagogical Research as Criticism (of Hegemony)

Up to this point, this paper has described migration pedagogy as a critical practice. What this means shall be briefly explained in the following with reference to the concept of criticism. Criticism, as it is understood here, is a social practice that should be differentiated, for example, from cavils or complaints. "The question of the conditions and the possibility of criticism appears wherever given conditions are analyzed, judged, or rejected as false. In this sense, criticism is a constitutive part of human practice" (Jaeggi and Wesche 2009, 7). This general human practice can be found in the sciences as well as in the arts or media. Scientific criticism does not simply present moral, aesthetic, or political judgments on societal structures and phenomena, as its primary concern is not normative judgment, but rather analysis and reflection on societal reality. At the same time, it is hardly convincing to describe every scientific statement on societal relations and developments as a critical practice, as this can contribute to the leveling or trivialization of those scientific practices that explicitly act as criticism. "Criticism cannot simply consist of stating how something *is*; it also has to take a position on the matter [...], how something *should be* or likewise *should not be*" (Jaeggi 2009, 279). Scientific criticism is normative in a very specific sense.

Criticism represents a non-affirmative mode of thought about established empirical structures and processes, and is disruptive in this regard. Thus, any reproach of criticism that suggests that criticism should be 'beneficial and constructive' misses the point and fails to grasp the nature of criticism. The argument that criticism is corrosive or not constructive hollows out and levels criticism's content. The aim of criticism is not to preserve and improve given circumstances. Criticism is ultimately a practice that must claim to (provisionally) set aside the law of the given without meeting any requirement to establish a new law. In this sense, criticism possesses a constitutive tendency toward destruction or non-constructiveness, making it suspicious and uncomfortable for a culture that is bent upon the endless perfection of what already exists.

The analysis of the effects of the supposedly legitimate (and allegedly without alternative) institutionalized asymmetric relation between a natio-racial-culturally coded 'us' and 'them' is of great importance for critical migration research. These relations are not simply the result of oppressive structures, as the subject is "bound to seek recognition of its own existence in categories, terms, and names that are not of its own making, the subject seeks the sign of its own existence outside itself, in a discourse that is at once dominant and indifferent" (Butler 1997, 20). Orders of belonging create inferior and superior subjects who remain tied to the logic of these orders.

Because dominance and hegemony cannot be reduced to force and oppression, there is a need for a concept of hegemony that includes the ambivalence of domination and the momentum of compulsion and hindrance, as well as concession and facilitation. In this sense, domination can be understood as an institutionalized, relatively permanent and temporarily solidified, structured and structuring social relationship, in which the opportunities for the mutual exertion of influence are asymmetrically allotted. In contrast to power relations based on force, relations of hegemony are characterized by an asymmetry that is taken for granted. As a reality that is built upon a solid history of individual experiences, such asymmetric relations appear to be implicit, unalterable, or putatively natural. Critical migration research examines the supposedly legitimate institutionalized asymmetric relations of difference, which are not only implicit, but represent, in their implicitness, impalpable asymmetric relations.

Against this backdrop, critical migration research has three main goals. In the sense in which Lawrence Grossberg (1997, 257) has described the central interest of cultural studies ("Cultural studies is always interested in how power infiltrates, contaminates, limits and empowers the possibilities that people have to live their lives in dignified and secure ways"), we can say that critical migration research is concerned with the analysis of those power structures of the migration society that hinder the opportunities of subjects to live a freer existence, and that limit their dignity. Second, critical migration research addresses subjectivization processes that take place within the context of these structures. In the frame of a non- or post-orthodox critical understanding, terms such as 'freer existence,' 'hinder,' and 'dignity' cannot and should not be conclusively determined, but instead need a constant conceptual reflection and empirical debate. If we are interested in analyzing the intrusion and penetration of power into the opportunities that "people have to live their lives in dignified and secure ways" (Grossberg 1997, 257), if we are interested in investigating the ways that power curtails people's options and opportunities, then we cannot avoid dealing empirically and conceptually with issues such as obstruction, limitation, and marginalization, as well as resistance. The fact that we are dealing empirically, here, with diverse forms of hindrance and facilitation that can be flexibly connected, formed into new constellations in specific contexts, decoupled, and reconnected, and thus inhibit simple analyses and suggestions for change, should not and cannot hamper efforts to deal with the basic fact of uneven relations of hindrance/facilitation. In a context marked by a commitment to criticism, it is impossible to abandon the motive to analyze the possibilities of people to live their lives in a more dignified and secure way. It is however necessary to keep the conception of what it means to live in a 'dignified and secure way' open, and to return to these conceptions and address these phenomena again and again in an interminable project of criticism. Third, critical migration research focuses on the analysis of possibilities for and forms of the shifting und transformation of orders of belonging and hegemonic structures of domination, as well as resistance to and within these orders and structures. Relations of domination are neither strictly determinative nor necessary; they feature spaces and latitude for action and are contingent. Critical migration research is very much concerned with the exploration of these spaces and options for contingency, because in this spaces may be researched practices that approach the comparative forms of 'freer' and 'more dignified.' The ways in which these may be 'freer' and 'more dignified' in any specific context is a topic for conceptual and empirical analysis. That which is 'freer' and 'more dignified' appears in varied contexts, diachronic and synchronic, differently. It is not fixed, and it is precisely this aspect of not being fixed, this modulation and variation, that must be understood.

Migration-pedagogical research explores orders of belonging as hegemonic forms of differentiation that allow members of a given society to make sense of societal phenomena in symbolic, material, institutional, and discursive terms, as well as to make sense of their own position in society. In other words, people experience, recognize, and understand societal reality and their own position within it with the help of natio-racial-culturally coded orders of belonging. In this context, the migration-pedagogic approach deals *first* with the examination of the situated or local generation of differences (for example in a school), and *second* with the analysis of general discursive practices, political or legal regulations (related to educational issues), and socio-economic conditions. Within the interrelation of these two 'analytical levels,' educational migration research concerns itself primarily with the investigation of the relationship between situated practices and the more general structures that generate relations of difference and orders of belonging. It thereby examines relations of belonging and difference as orders that display a certain inertia and inevitability, as well as relations that are historically and regionally determined and originated in specific contexts; it examines relations and orders of belonging not only with regard to the power that such relations exert over individuals, but also with regard to where and how subjects problematize, change, and shift these orders of belonging, giving them new meanings. Of particular interest here is the analysis of the contribution of educational institutions and pedagogical discourses concerning orders of belonging and opportunities to deal with or change these relations. Migration-pedagogical research examines societal structures, rules, and practices in formal education institutions and beyond these institutions in such a way that processes of the transformation of symbolic relations between the self and the world become the subject of investigation, with a particular focus on those processes of symbolic transformations that refer to the question of freer and more dignified lives.

The greater part of the German language original was translated into English by Michael Larsen.

References

Attia, Iman. *Die 'westliche Kultur' und ihr Anderes. Zur Dekonstruktion von Orientalismus und antimuslimischem Rassismus*. Bielefeld: transcript, 2009.

Bauman, Zygmunt. *Die Angst vor den Anderen*. Berlin: Suhrkamp, 2016.

Bublitz, Hannelore. *Diskurs*. Bielefeld: transcript, 2003.

Butler, Judith. *The Psychic Life of Power*. Stanford: Stanford University Press, 1997.

Castles, Stephen. "International Migration at the Beginning of the Twenty-First Century: Global Trends and Issues." *International Social Science Journal* 52.165 (2000): 269–281.

Castles, Stephen, and John Miller. *The Age of Migration: International Population Movements in the Modern World*. 4th ed. Basingstoke: Palgrave Macmillan, 2009.

Castro Varela, María do Mar, and Paul Mecheril. *Dämonisierung der Anderen. Rassismuskritik der Gegenwart*. 2nd ed. Bielefeld: transcript, 2015. 119–142.

Dahrendorf, Ralf. "Das Zerbrechen der Ligaturen und die Utopie der Weltbürgerschaft." *Riskante Freiheiten. Individualisierung in modernen Gesellschaften*. Eds. Ulrich Beck and Elisabeth Beck-Gernsheim. Frankfurt a.M.: Suhrkamp, 1994. 421–436.

Dhawan, Nikita, and Castro Varela, María do Mar. *Postkoloniale Theorie. Eine kritische Einführung*. 2nd ed. Bielefeld: transcript, 2015.

El-Tayeb, Fatima. *European Others: Queering Ethnicity in Postnational Europe*. Minneapolis/ London: University of Minnesota Press, 2011.

Fabian, Johannes. *Time and the Other: How Anthropology Makes Its Object*. New York: Columbia University Press, 1983.

Glick Schiller, Nina. "A Global Perspective on Transnational Migration: Theorising Migration Without Methodological Nationalism." *Diaspora and Transnationalism: Concepts, Theories and Methods*. Eds. Rainer Bauböck and Thomas Faist. Amsterdam: Amsterdam University Press, 2010. 109–129.

Gottuck, Susanne, and Paul Mecheril. "Einer Praxis einen Sinn zu verleihen, heißt sie zu kontextualisieren. Methodologie kulturwissenschaftlicher Bildungsforschung." *Bildung unter Bedingungen kultureller Pluralität*. Eds. Alexander Geimer and Florian von Rosenberg. Wiesbaden: Springer VS, 2014. 87–108.

Grossberg, Lawrence. "Cultural Studies: What's in a Name (One more Time)." *Bringing It All Back Home: Essays on Cultural Studies*. Durham, NC: Duke University Press, 1997. 245–271.

Hall, Stuart. “The West and the Rest: Discourse and Power.” *Modernity: An Introduction to Modern Societies*. Eds. Stuart Hall, David Held, Don Hubert and Kenneth Thompson. Malden, MA: Wiley-Blackwell, 1996. 184–228.

Höhne, Thomas, Thomas Kunz, and Frank-Olaf Radtke. *Bilder von Fremden. Was unsere Kinder aus Schulbüchern* über *Migranten lernen sollen*. Frankfurt a. M.: Johann Wolfgang Goethe-Universität, 2005.

Hormel, Ulrike, and Judith Jording. “Kultur/Nation.” *Handbuch Migrationspädagogik*. Ed. Paul Mecheril. Weinheim/Basel: Beltz, 2016. 211–225.

Jaeggi, Rahel, and Tilo Wesche. *Was ist Kritik?* Frankfurt a. M.: Suhrkamp, 2009.

Jaeggi, Rahel. “Was ist Ideologiekritik?” *Was ist Kritik?* Eds. Rahel Jaeggi and Tilo Wesche. Frankfurt a. M.: Suhrkamp, 2009.

Laclau, Ernesto, and Chantal Mouffe. *Hegemonie und radikale Demokratie. Zur Dekonstruktion des Marxismus*. Vienna: Passagen Verlag, 1991.

Lange, Dirk, and Sven Rößler. *Repräsentationen der Migrationsgesellschaft. Das Grenzdurchgangslager Friedland im historisch-politischen Schulbuch*. Hohengehren: Schneider, 2012.

Lücke, Martin. “Erinnerungsarbeit.” *Handbuch Migrationspädagogik*. Ed. Paul Mecheril. Weinheim/Basel: Beltz, 2016. 356–371.

Mecheril, Paul. *Prekäre Verhältnisse. Über natio-ethno-kulturelle (Mehrfach-)Zugehörigkeit*. Münster/New York: Waxmann, 2003.

Mecheril, Paul. “Kritik als Leitlinie (migrations)pädagogischer Forschung.” *Theoretische Perspektiven der modernen Pädagogik*. Eds. Albert Ziegler and Elisabeth Zwick. Münster: Lit-Verlag, 2014. 159–173.

Mecheril, Paul. “Das Anliegen der Migrationspädagogik.” *Schule in der Migrationsgesellschaft. Ein Handbuch*. Vol. 1: *Grundlagen – Diversität – Fachdidaktiken*. Eds. Rudolf Leiprecht and Antje Steinbach. Schwalbach/Ts.: Debus Pädagogik, 2015. 25–53.

Mecheril, Paul, María do Mar Castro Varela, İnci Dirim, Annita Kalpaka, and Claus Melter, eds. *Migrationspädagogik*. Weinheim/Basel: Beltz, 2010.

Mecheril, Paul, and Oscar Thomas-Olalde. “Die Religion der Anderen. Anmerkungen zu Subjektivierungspraxen der Gegenwart.” *Jugend, Migration und Religion. Interdisziplinäre Perspektiven*. Eds. Birgit Allenbach, Urmila Goel, Merle Hummrich and Cordula Weißköppel. Baden-Baden: Nomos-Verlag, 2011. 35–66.

Mecheril, Paul, and Monica van der Haagen-Wulff. “Bedroht, angstvoll, wütend. Affektlogik der Migrationsgesellschaft.” *Die Dämonisierung der Anderen. Rassismuskritik der Gegenwart*. Eds. María do Mar Castro Varela and Paul Mecheril. Bielefeld: transcript, 2016. 119–142.

Pöggeler, Franz. “Politik in Fibeln.” *Politik im Schulbuch*. Ed. Franz Pöggeler. Bonn: Bundeszentrale für politische Bildung, 1985. 21–50.

Pokos, Hugues Blaise Feret Muanza. *Schwarzsein im ‘Deutschsein’? Zu Vorstellungen vom Monovolk in der Schule und deren Auswirkungen auf die Schul- und Lebenserfahrungen von deutschen Jugendlichen mit schwarzer Hautfarbe. Handlungsorientierte Reflexionen zur interkulturellen Öffnung von Schule und zu rassismuskritischer Schulentwicklung*. Berlin: LIT Verlag, 2011.

Robertson, Roland. “Glocalization.” *The Post-Colonial Studies Reader*. Eds. Bill Ashcroft, Gareth Griffiths, and Helen Tiffin. Oxford: Routledge, 1995. 477–480.

Said, Edward William. *Orientalism*. 25th ed. New York: Vintage Books, 2003.

Sassen, Saskia. *Globalization and its Discontents: Essays on the New Mobility of People and Money*. New York: New Press, 1998.

Schissler, Hanna. "Toleranz ist nicht genug. Migration in Bildung und Unterricht." *Reflexion und Initiative* 4 (2003): 39–50.
Spivak, Gayatri Chakravorty. "The Rani of Sirmur: An Essay in Reading the Archives." *History and Theory* 24.3 (1985): 247–272.
Stöber, Georg. *Deutschland und Polen als Ostseeanrainer* (Studien zur internationalen Schulbuchforschung). Braunschweig: Georg Eckert Institut, 2006.
Stoler, Ann Laura. "Colonial Aphasia: Race and Disabled Histories in France." *Public Culture* 23.1 (2011): 121–156.
Wigger, Iris. "Die Schwarze Schmach am Rhein." *Rassische Diskriminierung zwischen Geschlecht, Klasse, Nation und Rasse.* Münster: Westfälisches Dampfboot, 2007.
Zebra. Lesebuch 2. Schuljahr. Stuttgart: Klett, 2010.

III: Stories, Histories, and Politics

Wulf Kansteiner

Unsettling Crime

Memory, Migration, and Prime Time Fiction

Generals always prepare for the most recent war and politicians always manage the most recent migration crisis – or what they and their voters remember as the most recent crisis. In this way collective memories of migration play a decisive role in policy decisions and may prevent societies from developing appropriate responses to resettlement processes.

That was the case in Europe in the year 2016. An estimated 2.2 million people arrived in the territory of the European Union during the so-called refugee crisis in 2015 and 2016 from Syria, Afghanistan, Kosovo, Eritrea, and many other countries (Connor 2017). That number represents merely 1.7 % of the 130.9 million people forcefully displaced worldwide in those two years (UNHCR 2016; UNHCR 2017). Housing, feeding, and educating the two million hardly constitutes a serious financial or material challenge for the EU where, in 2016, 511 million people accumulated a combined GDP of $ 16,4 trillion representing 21 % of global GDP (World Bank 2017). It would take a lot more refugees to overburden the resources of this area of extraordinary wealth. Moreover, in the long run, the refugees and their descendants are likely to become a demographic and financial asset rather than a burden to the EU (Kancs and Lecca 2017). We are clearly not dealing with a material problem but a problem of culture, mentalities, ethics – and memory. Some problems of misperception are short-term memory problems. The citizens of the EU remember the financial crisis of 2007 accompanied by economic contraction (for the first time since 1945), increased material inequality, and modest decreases in standards of living in some member countries (Magalhaes 2015). That experience might have negatively influenced member societies' willingness to welcome refugees into the EU. Other, more serious memory problems reflect long-term developments. Relevant segments of wealthy Northern European countries entertain decidedly negative views of the enduring consequences of post-WWII immigration events. They remember distinctly the labor migrations from Southern Europe in the 1960s and 1970s and from Eastern Europe after 1990 and the influx of political asylum seekers since the 1990s as having harmed their societies (Wilhelm 2017). Again, these negative assessments have little to no basis in fact. For most economists, the migrations in question have had positive rather than negative financial effects on host societies, and crime statistics reveal that only a few specific immigrant groups exhibit a clearly elevated propensity to commit

https://doi.org/9783110600483-009

crimes.[1] It is important to insist on these facts but that hardly suffices. Collective memories are not based in fact; they are based in stories, images, and feelings. Faced with incontrovertible evidence to the contrary, populist politicians across Europe have insisted that they and their voters feel threatened by immigrants. Unfortunately, their feelings are difficult to change. Dealing with dysfunctional memories requires effective memory analytics and memory interventions. We need new stories and feelings corresponding to the facts.[2]

The task of developing suitable memories and identities runs up against a veritable paradox. In principle, cultural memories and collective identities are fluid and malleable. Members of different collectives continuously reinvent themselves in symbolic and practical terms by differentiating individuals and groups that fall within their social contracts from those that fall outside of them or attain at best an ambivalent status. Plus, they constantly reshape the very nature of these contracts. For instance, scholars of nationalism have long abandoned the idea that cultural negotiations about the nation and its discontents unfold as stable processes and result in enduring notions of collective identity. They prefer to imagine nations as highly contingent cultural and psychological phenomena involving a wide range of heterogeneous and contradictory communications (Brubaker 1996). At the same time, imagined communities and their invented traditions, including myths of national belonging, seem to have significant staying power even if they are blurry around the edges. Malleability and persistence are not mutually exclusive characteristics in the politics of memory and identity. The figure of the migrant and stories about migration play a decisive role in Western post-colonial societies' efforts to imagine themselves as sedentary, stable, advanced, civilized, and deservedly affluent. Unfortunately, refugees are subjected to these powerful fantasies of Othering that come in many different flavors, from outright xenophobia to arrogant benevolence (Bauman 2016; Nail 2015).

The situation is analytically further complicated by the diversity of media arenas implicated in memory and identity negotiations. Nationally constituted, highly centralized, and transnationally networked old media landscapes like television co-mingle with social media environments with very different, multi-directional patterns of cultural communication and authorship and new forms

1 For a comprehensive assessment of the economic consequences of post-WWII and Post-Cold War immigration in (West) Germany, see Spencer 1994; Klingst and Venohr 2017.

2 On the importance of debate and messaging in counteracting xenophobic political actors, see Lochocki 2018. In this context it is important to consider how European national and transnational government institutions and media companies narrate and remember Europe and the EU, see Spohn and Eder 2016; and especially Bondebjerg et al. 2017; and avoid the danger of dysfunctional banal Europeanization (Trenz 2016).

of concentrated power over communication contents and structures (Hoskins 2018). Yet even if we focus on old media, which are our primary concern in this article, the task of cultural analysis remains complicated. There is ample material available to craft a compelling, robust story about the negative impact of TV content on migration memories. Media research, for example, has shown time and again that the narrative worlds of prime time television feature powerful negative stereotypes of foreign recalcitrance and criminality.[3] As prime time became ever more violent and crime-obsessed in the 1990s, hardly an evening went by in the big European TV markets without sleek, emotionally manipulative stories of Turkish sexism, Polish slyness, Russian brutality, and Arab terrorism, followed by exceedingly repetitive late night talk shows echoing similar views and possibly establishing them as widely shared social memories (Kretschmar 2002; Trebbe 2009; Goebel 2017). The story can be further elaborated by connecting key elements of popular culture to right wing radicalization. Effectively enhanced by racist propaganda, compacted collective symbols of alterity may have found an eager audience among the voters of right-wing populist parties who tend to be "male, young(ish), of moderate educational achievement and concerned about immigrants and immigration" (Arzheimer 2017, 287).[4] This stream of repetitive media images and its social base might be the collective memory one is up against in an attempt to build a more inclusive society.

At the same time, the story is hardly that simple. First of all, TV content about foreigners and migration has often not been dominated by negative stereotyping. To name just one example, after 9/11 US commercial television networks featured a great deal of Arab bashing but also a similarly sustained series of programs about good Muslims living the American dream while victimized by US xenophobia (Alsultany 2012). All of that raises the interesting question of how audiences react to the co-presence of positive and negative stereotyping and, more specifically, of how viewers respond to ironic representations of ethnic and racial prejudice designed to counteract bigotry (Keding and Struppert 2009). While it makes sense to assume that centralized cultural distribution systems like television offer important information about popular registers of collective identity, one should keep in mind that the carefully designed fantasies of national and ethnic belonging distributed via satellite and cable differ in form and content from the ways in which citizens think and feel about the burden and privileges of group affiliation.

---

3 On stereotyping of muslims after 9/11, see Morey and Yaqin 2011; Mertens and de Smaele 2016.

4 The storyline is consistent with hegemonic interpretations of right wing electoral success across Europe, see for instance the analysis of the 2017 federal election in Germany by *Spiegel Online* (Kwasnieski 2017).

After all, even after decades of TV research, we have frustratingly little insight into TV reception processes, although we know for instance that identical media texts may have very different effects in different settings (Lind 2010; Gorton 2009). Similar questions and lacunae arise when one directs one's analytical gaze at the assumed users of anti-migration memories, for example, the supporters of right wing political parties in Europe. Voter profiles generated by electoral sociologists only reflect a relative affinity between social background and voting decisions and even these affinities have already been subject to significant social mainstreaming (Rooduijn 2016). Moreover and more important, statements about voting propensities provide no empirical basis for connecting problematic media coverage to specific political phenomenon, for example, by linking the electoral success of the right wing political parties to prejudicial TV programs possibly consumed by their voters (Arzheimer 2018; Ellinas 2018).

With these words of caution in mind I will apply two conceptual perspectives to analyze the prehistory of the media event refugee crisis. First, a structuralist point of view well suited to highlight problematic dynamics of post-colonial Othering pervasive in Europe's mass media. This traditional approach captures layers and continuities of post-WWII and post-colonial xenophobia and racism, demonstrating that the cultural arena has never been a level playing field for new arrivals. Refugees always faced an uphill battle as they tried to insert themselves successfully into a cultural terrain systematically stacked against them. In addition, I apply a memory studies vantage point that paints a more complicated picture. As the collectively constructed crisis unfolded, two different memory regimes have struggled for supremacy. On the one hand, there were conventionally antagonistically structured memories confirming a simplistic us vs. them divide easily captured by structural analysis. These nationalistic memories contributed to the rise of the radical right across Europe – although it is remarkable how successfully the professed nationalists crafted transnational political and memory networks across the continent. On the other hand, some segments of Europe's societies deployed top-down cosmopolitan memories to frame the crisis within a humanitarian narrative trajectory tapping into human rights and Holocaust memories. Both approaches were successful in that they mobilized political actions, although hardly in any integrated fashion or with comparable degrees of sustainability. The countries following traditional antagonistic storylines far outnumbered the countries willing to give cosmopolitanism a try, as for example Sweden and Germany did. Moreover, in the end, neither memory regime succeeded in providing compelling long-term perspectives and policy solutions. The antagonistic narrative collides with the simple fact that Europe is a site of irreversible, large-scale migration whereas the cosmopolitan utopia fails to provide practical guidelines for generous yet also clearly circumscribed immigration policies. As we will see, Europe needs a third memory culture that can extend welcome to many more refugees and help with the difficult task of developing ethics of material and legal global inequality.

After that brief conceptual contextualization we now turn to a layer of media history, and arguably also a powerful source of collective memories of migration that predates the "refugee crisis" by over a decade and thus represents exactly the kind of media fare the witnesses, bystanders, and protagonists of the "migration crisis" of 2015/2016 have consumed over the course of the last decades. Our case study concerns the 2003–2004 season of the German TV series *Scene of the Crime* (*Tatort*), which is Germany's Sunday evening prime time flagship program and whose 2003–2004 season featured 34 broadcasts and a number of in-depth engagements with the topic of migration.

1 Whose Jungle Is It Anyway?

On 26 October 2003, 6.5 million German TV viewers watched Berlin detectives Ritter and Stark solve the murder of an illegal immigrant from Nigeria whose corpse had been dumped at the airport. *Jungle Brothers* (*Dschungelbrüder*), as the program is provocatively entitled, does not spare any didactic efforts but suffers from a number of shortcomings.

Right at the start, Ritter delivers an impressive yet also improbable impromptu lecture about the pieces of West African jewellery found on the dead man. Further along in the investigation, the Berlin cops dutifully, albeit somewhat laconically, inform the viewers about poverty and corruption in a long list of West African countries, criticize the exploitation of illegal immigrants as a source of cheap labor, and acknowledge that the majority of African immigrants, including people who have lived in Germany for most of their lives, have no chance to attain refugee status. In passing, Ritter and Stark also perfunctorily deplore racism in Germany. But the program does not show vivid instances of racism with the exception of the occasional use of the word darky ("Bimbo") by one of the shadier characters of the story as he addresses his illegal immigrant workers. The program offers theoretical information about xenophobia but does not render it tangible in any meaningful way.

Jungle Brothers has other weaknesses. Perhaps in an attempt at realism in the depiction of police–immigrant relations, the creators of the film have the detectives treat black witnesses and suspects with considerable indifference. Thus the very carriers of PC-consciousness in the story, whom regular TV viewers know as compassionate figures, are reinforcing the symbolic divide between Germans and others – notwithstanding the fact that some of the others are German citizens (Wallnöfer 2003). Moreover, in order to satisfy the requirements of prime time and in an effort to illustrate that immigrants are human beings just like the viewers in front of the screen, the director has all black actors deliver their lines in flawless German. Thus characters in the story with non-German backgrounds, including recent arrivals, display stunning linguistic capabilities that simply do not reflect everyday life in multi-lingual German cities (Bonner Generalanzeiger 2003). This feat of foreign language acquisition bestows a cyborgian quality on some immigrant figures, inadvertently turning them into a specially marked group of people within the narrative universe on the screen.

However, the most decisive flaw of *Jungle Brothers* concerns the depiction of the crime itself. In the first scene of the film, the refugee Koffi is murdered by the black father of his girlfriend. The father is a successful business tycoon who immigrated to Germany from Cameroon in the late 1960s. He now owns commercial property all over Berlin and moves in the highest circles of the city, where he is almost but not completely accepted as an equal by his urban jungle brothers. The father finds Koffi unacceptable as a son-in-law, offers him money if he promises to leave his daughter alone, and accidentally kills him in the fight that ensues after Koffi rejects the bribe. Thus, the first-generation black immigrant turns out to be the biggest bigot in the world of *Jungle Brothers*.

The brief look at the plot structure of *Jungle Brothers* already hints at a fundamental dilemma that lurks at the intersection of the crime genre and the topic of migration. The director and screenwriter of *Jungle Brothers*, Lars Becker, certainly had the best of intentions. He himself is married to a woman of color and planned to make viewers aware of the problems immigrants face in German society (Gehrmann 2003). But within the current genre setting a crime show about migration requires migrants in the roles of victims and/or perpetrators – in addition to a set of more or less heroic non-migrant detectives representing the host society. In this narrative terrain plot options are clearly limited and the creative staff will have to engage with migrant stereotypes in one way or another. That raises tough questions about how much (violent) racism one can responsibly depict on the screen. The problems are exacerbated by Becker's script. How does one turn foreigner-on-foreigner violence into a self-reflexive study of German racism, especially if the viewers learn the identity of the murderer right after the opening credits? Finally, Becker faced even more of an uphill battle because *Jungle Brothers* is part of the prestigious series *Scene of the Crime* that restricted his aesthetic options. The German detective team Ritter/Stark were established prime time figures combining cool with conscience and leaving little space for alternative heroes. Under these circumstances it is not surprising that Becker did not succeed in subverting powerful prejudices. The film follows the conventional rules of the genre and validates the superior sceptical gaze of the enlightened German law-enforcement officers who bring to light the shady dealings and psychological hang-ups of ethnic minorities living in the German capital. *Jungle Brothers* thus sends a devastating message about German society that probably only partly reflect Becker's political intentions: the integration of black immigrants in Germany is doomed to fail even under the best of circumstances. The immigrants might be economically successful and seemingly well-adjusted but the combination of traditional tribal prejudice, immigrant insecurities, and entrenched German racism may sooner or later turn them into dangerous xenophobes, especially if they successfully emulate German values (Wallnöfer 2003; Bohn 2003b). One can easily imagine that these kinds of stories, reproduced in the media on a daily basis, have dire effects. When they intersect with a fear of migration on the part of viewers and the arrival of real migrants, especially migrants perceived as poor and dark-skinned, the stories likely contribute to the collective construction of a 'refugee crisis,' if the perceived risks and fears of migration are not held in check by countervailing memories.

2 A Memorial to Whiteness

As an installment of the series *Scene of the Crime*, *Jungle Brothers* is part of a venerable German media tradition. *Scene of the Crime* has been produced for over four decades and, by the end of 2017, had a record of 1040 original programs, not to mention thousands of reruns (Tatort-Fundus 2018a; Brück et al. 2003, 159–160; Hickethier 1998, 237). It might very well be the longest-running prime time fiction program in the world. Yet while *Scene of the Crime* may rightfully lay claim to global significance in statistical terms and has been broadcast in over 50 countries, it remains a thoroughly German phenomenon that thrives in Germany's peculiar cultural and media landscape (Hamburger Abendblatt 2016). The 48-year history of *Scene of the Crime* comprises over seventy-five different investigative teams located in cities all across the Federal Republic. Each team represents a different region with its landmarks and cultural and linguistic peculiarities. *Scene of the Crime* renders *Heimat* both regionally specific and nationally recognizable (Bollhöfer 2007; see Mously 2007). In this vein, *Scene of the Crime* has been a cultural vehicle of national integration in West, East, and unified Germany. Consequently, *Scene of the Crime* has a large, dedicated, multi-generational cult following that turns the broadcast of new installments into media events.[5] These audiences expect entertainment and suspense but also social commentary. As a result, *Scene of the Crime* provides an important platform for political self-reflection with each investigation carefully calibrated to navigate, manipulate, and occasionally challenge the perceived limits of popular taste. In this way, *Scene of the Crime* presents interesting interpretations of the social and political status quo in Germany (Hißnauer et al. 2014). Every important political issue has been addressed over the years, including German–German relations and unification, neo-Nazis and the Nazi past, feminism, social inequality, unemployment, abortion, xenophobia, AIDS, religious sects, euthanasia, prostitution, migration, child labor and abuse, globalization, and genocide. The people behind *Scene of the Crime*, including many first-rate directors, screenwriters, and actors, are very aware of the fact that they are managing a national label that offers them extraordinary career opportunities.[6]

There are other developments that have turned *Scene of the Crime* and similar programs into central sites of cultural memory. Germany's media landscape changed decisively in the early 1990s when commercial stations, which had first been licensed in West Germany in 1984, started to match their public service rivals in program diversity and audience appeal. The ensuing competition transformed television fare in Germany, especially during prime time, killing off a number of venerable public service series and formats (Krüger 2002). But crime dramas in general and *Scene of the Crime* in particular have thrived in the new TV environment. Between the mid-1980s and the mid-1990s, airtime for prime time cop shows increased more than fivefold (Wehn 2002, 7–8, 194). *Scene of the Crime* showings jumped from twelve broadcasts in 1989 to over thirty per year since 2002. There is a simple explanation for this development. Crime dramas constitute one of the few reliable recipes for successful ratings in a crowded field of TV content providers. *Scene of the Crime* is again a case in point. The series no longer reaches audience shares of over 50 percent, which was not unusual during the era of the public service monopoly of ARD and ZDF. But with an average of 8 million viewers and 23% market shares per prime time broadcast, *Scene of*

5 Some *Scene of the Crime* investigators have become celebrities – most notably Götz George aka Schimanski – and the introduction of new detectives is vigorously discussed in national tabloids and TV guides; on the Schimanski phenomenon see Harzenetter 1996.

6 *Scene of the Crime* was for instance the springboard for Wolfgang Petersen's career. He directed a number of episodes in the 1970s, including the famous *Reifezeugnis* of 1977; Tatort-Fundus 2018b.

the Crime has generally remained the most popular program of the prime time line-up on Sunday evening (Tatort-Fundus 2018a).

At the precise moment when crime shows in general and *Scene of the Crime* in particular filled ever more airtime, unified Germany experienced increased immigration and a wave of right-wing violence against foreigners in the early 1990s and, partly in response to these events, drastically curbed non-EU immigration by negotiating a system of EU controls (Geddes and Scholten 2016, 95). Yet despite anti-immigration legislation prompted by popular xenophobic resentment, Germany relies on immigrant labor for its economic wellbeing and by 2017 the percentage of foreign citizens and people of so-called non-German background living in Germany had risen to 11,2 % and 22,5 % respectively (Deutsche Wirtschaftsnachrichten 2017). In this heterogeneous group of approximately 18 million, three million people of Turkish descent represent the largest subsection and, until the events of 2015/2016, were considered the least well-integrated foreigners in Germany. As a result of recent developments, hierarchies of belonging are somewhat in flux in the Federal Republic although there have always been various cultural constructs of migrating Muslims that served as the symbolic antipodes of the constructs 'German' and 'European' in post-Cold War Germany (Dietrich and Frindte 2017, 112).

German television has played an important role in the remapping of collective national-ethnic identities after the fall of the Berlin Wall. A study from 2007 revealed that 81 % of the coverage about Muslims on the public stations ARD and ZDF presented Islam in an overwhelmingly negative light (Hafez and Richter 2007). That figure neatly corresponds to the 83 % of the German population who held decidedly unfavorable opinions about Islam at that point in time (Noelle and Petersen 2006). Both trends have continued unabated in the last ten years (Focus-Online 2017; Hafez 2017). However, while these statistics seem to speak a clear language, German television's encounter with Islam has been quite complex. TV executives have tried to deal with the problem of racial prejudice by offering viewers a number of critically acclaimed programs calling into question ethnic stereotypes, especially regarding Turks. The programs include the docusoap *Die Özdags* (WDR 2007–2008) and the award-winning sitcom *Alle lieben Jimmy* (RTL 2005–2007) (Domaratius 2009). *Scene of the Crime* itself may serve as another, on first sight positive example. Between 1991 and 2003, partly in response to the wave of xenophobic violence mentioned above, the topic of migration has been addressed, in greater or lesser detail, in almost 20 % of all *Scene of the Crime* episodes (Ortner 2007, 85; Buhl 2014, 76). Thus German public television has tried to step up to the plate and media scholars have generally acknowledged these efforts (Wellgraf 2008; Ortner 2007; Thiele 2005), although, as we have seen above, engaging with the topic is not synonymous with effectively counteracting xenophobic sentiments. Consider in this context the first (!) German-Turkish detective, Cenk Batu, whom ARD *Scene of the Crime* executives sent on crime-fighting missions in the gritty urban setting of Hamburg from 2008 to 2012 (Tatort-Fundus 2018c). Batu retired after only 6 episodes and the figure assumed a very ambivalent profile right from the start. Batu was the only undercover detective in *Scene of the Crime*. In that role he assumed the personae of Turkish criminals and Islamist terrorists, and, on his last case, went on a vigilante rampage. Consequently, especially on a visual level, Batu often performed rather deconstructed hegemonic cultural patterns of racial prejudice and Muslim migratory threats (Spielberger 2012).[7] 48 years after its inception, *Scene of the Crime* remains a thoroughly racist narrative universe in which white detectives chase white, brown, and black criminals.[8]

TV networks and critics have simply not paid enough attention to the problematic narrative and aesthetic strategies with which crime shows influence perceptions of foreigners and processes of social integration. Cop shows like *Scene of the Crime* focus on deviancy and symbolically adjudicate social behavior and in this fashion shape the everyday performance of collective memory. According to Jack Katz, crime shows are so successful and such a great source of historical analysis because they appeal to the audience's strongest-felt positive and negative emotions. By watching crimes unfold on the screen, viewers collectively partake in a daily moral workout that reflects their passionate desire for safety and raises fundamental questions about the reproduction of social order (Katz 1987). Thus popular television offers intriguing interpretations of fundamental fears and hopes, especially with regard to the cultural construction of others. After all, few topics are more emotionally and politically engaging than the question of who does and who does not belong to one's society and how people aspiring to membership should conduct themselves. In this sense, the communications surrounding crime shows about migration touch upon the very essence of human self-consciousness (Rudolph 2007, 49–51). Put differently, the serial production of xenophobically structured programs is not just a likely risk in an innovation adverse genre setting like prime time crime but should be recognized as one of the genre's key functions at a time when positive collective self-images in the West appear to be inextricably linked to negative constructions of foreignness

---

7 Similar problems arise with regard to the longest serving immigrant *Scene of the Crime* detective, i.e., Munich's Ivo Batic with Yugoslavian roots who also suffers from an excess of Mediterranean temper (Thiele 2005, 197).

8 See the gallery of white and predominantly male detectives at Tatort-Fundus 2018d.

and migration. Needless to say, these imagined worlds come with a hefty price tag, first and foremost for migrants but also for host societies.

3 Orientalizing Crime

Lars Becker was not the only creative mind at ARD who used *Scene of the Crime* in the 2003–2004 season to address the topic of migration. In August 2003 screenwriter Harald Göckeritz and director Martin Weinhart tackled a particularly thorny issue when they explored the interdependence between illegal migration and the international black market for organ transplants. Their film focuses on the plight of migrants who have been smuggled into Germany under false pretenses and are now pressured to serve as organ donors. The refugees have found temporary shelter in a dilapidated public-housing development in Ludwigshafen where they are eyed with considerable suspicion by the authorities and left unprotected against assaults from an international organ transplant mafia. Detectives Odenthal and Kopper take a close look at the migrants when a refugee headed for the housing development is found dead in a nearby gravel pit. The film quickly focuses on the anguish of a 12-year old Kurdish girl who survived persecution in Turkey and Iraq and now has to fend for herself because her brother, her only remaining relative, has not returned from a mission as an organ donor in Ukraine (Frohn 2003).

Leyla, broadcast on 31 August 2003 to an audience of 6.9 million viewers, has self-critical potential. Göckeritz and Weinhart forgo any simple happy end. As the detectives are successfully wrapping up their investigation, the camera focuses on another helpless migrant on his way to yet another risky surgery abroad (Bohn 2003a). In other compelling moments of the film, viewers get a palpable sense of the despair that makes parents of chronically sick children participate in a perfidious system of exchange in which the desperate exploit the poor. Last but not least, the script occasionally assumes the perspective of the refugees, especially by offering a detailed account of the growing panic of 12-year-old Leyla. But the figure of the girl also proves Göckeritz' and Weinhart's downfall. They string together scene after scene of Leyla mournfully inspecting the empty mailbox, placing yet another unanswered call from a public phone booth, or patiently waiting by the window, illuminated by candlelight, with vaguely Middle Eastern music playing in background. The stereotypical spectacle of childlike innocence, flavored with a dash of exoticism, robs the character of complexity and agency, rendering Leyla and her peers defenseless against organ smugglers as well as self-indulgent projections of munificence on the part of the viewers. *Leyla* is a picture-perfect example of orientalism.[9]

Moreover, the implicit audience constructed in the televisual universe of the film, which is treated to the spectacle of exotic innocence and seemingly invited to identify with Leyla, is defined along narrow ethnic and linguistic lines that undercut empathetic engagement with the plight of the refugees. The team of investigators and medical support staff, i.e., the heroes of the film, only consists of white, middle-class German native speakers.[10] This homogeneous group of well-adjusted individuals enlightens the audience about a thriving international business of illegal organ transplants in China, Russia, Ukraine, India, Brazil, and the Czech Republic. Through their profile and their actions they draw a line in the sand, somewhere east of Berlin. After all, as Odenthal puts it succinctly on one occasion: "This is Germany." Consequently, all perpetrators are clearly marked as outsiders. The leader of the smuggling enterprise is a stereotypically stoic Russian who kills people without the slightest hesitation; the physician examining the refugees is an ethnic German who grew up in Siberia and still mutters Russian under her breath; the thug who is sent after the victims speaks with the thickest of Bavarian accents, marking him with a powerful sign of alterity according to the cultural codes of the German media.

There are two signs of ambivalence in the neat narrative world of *Leyla*. The first unpersuasive marker of hybridity is Leyla herself. After only two months in Germany she speaks excellent German. The second character caught between the lines is the social worker Marler, portrayed masterfully by veteran actor and *Scene of the Crime*-regular Jürgen Tarrach. Marler frequently helps the migrants with their everyday problems but also advises them to accept the deals offered by the smugglers. When confronted by Odenthal and Kopper with the consequences of his corrupt dealings, Marler deeply regrets his actions and contemplates suicide. But he remains the only wrinkle in the tidy national universe of *Leyla* (Bohn 2003a). Overall, the film depicts crimes committed by foreigners against foreigners with the former complaining about German laws that complicate their business and about German welfare payments that drive up the prices for organ 'donations.' The viewers can hardly avoid the conclusion that more rules

---

9 A similar process of balkanization occurs in *Scene of the Crime* installments dealing with crimes linked to the wars in the former Yugoslavia (Gladis 2016). On Orientalization as a widespread pattern of perception in contexts of migration see Evangelos Karagiannis and Shalini Randeria's contribution to this volume.

10 The picture is complicated by the fact that *Tatort* aficionados know Kopper to be of Italian-German descent (although that does not play a role in *Leyla*). In addition, the actress Ulrike Folkerts, who plays Odenthal, is widely known as one of a few openly gay German actresses.

and more aggressive policing are needed in order to force the thugs and their victims to conduct their unappetizing business elsewhere – without the benefit of German infrastructure and taxpayer support.

4 Into the Holocaust Ghetto

In addressing the problems of multiculturalism in Western societies, the makers of *Scene of the Crime* are clearly handicapped. The crime-drama genre provides only limited opportunities to highlight the positive side of migration. A murder mystery needs a corpse, a murderer, and a compelling motive; a clash of cultures is an excellent plot structure that offers all of the above. In the 2003–2004 season, *Scene of the Crime* addressed questions of migration and integration on a routine basis and each time the gripping storylines made a powerful case for cultural homogeneity. In fact, the message could not have been stated more bluntly: the West is better off if it does not get involved in the shady dealings, tribal tensions, and reactionary values of migrant communities. None of the regretful deaths of migrants would have occurred – or at least would not have occurred on Western soil – if the foreigners had simply stayed put. *Leyla* und *Jungle Brothers* convey attractive and highly antagonistic memories of migration.

A closer look at the cross-section of crime dramas focusing on migrants and foreigners has revealed rather disappointing results. In each case, the murder is committed by a migrant/foreigner and national identity and mainstream values are affirmed in simplistic ways that offer few opportunities for self-reflection. The programs are obviously not dealing well with the multicultural challenges of the 21st century. Under the circumstances it might be helpful to broaden the scope of analysis and gain some comparative perspective. The topic of contemporary migrations seems to overtax the didactic skills and political imagination of the creative staff in charge of crime-drama production. But perhaps screenwriters, directors, and TV executives are more astute in their representation of established and well-integrated minority communities like German Jewry. After all, Western culture has had a lot of opportunities to think about Jewish–Gentile relations and after 1945 these reflections have often taken a decidedly self-critical turn, especially in Germany. Moreover, especially since the 1970s, television has played a particularly important and often constructive role in Holocaust education (Shandler 1999; Kansteiner 2006; Kansteiner 2011).

On 7 December 2003, over 7 million viewers watched an aesthetically and politically ambitious episode of *Scene of the Crime* that painted a critical and discouraging picture of Jewish–Gentile relations in contemporary Germany. The events in *The Shochet* (*Der Schächter*) are set into motion by a mentally handicapped young man who, in a state of panic, cuts the throat of a homeless boy after having been threatened by the boy's attack dog. The brother of the perpetrator covers up the crime by planting corpse and murder weapon in the house of a shochet, who is subsequently vilified in the press and relentlessly pursued by an anti-Semitic district attorney. The actual crime and the pursuit of the murderer are not center stage in the program. Instead, the veteran screenwriter Fred Breinersdorfer offers a morality play about persistent anti-Semitic prejudice in the German provinces (Sauerwein 2003). The convoluted storyline, lacking suspense and veracity but overflowing with didactic ambition, is delivered in carefully constructed, increasingly more disturbing images and combined with an evocative, highly symbolic soundtrack. The program begins with idyllic scenes of the shochet, the lead detective, and the handicapped man playing boule along the banks of Lake Constance. As the case unfolds, the viewers witness the repeated, aggressive interrogation of the shochet staged with Brechtian techniques in sterile early 20th-century institutional architecture. Apart from the interrogations, the district attorney calmly and seemingly rationally invokes all kinds of anti-Semitic stereotypes, from greed to blood sacrifice, to bolster his case. Shortly before the climax of the story, director Jobst Oetzmann shifts aesthetic gears and replaces the distance-inducing alienation techniques with subjective camera angles and unsteady camera movements designed to make the viewers sense the rising panic that the shochet, a survivor of Treblinka, experiences as he is once more victimized by German officials (Heinen 2003).

The director of *The Shochet* develops a lot of cinematic ambition in an effort to induce empathy with the accused and instill properly anti-anti-Semitic values in his audience. But the program also contains a number of contradictory plot elements that insert an interesting level of narrative ambivalence and raise doubts about the philosemitic credentials of the lead detective and possibly the program as a whole (Gangloff 2003; Anders 2003). On the one hand, the proceedings against the shochet are always conducted according to the rule of law, despite the crude anti-Semitic slogans that the prosecutor delivers outside the courtroom. The lead detective and an impartial judge make sure that justice is done. So the fears of the shochet, especially his growing suspicion of his detective friend, appear increasingly paranoid. On the other hand, a strangely passive lead detective does not seem to resist the anti-Semitic onslaught as aggressively as one might expect. Moreover and most importantly, viewers are told several times over the course of the film that the shochet spent his childhood in Konstanz, became a well-respected member of the Strasbourg Jewish community after 1945, and only returned to Konstanz on a part-time basis in recent years because he had inherited a beautiful and valuable house on the shore of Lake Constance. Now that he is harassed by law enforcement, the shochet plans his illegal departure from Konstanz and permanent escape to France.

Thus Breinersdorfer invokes anti-Semitic stereotypes of wandering Jews with shifting loyalties in pursuit of material gain and illustrates how difficult it is to find one's way out of the maze of philo- and anti-Semitic traditions that is contemporary German culture. In this way, he and Oetzmann open up the film for a wide range of possible readings. Some viewers might have felt encouraged to reflect about their own latent fears of ethnic-cultural alterity and moments of disturbing transference (as Breinersdorfer and Oetzmann probably hoped they would). Others might have come to the conclusion that disengagement seems like the best solution and that the local German–Jewish symbiosis should be ended before worse things happen. In fact, until the very last seconds of the show, the plot and the camerawork squarely point towards segregation and the benefits of ethnic homogeneity (Winter 2015). Finally and most disturbingly, the ambivalences and performances of anti-Semitism might have prompted some members of the audience to revel quietly in anti-Semitic prejudice.[11]

Crime shows with Jewish and Holocaust subject matter convey stern warnings about the need for political vigilance and ethnic tolerance, in the case of *The Shochet* cast into an ambitiously complex and self-reflexive TV story. Everybody involved in television, from producers to consumers, seems to recognize the Holocaust as a site of humility and earnest historical exploration. It is very possible, however, that decades of Holocaust memory work have simply resulted in the creation of narrowly circumscribed cultural ghettoes of cosmopolitan self-reflexivity. The habitual, self-critical impetus noticeable in TV programs with Jewish subject matter does not seem to be transferred to televisual engagements with other contemporary problems of memory whose solutions depend on a similar combination of self-reflexive resolve and innovative zeal. In fact, from a structuralist perspective, the situation appears to be even more sinister. What if the belated, proudly performed critical engagements with the legacy of Nazism play an important role in the thoughtless reproduction of contemporary patterns of anti-migration prejudice? Does Holocaust memory facilitate contemporary racism and nationalism? Or have the rituals of German Holocaust culture effectively curbed xenophobia, for instance by helping to create the welcoming culture for refugees that briefly flourished in Germany in 2015/2016? Is chancellor Merkel's

---

11 There are good reasons to assume that a significant share of the audience might have indeed favored this last reading. In December 2003, a month before the broadcast of *Der Schächter*, a high-profile political scandal broke in Germany that revealed the persistent anti-Semitic sentiments of conservative politicians and their voters; Herzinger 2003; Kansteiner 2006, 308–312.

famous sentence "We'll manage that" ("Wir schaffen das") a concrete application of the ethical demands of Holocaust memory as suggested by the Holocaust survivor Ruth Klüger in her speech to the German parliament in January 2016 (Galaktionow 2016)? Maybe 2015 represents a brief, unusual moment in which collective memories of past mass crimes intersected with resettlement processes for the benefit of refugees? Either way, the brief look at the symbolic-ideological dynamics of the media world of primetime crime highlights the difficulties involved in any attempts to brush the genre against the grain and engage in truly self-reflexive communication with the audience.

5 Leveling the Playing Field

The culture of prejudice, which is a constituent element of primetime TV, is not representative of all mainstream visual culture. German film and TV culture, for instance, features a long and very successful tradition of anti-xenophobic media stories exemplified by such different media texts as the documentaries of Hans Dieter Grabe from the 1980s (Hißnauer 2009) or the successful feature films of Fatih Akin produced since 1998 (Mackuth 2007; Rings 2016). Even the annals of *Scene of the Crime* include an exemplary 1998 episode called *Trapped* (*In der Falle*), which effectively criticizes the campaign of structural violence waged by the German state against Turkish immigrants (Ortner 2007, 139–160). But, at the same time, the low profile, everyday cultural reproduction of structural racism and nationalism continues unabated.[12]

The above findings offer an opportunity to develop a systematic, structural critique of the crime drama with regard to its construction of alterity. The overwhelming presence and popularity of cop shows give rise to the suspicion that law and order programs have emerged as one of the primary cultural sites of collective and national identity in a post-Cold-War era in which the media have sworn off the use of overtly nationalistic and xenophobic strategies of representation. In the absence of the kind of clear cut visual and narrative codes circulating at the height of colonialism, fascism, and communism, television producers and their audiences in the West have settled on new master scripts to regain stable Western, European, and national bearings. In this fashion, the crime drama has become a powerful contemporary master symbol whose language, metaphors, and nar-

12 See for instance the more recent *Scene of the Crime* installments *The Saint* (*Die Heilige*), ARD, 10 March 2010; and *War Fragments* (*Kriegssplitter*), ARD, 3 May 2017.

rative designs are well suited for boosting morale in the respective in-group at the expense of various symbolic others along shifting lines of demarcation. The crime drama has emerged as a site of troublesome historical disinformation and self-fulfilling analytical prophecies. In the course of constantly rewriting and symbolically adjudicating the history of ethnic relations in the West and beyond, popular TV shows provide viewers with powerful tools for the interpretation of contemporary debates about migration, which make them all the more likely to adopt nationalistic and racist points of view. All these factors explain why producing a truly enlightened crime drama about migration seems to require nothing short of subverting the essence of a genre thriving on passive-aggressive politics of exclusion and stereotyping.

In the current cultural context, educational campaigns engaging with the topic of migration face a number of obstacles. The fight against prejudice often seeks to criticize ethnic stereotypes by presenting them in an exaggerated form whose political effects are difficult to gauge (Wellgraf 2008, 39). Ironically, the exaggerations are linked to collectively constructed standards of veracity. Viewers expect from television a realistic engagement with contemporary issues, including realistic representations of the kinds of problems they assume exist as a result of migration. But the producers' and viewers' perceptions of reality, which play an important role in successful communication via television, exist largely independent of the type of realism constructed in crime statistics, scholarly papers, or other scientific data. In fact, prime time fiction and non-fiction create their own standards of truth (Enli 2015). The exploration of minority communities from the vantage point of law enforcement for instance creates a seemingly compelling causal link between migration and crime and in that respect systematically misrepresents the basic characteristics of actual crimes as well as the efficacy of law enforcement agencies (Ortner 2007, 176). Needless to say, the actual conduct of migrants is less important for their perception by host communities than the characteristics attributed to migrants in the media.

In light of the high stakes involved in this media feedback loop, it is useful to recall guidelines for the representation of migrants on TV developed by the media scholar Christina Ortner. Empathy with members of minority groups is more easily induced by TV stories that feature migrants as main protagonists, present events from the migrants' point of view, and depict the private and professional complexities of their lives. All too often, even in shows designed to criticize xenophobic violence, camera and script linger on the non-migrant perspective in an attempt to understand and condemn anti-migrant sentiments. In addition, the cause of cross-ethnic peace and understanding is more successfully championed by TV programs that emphasize the heterogeneity of in-groups and out-groups and delineate generous criteria for membership in in-groups (Ortner 2007, 37–46).

An implementation of these guidelines might indeed result in politically more desirable television programming although the shows would transgress the limits of the genre and no longer meet the expectations of many audience members who take pleasure in watching upstanding white police officers solving the problems of a globalized world (Süss 1993, 217, 223). Ortner's well-intended cosmopolitical guidelines thus inadvertently highlight the structural dilemma inherent in televisual communication and the need for more radical cultural change. Any coverage of a politically sensitive and complex topic like migration, especially in prime time, will *nolens volens* convey the kind of messages that Roland Barthes has defined as myth, i.e., fairly simplistically structured, easily understood, and semantically over-determined arguments concerning the dangers or advantages of diversity and migration that are difficult to replace by other, ethically, politically, and economically more suitable myths (Barthes 1972, 116). Reforming myths about migration amounts to an uphill battle because the cultural terrain in which these negotiations take place does not constitute a level playing field. As Jürgen Link has shown, most collective symbols and metaphors, which have traditionally been used in discussions about difference and migration – and that are thus almost indispensable for effective mass-media communication about the topic – carry a clear bias against many forms of alterity. For instance, the moment any speaker deploys weather, body, house/home, or other territorial metaphors in the context of talk about migrants, the audience is encouraged to wonder about ways to stop the 'flood,' close 'leaks' and 'loopholes,' or prevent 'invasions' and 'injuries.' In this way foreigners are symbolically burdened with a wide range of negative outcomes, which are not counterbalanced by similarly compelling and extensively used positive symbols (Link 1978, 184–222; Link 2006; Link and Jäger 1993; Thiele 2005).

Obviously, well-meant cosmopolitan memories of migration are a useful point of departure but will not suffice. We need more radically innovative, agonistically structured memories of mobility to bridge the gap between fact and perception and create political options for inclusivity (Bull and Hansen 2016). How can we better imagine the migration experiences of the past to help solve perceived migration crises today? How can we rethink the population movements of the world wars, the political migrations of the Cold War, and the labor migrations of the age of European affluence to shape the kind of memories that help us foster rather than derail integration processes? How can we lay the foundations for helpful memories of the war in Syria, the destruction of Iraqi society, the misery of continuous warfare in Afghanistan, and the climate migrations of the future (Mayer 2016) to help migrants and their reluctant hosts craft a productive, sustainable life together?

6 Memory Agonistics: Honoring the Refugee & Forgetting the Migrant

The interpretive feedback loop is all too real: the memory of past mobility, or lack thereof, does amount to a decisive factor in a given society's ability to welcome newcomers as equals. If you conceive of yourself as being part of a tradition of people on the move, it is much easier to relate to people moving towards you. The plasticity of social memories is similarly real. The field of memory studies has proven time and again that what counts in the realm of collective remembrance is not what actually happened but what people feel and think happened. In this fashion, 20th century culture has moved mountains. Without having had much help in terms of cultural precedents the mass media turned the most brutal and depressing chapters of World War II, Nazi rule, and forced migration into uplifting stories of heroism, innocence, victimhood, and justice accomplished (Fogu et al. 2016). In a kind of secular reformation, the survivors of Auschwitz became figures of public veneration in the arena of popular Holocaust memory (Shenker 2015). That example illustrates that the plasticity of collective memory defines an important ethical responsibility. As a society we are accountable for our memories and should choose wisely when it comes to remembering the events and stories that define our collective selves. Unfortunately, remembering responsibly is a complicated affair leading invariably to a troublesome paradox. Migration is an excellent case in point. On the one hand, we should do all we can to remember the plight of people who have escaped hardship. They deserve our respect and empathy. A widely shared memory of their suffering and resolve is a precondition for justice and humanitarian action. On the other hand, migrants are not citizens. For instance, the word "migrant" is hardly a neutral term. By designating a person as deserving of special treatment, the word turns that person into a target of discrimination, pity, and arrogance. So we have to forget the migrant for the purpose of extending full-term membership in society to new arrivals. To sideline dysfunctional memories, Europe needs to embrace the figure of the refugee as memory icon and forget the migrant's distinctiveness – all at the same time. How do we reconcile these two sound and seemingly diametrically opposed ethical principles?

Societies have different levels of tolerance for moving people and different capabilities of integrating them after their arrival, among other reasons because they develop different memories of migration. In classic immigration societies like the US and Australia, the troublesome and shameful details of mass exodus and deportation from Europe were swiftly pushed aside by uplifting stories of liberty and self-made prosperity at the expense of indigenous populations. Past suffering and dashed hopes were covered up by inauthentic charades of ethnic folklore and fantasies of national unity, which in turn became a solid foundation for democracy and the rule of law (Bedad 2005). European societies handled the large population displacements caused by World War II in similar ways. All across Eastern Europe and Germany, refugees, displaced persons, and the many ideologically uprooted sought coverage under fictions of ethnic homogeneity and political consensus. The forces of integration exerted a great deal of structural violence. Languages, heritage, political discontent, and many unwelcome memories disappeared in relentlessly selective invented traditions. The cultural violence designed to hide past enthusiasm for Nazism, extensive collaboration, and a great number of Nazi perpetrators also helped turn refugees into citizens. An impoverished and destroyed continent managed to find homes for at least 15 million migrants, especially in West Germany, and in later years continued that tradition by welcoming millions of people eager to leave Communist Eastern Europe. Ironically, West Germany and many of its neighbors meanwhile maintained the fiction that their countries were not immigration societies (Wilhelm 2017). For that purpose they invented new terms for people on the move, including such creative designations as "belated re-settlers" (*Spätaussiedler*; Menzel and Engel 2014). The strategies intended to forget the stigma of migration worked wonders for the cause of integration but turned into memory liabilities when migrants from Southern Europe were encouraged to move to the North and did not receive similarly optimistic long-term perspectives and fanciful designations. One is tempted to conclude that the integration of the 15 million succeeded because they were Germans and therefore naturally belonged to Germany, whereas the much smaller group of Southern European arrivals hailed from another culture and could not be integrated as easily. That explanation underestimates the cultural distance between Königsberg and Bavaria and the cultural diversity among the expellees (Rock and Wolff 2002); the explanation also underestimates the importance of social scripts and social memory for processes of social integration. If the movement of Turks to West Germany in the 1960s had been designated as permanent, inevitable, and based on traditions of long-term German-Turkish economic collaboration, the descendants of Turkish immigrants would find themselves in a much different position in German society today.

Honoring the refugee and forgetting the migrant does not have to be a contradiction. Cultural memories are never consistent; they are dynamic, multi-faceted, and contradictory. Official institutional memories of migration can celebrate the refugee while more fluid everyday media memories help deconstruct the stark artificial divide between migrants and non-migrants. Now that the survivors of the Shoah are leaving the stage, the refugees should take their place. Having overcome war, human trafficking, and persistent prejudice, they deserve their own memorial sites and prominent didactic presence in schools, museums, and the media. Permanently inscribed in our cultural memories, the figure of the refugee could make sure that we remember their woes and accomplishments and strive to live up to the dictum of "never again!" So we are advocating here for a mobility turn in memory studies and the politics of memory, conceptualizing moving people not as deficient vis-à-vis an allegedly static location from which they emerge (emigration) or to which they aspire (immigration). Instead, the process of movement itself becomes the focus of analytical and narrative interest resulting in a wide spectrum of multi-directional memories reflecting a range of nomadic experiences and diasporic feelings that include such different perspectives as that of the exchange student and civil war refugee without obliterating the dissimilarities between them (Nail 2015; Sheller and Urry 2006). Social and media scripts that acknowledge differences of opinion and interests about people movements while de-essentializing these differences could lay the foundation for agonistic memories of migration and agonistic debates about the privileges of affluent birth and the limits of human solidarity. What difference should this perspective make for prime time crime? A lot and very little. Television is hardly at the cutting edge of 21st centuries media technologies. Future memories of flight and migration will be negotiated in immersive gaming and AI environments (de Smale et al. 2017). But let us pretend for a moment that we could rewrite media history. In *Leyla* and *Jungle Brothers* script and camera could for instance have stayed with the moving victims and moving perpetrators, rendering their decisions and behavior complex, intriguing, and plausible. The detectives should have been much more diverse and sported a moving background easy to relate to. Itinerancy, deviancy, and helplessness, on the one hand, and settledness, integrity, and power, on the other hand, should not have formed such a rigid semantic field. If more fluid stories of people movements had been widely spread in prime time crime at the beginning of the new millennium, perhaps the welcoming culture of 2015/2016 might have lasted just a little bit longer.

The scripts for better stories of movement can be found in the real world. Between 2013 and 2017, over 24 % of newly hired police officers in Berlin hailed from a minority background but the *Scene of the Crime* detectives serving in Berlin, who were hired in 2015, still exude settled whiteness (Mediendienst-Integration 2017; Tatort-Fundus 2018e). The toughest challenge in building new collective memories of movement remains the deconstruction of racism.

References

Alsultany, Evelyn. *Arabs and Muslims in the Media: Race and Representation after 9/11*. New York: NYU Press, 2012.

Anders, Manfred. "Film verwechselt." *Sächsische Zeitung* (9 December 2003).

Arzheimer, Kai. "Electoral Sociology: Who Votes for the Extreme Right and Why and When." *The Populist Radical Right*. Ed. Cas Mudde. London: Routledge, 2017. 277–289.

Arzheimer, Kai. "Explaining Electoral Support for the Radical Right." *The Oxford Handbook of the Radical Right*. Ed. Jens Rydgren. Oxford: Oxford University Press, 2018. 143–165.

Barthes, Roland. *Mythologies*. London: Cape, 1972.

Bauman, Zygmunt. *Strangers at Our Door*. Cambridge: Polity, 2016.

Bedad, Ali. *A Forgetful Nation: On Immigration and Cultural Identity in the United States*. Durham, NC: Duke University Press, 2005.

Bohn, Angelika. "Das Organ-Geschäft. Nur ein Teilerfolg erzielt der Tatort Leyla." *Ostthüringer Zeitung* (2 September 2003a).

Bohn, Angelika. "Spannender Wettlauf." *Ostthüringer Zeitung* (28 October 2003b).

Bollhöfer, Bjørn. *Geographien des Fernsehens. Der Kölner 'Tatort' als mediale Verortung kultureller Praktiken*. Bielefeld: transcript, 2007.

Bondebjerg, Ib, et al., eds. *Transnational European Television Drama: Production, Genre and Audiences*. Cham: Palgrave Macmillan, 2017.

Bonner Generalanzeiger. "Tödliche Langeweile." (28 October 2003).

Bull, Anna, and Hans Hansen. "On Agonistic Memory." *Memory Studies* 9.4 (2016): 390–404.

Brubaker, Rogers. *Nationalism Reframed: Nationhood and the National Question in the New Europe*. Cambridge: Cambridge University Press, 1996.

Brück, Ingrid, Andrea Guder, Reinhold Viehoff, and Karin Wehn. *Der deutsche Fernsehkrimi. Eine Programm- und Produktionsgeschichte von den Anfängen bis heute*. Stuttgart: Metzler, 2003.

Buhl, Hendrik. "Zwischen Fakten und Fiktionen." *Zwischen Serie und Werk. Fernseh- und Gesellschaftsgeschichte im "Tatort."* Eds. Christian Hißnauer, Stefan Scherer, and Claudia Stockinger. Bielefeld: transcript, 2014. 67–87.

Connor, Phillip. "Still in Limbo: About a Million Asylum Seekers Await Word on Whether They Can Call Europe Home." *Pew Research Center Global Attitudes & Trends* (20 September 2017. <http://www.pewglobal.org/2017/09/20/a-million-asylum-seekers-await-word-on-whether-they-can-call-europe-home/> [accessed: 23 March 2018].

de Smale, Stephanie, et al. "The Case of *This War of Mine*: A Production Studies Perspective on Moral Game Design." *Games and Culture* (29 August 2017): 1–23.

"Bundesamt: Ausländeranteil in Deutschland bei 22,5 Prozent." *Deutsche Wirtschaftsnachrichten*. (17 August 2017). <https://deutsche-wirtschafts-nachrichten.de/2017/08/05/deutschland-anteil-der-einwohner-mit-auslaendischen-wurzeln-stark-gestiegen/> [accessed: 23 March 2018].

Dietrich, Nico, and Wolfgang Frindte. "Einstellungen zu Muslimen und zum Islam II und der Terrorismus." *Muslime, Flüchtlinge und Pegida. Sozialpsychologische und kommunikationswissenschaftliche Studien in Zeiten globaler Bedrohungen*. Eds. Wolfgang Frindte and Nico Dietrich. Wiesbaden: Springer, 2017. 89–137.

Domaratius, Jana. "Cultural Diversity Mainstreaming in *Türkisch für Anfänger* und *Alle lieben Jimmy*." *Heimat und Fremde. Selbst-, Fremd- und Leitbilder in Film und Fernsehen*. Eds.

Claudia Böttcher, Judith Kretzschmar, and Markus Schubert. Munich: Peter Lang, 2009. 199–214.

Ellinas, Antonis. "Media and the Radical Right." *The Oxford Handbook of the Radical Right*. Ed. Jens Rydgren. Oxford: Oxford University Press, 2018. 269–284.

Enli, Gunn. *Mediated Authenticity: How the Media Constructs Reality*. New York: Peter Lang, 2015.

"70 Prozent der Deutschen finden, dass der Islam nicht zu Deutschland gehört." *Focus-Online* (4 October 2017). <https://www.focus.de/politik/videos/repraesentative-umfrage-70-prozent-der-deutschen-finden-dass-der-islam-nicht-zu-deutschland-gehoert_id_6027429.html> [accessed: 23 March 2018].

Fogu, Claudio, Wulf Kansteiner, and Todd Presner, eds. *Probing the Ethics of Holocaust Culture*. Cambridge, MA: Harvard University Press, 2016.

Frohn, Axel. "Ulrike Folkerts. Die dienstälteste TV-Kommissarin feiert heute Jubiläum im Tatort Leyla." *BZ* (31 August 2003).

Galaktionow, Barbara. "Holocaust-Überlebende nennt 'Wir schaffen das' einen heroischen Slogan." *Süddeutsche Zeitung* (27 January 2016).

Gangloff, Tilmann. "Irgendwann rächt sich alles." *Stuttgarter Zeitung* (6 December 2003).

Geddes, Andrew, and Peter Scholten. *The Politics of Migration and Immigration in Europe*. 2nd ed. Los Angeles: Sage, 2016.

Gehrmann, Alva. "Plastikleichen pflastern seinen Weg." *Der Tagesspiegel* (15 June 2003).

Gladis, Lea. "Mapping Stereotypes und Tatort: Aspekte stereotyper Perzeptionen Südost-europas im 21. Jahrhundert." *Grenzräume – Grenzbewegungen*. Eds. Nina Frieß et al. Potsdam: Universitätsverlag, 2016. 49–62.

Goebel, Simon. *Politische Talkshows über Flucht. Wirklichkeitskonstruktionen und Diskurse. Eine kritische Analyse*. Bielefeld: transcript, 2017.

Gorton, Kristyn. *Media Audiences: Television, Meaning and Emotion*. Edinburgh: Edinburgh University Press, 2009.

Hafez, Kai. "Der Islam in den Medien. Der Islam hat eine schlechte Presse." *Zeit Online* (21 February 2017). <http://www.zeit.de/gesellschaft/zeitgeschehen/2016-12/islam-verstaendnis-medien-berichterstattung-populismus-gefahr> [accessed: 23 March 2018].

Hafez, Kai, and Carola Richter. "Das Islambild von ARD und ZDF." *Aus Politik und Zeitgeschichte* 26–27 (2007): 40–46.

"Der 'Tatort' wird in rund 50 Ländern ausgestrahlt." *Hamburger Abendblatt* (12 November 2016). <https://www.abendblatt.de/wirtschaft/article208704647/Der-Tatort-wird-in-rund-50-Laendern-ausgestrahlt.html> [accessed: 23 March 2018].

Harzenetter, Wilma. *Der Held 'Schimanski' in den 'Tatort'-Folgen des WDR. Ein Protagonist der achtziger Jahre*. Alfeld: Coppi, 1996.

Heinen, Christina. "Blut und Boule. Tatort über Antisemitismus." *Frankfurter Rundschau* (6 December 2003).

Herzinger, Richard. "Der Fall Hohmann. Raunen, Angst und Haß." *Die Zeit* (13 November 2003).

Hickethier, Knut. *Geschichte des Deutschen Fernsehens*. Stuttgart: Metzler, 1998.

Hißnauer, Christian. "Fremdes Deutschland. Heimat und Fremde aus der Sicht von Migranten. Hans-Dieter Grabes Dokumentarfilme der 1980er Jahre." *Heimat und Fremde. Selbst-, Fremd und Leitbilder in Film und Fernsehen*. Eds. Claudia Böttcher et al. Munich: Peter Lang, 2009. 35–46.

Hißnauer, Christian, Stefan Scherer, and Claudia Stockinger, eds. *Zwischen Serie und Werk. Fernseh- und Gesellschaftsgeschichte im 'Tatort.'* Bielefeld: transcript, 2014.

Hoskins, Andrew, ed. *Digital Memory Studies: Media Pasts in Transition*. New York: Routledge, 2018.

Kancs, d'Artis, and Patrizio Lecca. "Long-term Social, Economic and Fiscal Effects of Immigration into the EU: The Role of the Integration Policy." *JRC Working Papers in Economics and Finance* 4 (2017).

Kansteiner, Wulf. *In Pursuit of German Memory: History, Television, and Politics after Auschwitz.* Athens, OH: Ohio University Press, 2006.

Kansteiner, Wulf. "What is the Opposite of Genocide? Philosemitic Television in Germany, 1963–1995." *Philosemitism in History.* Eds. Jonathan Karp and Adam Sutcliffe. Cambridge: Cambridge University Press, 2011. 289–313.

Katz, Jack. "What Makes Crime 'News'?" *Media, Culture, and Society* 9 (1987): 47–75.

Keding, Karin, and Annika Struppert. *Ethno-Comedy im deutschen Fernsehen*. Leipzig: Frank & Timme, 2009.

Klingst, Martin, and Sascha Venohr. "Wie kriminell sind Flüchtlinge? Was die Kriminalstatistiken der Bundesländer über die Zunahme von Gewalttaten seit 2015 verraten: Sechs Trendmeldungen zur Zuwanderungskriminalität." *Zeit Online* (19 April 2017). <http://www.zeit.de/2017/17/kriminalitaet-fluechtlinge-zunahme-gewalttaten-statistik> [accessed: 23 March 2018].

Kretzschmar, Sonja. *Fremde Kulturen im europäischen Fernsehen. Zur Thematik der fremden Kulturen in den Fernsehprogrammen von Deutschland, Frankreich und Großbritannien.* Wiesbaden: Westdeutscher Verlag, 2002.

Krüger, Udo. *Programmprofile im dualen Fernsehsystem 1991–2000.* Baden-Baden: Nomos, 2002.

Kwasnieski, Nicolai. "Wie Gehalt, Beruf und Wohnort die Wahlentscheidung prägen." *Spiegel Online* (19 July 2017). <http://www.spiegel.de/wirtschaft/soziales/bundestagswahl-2017-wer-waehlt-cdu-csu-spd-fdp-gruene-linke-afd-a-1158543.html> [accessed: 23 March 2018].

Lind, Rebecca Ann. *Race/Gender/Media: Considering Diversity Across Audiences, Content, and Producers.* 2nd ed. Boston: Pearson, 2010.

Link, Jürgen. *Die Struktur des Symbols in der Sprache des Journalismus. Zum Verhältnis literarischer und pragmatischer Symbole.* Munich: Fink, 1978.

Link, Jürgen. *Versuch über den Normalismus. Wie Normalität produziert wird.* 3rd ed. Göttingen: Vandenhoeck & Ruprecht, 2006.

Link, Jürgen, and Siegfried Jäger, eds. *Die vierte Gewalt. Rassismus und die Medien.* Duisburg: Duisburger Institut für Sprach- und Sozialforschung, 1993.

Lochocki, Timo. *The Rise of Populism in Western Europe: A Media Analysis of Failed Political Messaging.* Cham: Springer, 2018.

Mackuth, Margret. *Es geht um Freiheit. Interkulturelle Motive in den Spielfilmen Fatih Akins.* Saarbrücken: Akademikerverlag, 2007.

Magalhaes, Pedro, ed. *Financial Crisis, Austerity, and Electoral Politics: European Voter Responses to the Global Economic Collapse 2009–2013.* New York: Routledge, 2015.

Mayer, Benoît. *The Concept of Climate Migration: Advocacy and Its Prospects.* Cheltenham: Elgar, 2016.

"Beamte mit Migrationshintergrund: Wie entwickelt sich die Vielfalt bei der Polizei?" *Mediendienst-Integration.* January 2017. <https://mediendienst-integration.de/fileadmin/Dateien/Polizisten_mit_Migrationshintergrund_2017.pdf> [accessed: 05 February 2018]

Menzel, Birgit, and Christine Engel, eds. *Rückkehr in die Fremde. Ethnische Remigration russlanddeutscher Spätaussiedler.* Berlin: Frank & Timme, 2014.

Mertens, Stefan, and Hedwig de Smaele, eds. *Representations of Islam in the News: A Cross-Cultural Analysis*. Lanham, MD: Lexington Books, 2016.
Morey, Peter, and Amina Yagin. *Framing Muslims: Stereotyping and Representation after 9/11.* Cambridge, MA/London: Harvard University Press, 2011.
Mously, Sara. *Heimat im Fernsehen. Eine medienpsychologische Untersuchung am Beispiel des 'Tatort.'* Saarbrücken: VDM, 2007.
Nail, Thomas. *The Figure of the Migrant.* Stanford: Stanford University Press, 2015.
Noelle, Elisabeth, and Thomas Petersen. "Allensbach-Analyse. Eine fremde, bedrohliche Welt." *Frankfurter Allgemeine Zeitung* (17 May 2006).
Ortner, Christina. *Migranten im Tatort. Das Thema Einwanderung im beliebtesten deutschen TV-Krimi.* Marburg: Tectum, 2007.
Rings, Guido. *The Other in Contemporary Migrant Cinema: Imagining a New Europe?* New York: Routledge, 2016.
Rock, David, and Stefan Wolff, eds. *Coming Home to Germany? The Integration of Ethnic Germans from Central and Eastern Europe in the Federal Republic since 1945.* New York: Berghahn, 2002.
Rooduijn, Matthijs. "Closing the Gap? A Comparison of Voters for Radical Right-wing Populist Parties and Mainstream Parties over Time." *Radical Right-Wing Populist Parties in Western Europe: Into the Mainstream?* Eds. Tjitske Akkermann et al. New York: Routledge, 2016. 53–69.
Rudolph, Ulrich. *Die Visualität der Teilsysteme. Intersubjektivität der Wahrnehmung visueller Symbole am Beispiel einer TATORT-Filmreihe.* Marburg: tectum 2007.
Sauerwein, Uwe. "Tatort-Krimi gegen Antisemitismus. Der Schächter dreht sich um die uralte Legende vom Ritualmord." *Berliner Morgenpost* (7 December 2003).
Shandler, Jeffrey. *While America Watches: Televising the Holocaust.* New York: Oxford University Press, 1999.
Sheller, Mimi, and John Urry. "The New Mobilities Paradigm." *Environment and Planning* 38 (2006): 207–226.
Shenker, Noah. *Reframing Holocaust Testimony.* Bloomington, IN: Indiana University Press, 2015.
Spencer, Sarah, ed. *Immigration as an Economic Asset: The German Experience.* London: Trentham, 1994.
Spielberger, Christoph. "Befehl des NDR. Cenk Batu muss erschossen werden." *achgut.com* (7 May 2012). <http://www.achgut.com/artikel/befehl_des_ndr_cenk_batu_muss_erschossen_werden> [accessed: 23 March 2018].
Spohn, Willfried, and Klaus Eder, eds. *Collective Memory and European Identity: The Effects of Integration and Enlargement* (2005). London/New York: Routledge, 2016.
Süss, Daniel. *Der Fernsehkrimi, sein Autor und die jugendlichen Zuschauer.* Bern: Huber, 1993.
Tatort-Fundus. "Alle Folgen im Überblick." (2018a). <https://www.tatort-fundus.de/web/folgen/chrono/ab-2010/2018.html> [accessed: 03 April 2018].
Tatort-Fundus. "Reifezeugnis." (2018b). <https://www.tatort-fundus.de/web/folgen/chrono/1970-bis-1979/1977/073-reifezeugnis.html> [accessed: 03 April 2018].
Tatort-Fundus. "Cenk Batu." (2018c). <https://www.tatort-fundus.de/web/ermittler/sender/ndr-norddeutscher-rundfunk/batu.html> [accessed: 03 April 2018].
Tatort-Fundus. "Die Ermittler – alphabetisch sortiert." (2018d). <https://www.tatort-fundus.de/web/ermittler/alphabetisch-sortiert.html> [accessed: 03 April 2018].

Tatort-Fundus. “Nina Rubin und Robert Karow.” (2018e). <https://www.tatort-fundus.de/web/ermittler/sender/rbb-rundfunk-berlin-brandenburg/karow-rubin.html> [accessed: 02 February 2018].

Thiele, Matthias. *Flucht, Asyl und Einwanderung im Fernsehen*. Konstanz: UVK Verlagsgesellschaft, 2005.

Trebbe, Joachim. *Ethnische Minderheiten, Massenmedien und Integration. Eine Untersuchung zu massenmedialer Repräsentation und Medienwirkungen*. Wiesbaden: Springer VS, 2009.

Trenz, Hans-Jörg. *Narrating European Society: Towards a Sociology of European Integration*. Lanham, MD: Lexington Books, 2016.

UNHCR: The UN Refugee Agency, ed. *Global Trends: Forced Displacements 2015*. Geneva: UNHCR, 2016.

UNHCR: The UN Refugee Agency, ed. *Global Trends: Forced Displacements 2016*. Geneva: UNHCR, 2017.

Wallnöfer, Pierre. “Lügen wie gedruckt.” *Salzburger Nachrichten* 10/28 (2003).

Wehn, Karin. *‘Crime-Time’ im Wandel. Produktion, Vermittlung und Genreentwicklung des west- und ostdeutschen Fernsehkrimis im Dualen Rundfunksystem*. Bonn: ARCult-Media, 2002.

Wellgraf, Stefan. *Migration und Medien. Wie Fernsehen, Radio und Print auf die Anderen blicken*. Berlin: LIT Verlag, 2008.

Wilhelm, Cornelia, ed. *Migration, Memory, and Diversity: Germany from 1945 to the Present*. New York: Berghahn, 2017.

Winter, Renee. “Tatort.” *Handbuch des Antisemitismus* 7 (2015): 484–486.

World Bank. “Data European Union.” <https://data.worldbank.org/region/european-union> [accessed: 23 March 2018].

Friederike Eigler

Post/Memories of Forced Migration at the End of the Second World War

Novels by Walter Kempowski and Ulrike Draesner

For the past 25 years, scholarship on the forced migration of approximately 12 million ethnic Germans at the end of World War II and the lasting effects of these massive population movements on the two post-war German states has been on the upswing. This renewed interest contrasts sharply with the latter part of the long post-war period (1970s and 1980s), when most historians and literary scholars shied away from these ideologically fraught historical events. The situation only began to change with the end of the Cold War, when the German-Polish post-war borders were finalized and, more noticeably, in the new millennium.[1] This delayed scholarly attention has to be seen in the larger context of shifting public discourses on the "German past" in general – WWII, National Socialism, the Holocaust – and on "German wartime suffering"[2] in particular. The broad scope of many recent studies[3] is testimony to the dearth of research in the preceding periods. This renewed scholarly interest coincides with and in some cases responds to increased media attention to German wartime suffering that started in the 1990s and intensified in the early 2000s.

In light of the history of contested public discourses on flight and expulsion it is not surprising that a considerable part of the new scholarship draws on memory studies as methodological framework. Across the disciplines, scholars consider the changing discourses on forced migration in the public sphere and the media (Traba and Zurek 2011; Hahn and Hahn 2010; Röger 2011, among others), in cultural representations including film and literature (Kopp and Niżyńska 2012; Eigler 2014; Niven 2014; Berger 2014), and, last but not least, in academia itself (Beer 2011a; Röger 2014). What is at stake are conflicted and conflicting memories of a decisive period of 20th-century German and European history and their

1 Polish authors and local initiatives in Poland began to explore the histories of border regions, including the forced migration of Germans, as early as the late 1980s, yet these trends received little attention in Germany (see Traba and Zurek 2011, 369; Sywenky 2013, 49–84).

2 This umbrella term also includes Allied bombing of German cities and the mass rapes of German women by Soviet soldiers at the end of the war (Assmann 2006, 184). For a discussion of the reasons for the delay in scholarship, see Eigler 2014, 51–60.

3 For a review of these broadly conceptualized studies, see Röger 2014, 50–55.

https://doi.org/9783110600483-010

implications not only for the two post-war Germanies but also for contemporary Germany and its relations with Eastern European countries, especially Poland. As Robert Traba and Robert Zurek point out in a comprehensive article titled "'Expulsion' or 'Forced Resettlement'? The Polish-German Dispute about Notions and Memory," the collective memory of forced migration continues to be a major sticking point in German-Polish relations because they belong to competing identity narratives in the respective countries (Traba and Zurek 2011, 400). From the Polish perspective, the expulsion of Germans under often inhuman conditions from territories in the East that became part of Poland cannot be seen in isolation but is placed in the context of Germany's role as aggressor in Eastern Europe, most prominently the murder and victimization of Poles for the entire duration of the war; from the German perspective, the awareness of guilt tends to focus on the Holocaust and less so on the occupation in the East – thus promoting narratives of German suffering and expulsions that are perceived as fundamentally unjust (Traba and Zurek 2011, 397).[4]

According to Traba and Zurek, attempts to establish a common European memory site via the federally funded German foundation "Flucht Vertreibung Versöhnung" (with a planned permanent exhibit in Berlin) have largely ignored diverging European memory discourses (esp. in Germany, Poland, and the Czech Republic). They argue convincingly that it would be more constructive to foster awareness of and respect for competing national memory discourses.[5] The difficulties the foundation continued to face until the foundation agreed on a revised and expanded plan for the permanent exhibit in 2017 provided further evidence for the inherent tension between claims for a European project and the actual dominance of a particular national (German) perspective.[6] By contrast, some recent scholarship examines German and Polish memory discourses by adopting comparative or transnational approaches. Examples are the comprehensive transnational project *Deutsch-Polnische Erinnerungsorte* co-edited by Robert Traba and Hans Henning Hahn that appeared in both German and Polish and the

4 This summary cannot do justice to the account of Traba and Zurek 2011, who also show the discursive fault lines within and across both countries. [Please note: The present article was completed before the publication of a revised exhibition plan in June of 2017]

5 Models for such an approach exist, for instance the German-Polish textbook commission (Traba and Zurek 2011, 394).

6 See <https://www.sfvv.de/en/foundation/founding-documents> for the expanded and revised outline of the permanent exhibition (to be opened at the end of 2018, ten years after the foundation's creation and original charge). The revised plan explicitly addresses some of the competing (national) memory discourses discussed by Traba and Zurek (2011) among others.

volume *Germany, Poland and Postmemorial Relations* (Kopp and Niżyńska 2012).[7] Much of this research espouses a high degree of self-reflection that examines not only the subject matter at hand but also how the history of their memory (*Erinnerungsgeschichte*) has become an intricate dimension of the ways we think about the events themselves (*Ereignisgeschichte*).[8]

Against this backdrop, this contribution pursues the following interlinked objectives: first, a discussion of the central role of memory studies in recent scholarship on flight and expulsion.[9] Beyond a consideration of collective memory discourses, I examine in what ways scholars draw on insights on the role of trauma and postmemory that initially emerged in the context of Holocaust studies. Second, a discussion of two literary examples that aims to show the benefits of such a memory studies approach for a nuanced assessment of novels on flight and expulsion; conversely, I will ask how literary texts referencing national and transnational contexts can foster our understanding of particular aspects of memory and postmemory and thus contribute to a further refinement of methodological approaches and theoretical frameworks.

1 Memory Studies as Conceptual Framework for the Study of Forced Migration

The field of memory studies straddles multiple disciplines and includes a wide range of approaches and terminologies. Regarding the subject matter of this contribution – literary representations of forced migration – two strands of scholarship are most pertinent: studies of collective and communicative memory, as well as the role of remediation, informed by the work of Aleida Assmann and Astrid Erll respectively; and approaches that are closely linked with the study of the Holocaust and its lasting effects across generations and cultures, informed by the work of Marianne Hirsch and Michael Rothberg, among others.

7 Other publications that combine memory studies with a comparative approach include: Möller 2016; Breuer 2015; Stickler 2014; Röger 2011.

8 For a helpful review of recent scholarship on forced migration with reference to these two approaches to historiography, see Röger 2014.

9 My own research on this topic has to date focused primarily on notions of belonging in the context of spatial theories (Eigler 2014). The present focus on memory, while often linked to issues of space and place, grew out of this contribution's objective to provide a meta-critical assessment of the changing scholarly and literary discourses on forced migration.

A good example for the prominent role of memory studies for recent scholarship on flight and expulsion is a handbook published in 2015 by Schöningh. As suggested by the title, *Die Erinnerung an Flucht und Vertreibung. Ein Handbuch der Medien und Praktiken*, it provides succinct information on a large range of 'media' and 'practices' related to flight and expulsion. Since this handbook exemplifies the larger trend in scholarship outlined above, I will comment on the approach it takes in more detail. At the most fundamental level, the handbook's main title, *Die Erinnerung an Flucht und Vertreibung*, is noteworthy. Despite its problematic history, the term 'Flucht und Vertreibung' (flight and expulsion) continues to be widely used in German public and academic discourses as shorthand for a broad and heterogeneous set of historical events.[10] Using the term as part of the title – and then commenting on its meaning and usage at the very beginning of the introduction (Scholz et al. 2015, 9) – exemplifies the handbook's overall approach: it takes on popular topoi, representations, and practices associated with flight and expulsion and subjects them to nuanced analyses.

In the introduction, the editors Stephan Scholz, Maren Röger, and Bill Niven also elaborate on the second part of the title, *Ein Handbuch der Medien und Praktiken*, especially on the central role of the media of memory, including memorials, TV series, oral and written accounts, photography, among others. Drawing on Erll's work, they underscore the constitutive role of mediation, that is, the insight that any particular medium does not only transmit or preserve the memory of a given event or phenomenon but co-constructs its very meaning and thus shapes its reception (Scholz et al. 2015, 10). Examples are images of the flight from East Prussia in the winter of 1945 that have attained iconic status: long treks of mostly women and children crossing the frozen Haff by foot or on overloaded wagons pulled by horses. As Beata Halicka notes in her contribution to the Handbook, these images were widely circulated in documentaries and docu-fictions and have thus shaped the public imagination and collective memory of the flight – even though the overall conditions and extreme hardship suffered by those who fled from East Prussia[11] are not representative of the flight from most other regions (2015, 96–97).

Beyond this role of specific media, the editors discuss what they call, with reference to Astrid Erll and Patrick Schmidt, the "Plurimedialität" ("plurimedi-

---

10 For more detailed commentaries on the term, see Beer 2011, 13–22 as well as Traba and Zurek 2011, 389. Like most scholars, I continue to use the term due to its wide currency – but with awareness of its problematic history.

11 East Prussia was the most Eastern region in the pre-war German Reich; the advancing Soviet army in the winter of 1945 led to mass flights among German civilians.

ality") of cultural memory, that is, the insight that different media interact with one another, shaping, reinforcing, or complicating particular memory discourses (Scholz et al. 2015, 12). An example is the powerful impact of Günter Grass's novel *Im Krebsgang* (*Crabwalk*) from 2002. The public attention resulted in part from the prominence and liberal outlook of its author, a political stance that was at the time perceived as precluding an engagement with 'flight and expulsion' (topics that instead were associated with the political right). But the public persona of Grass only partially explains the attention the novel received. As mentioned in the introduction to the Handbook, the book's success was prepared by TV-documentaries on flight and expulsion in the preceding year; its impact was then amplified and shaped by multiple reviews in *Der Spiegel*, among other popular news media, as well as by subsequent publications on the historical events portrayed in the novel. Ironically, the multiple media that reinforced one another and contributed to the novel's prominence were also responsible for erasing their own role, that is, the 'plurimediality' that helped create its success. Instead of looking at the surging public interest in 'German wartime suffering' as the result of a confluence of political and cultural factors, intertwined with ongoing mediation and remediation, the author Grass was singled out and credited with breaking a taboo regarding the representation of the flight. To this day his novel is mentioned as the first to address these events, even though other writers, including Walter Kempowski and Hans-Ulrich Treichel, had done so prior to Günter Grass.

In light of the meta-critical objectives outlined above, this contribution focuses primarily on literature's response to and participation in changing discourses on flight and expulsion. Literature, itself a powerful medium of memory, often references communicative and collective memory as part of the plot and its narrative organization. Grass's *Im Krebsgang* is a case in point. The novel is structured around diverging post/memories of three generations: the war generation (Tulla Prokriefke and "der Alte," alter ego of the author), the postwar generation that grew up in the shadow of the war (Tulla's son Paul, the narrator), and the subsequent generation (Paul's son Konny and Wolfgang who poses as a Jew and is ultimately killed by Konny). The historical events that make up the traumatic core of the novel (the Soviet attack of the ship Gustloff in January 1945 and the drowning of thousands of civilians, mostly women and children next to some members of the military) are narrated in a highly mediated fashion, that is, via references to a feature film about the sinking of the Gustloff and multiple textual sources. Ironically, the accounts of Tulla, eyewitness and survivor, are presented as unreliable. As Aleida Assmann maintains in her insightful comments, the novel thus marks the traumatic experience of the Gustloff sinking as a void ("Leerstelle;" 2006, 198).

As the example of *Im Krebsgang* illustrates, literary representations provide insights into the subjective, human responses to the violent history of WWII and its aftermath. Furthermore, literature opens up opportunities for exploring multiple individual and generational perspectives not only on the events themselves but also on their continued effects across several generations. Lastly, literature provides discursive spaces for the critical reflection on these memories and postmemories. For instance, *Im Krebsgang* draws attention to the mediated and constructed character of all memories.

Scholarship, informed by memory studies, has in the past few decades refined its approach to and assessment of literary texts that engage with aspects of memory and postmemory. Considering the emergence of memory studies in the US it would be difficult to underestimate the connection to the study of the Holocaust and its effects. Importantly, it was the convergence of memory studies with Holocaust studies that has produced new insights into the functioning and long-term effects of trauma and memory both in the individual and in the collective realms. The concept of "postmemory" as developed by Marianne Hirsch and the notion of "multidirectional memory" as proposed by Michael Rothberg are two cases in point. Recognizing the broad effects of the Holocaust at the individual and familial level (Hirsch) – roughly corresponding with Assmann's "communicative memory" – and at the collective level (Rothberg), these and other scholars have developed new approaches for the study of memory and trauma.

Arguably advances in memory studies in conjunction with Holocaust studies have also changed the ways in which we examine the legacies of other violent histories, including the forced migration of millions of Germans at the end of the war. It is this shift and the insights this theoretical framework affords that will be explored in the remainder of this section.

The concepts of 'postmemory' and 'transgenerational trauma' originated in the context of the study of documents and art by Holocaust survivors and their descendants. According to Hirsch, who first introduced the term in a 1992 article on Art Spiegelman's *Maus* (*Mouse*),

> 'Postmemory' describes the relationship that the 'generation after' bears to the personal, collective, and cultural trauma of those who came before – to experiences they 'remember' only by means of the stories, images, and behaviors among which they grew up. But these experiences were transmitted to them so deeply and affectively as to *seem* to constitute memories in their own right. Postmemory's connection to the past is thus actually mediated not by recall but by imaginative investment, projection, and creation. (Hirsch 2012, 5)

The references to "imaginative investment [...] and creation" point to the significance of literature (as well as graphic novels, photography, film, multi-media installations, and other artistic approaches) for representing and working through postmemorial constellations. In her recent monograph, Hirsch explicitly acknowledges the broad significance of postmemorial constellations. Responding to the growing field of memory studies in *The Generation of Postmemory* (Hirsch 2012), she sees her own work now as part of efforts among scholars to establish "connective" approaches to memory and considers her analysis of post-Holocaust art "in dialogue with numerous other contexts of traumatic transfer that can be understood as postmemory" (Hirsch 2012, 18). Among many other phenomena, she lists American slavery, the Vietnam War, the Dirty War in Argentina, and Communist terror as examples for sites where the notion of "postmemory" has become an important explanatory vehicle (Hirsch 2012, 19).

Regarding scholarship on the 'German past,' one can observe a broader use of concepts and approaches developed in Holocaust studies as well. For instance, in an article titled "*Flucht und Vertreibung* and the Difficult Work of Memory," Linda Warley (2013) explores how experiences not one's own can form part of one's identity. Combining aspects from her own family history with academic discourse on memory and life writing, she links her mother's individual story not only to the historical events of forced migration but also to her own identity formation. Reflecting on the usefulness of memory studies for understanding the transgenerational effects of forced migration, Warley draws on Hirsch's work on "postmemory." Yet in a surprising move, she introduces the notion of "acquired memory" (Warley 2013, 329) in explicit juxtaposition to Hirsch's term. By employing a new term she might try to dispel here any suggestion that transgenerational effects of the Holocaust are comparable to those of German wartime suffering.

Warley's use of this terminology raises the larger question of the relationship between the Holocaust and forced migration in collective memory as well as in academic discourse. While this is not the place to explore this complex constellation in any detail, suffice is to say that some scholars are highly critical of attempts to see both the Holocaust and forced migration in global contexts of ethnic cleansing.[12] Eva and Hans Henning Hahn have provocatively referred to the "Holocaustization of the memory of flight and expulsion" that results in dehistorizing flight and expulsion and absolving Germans of responsibility for the Holocaust and National Socialism (2008). They identify these revisionist trends not only in historiography but also in collective memory discourses shaping German national identity.[13] While Hahn and Hahn raise important concerns regarding historiographical research on the forced migration of Germans in

12 For transnational studies of ethnic cleansing, see Aly 2005; Naimark 2001; Schwartz 2013.

13 The phrase was first used by Andreas Kelletat 2005. For a more comprehensive argument along similar lines, see also the monograph by Hahn and Hahn 2010 and Lange's 2016 study.

the context of nationalism and ethnic cleansing, their competitive approach to memories of violent histories risks ignoring the benefits of transnational scholarship on forced migration.[14] Briefly put, transnational approaches do not preclude historical contextualization, as the work of Michael Schwartz and Jan Maria Piskorski illustrates (Schwartz 2013; Piskorski 2013).

Returning to the issue of methodology and terminology, Warley's adoption of a new term reminds us to take extra care when employing theoretical concepts like 'postmemory' in new scholarly contexts like forced migration. Arguably, the borrowing of theoretical concepts is a process of translation, that is, an interactive and dynamic process that affects both a particular field of research and the respective concept or approach.[15] From this vantage point, Warley's adoption of the term 'acquired memory' (related to but not identical to 'postmemory') is as relevant as the employment of the term 'postmemory' in interaction with new areas of research. In both cases, a meta-critical awareness of the contextual origins of specific concepts is important as it counteracts the uncritical appropriation (or naïve translation) of theoretical terms.

An excellent example for broadening the scholarship on postmemorial constellations in such a self-reflective manner is Gabriele Schwab's monograph *Haunting Legacies: Violent Histories and Transgenerational Trauma* (2010). Building on the work of Marianne Hirsch, Judith Butler, and Michael Rothberg, she argues convincingly for including effects of violent histories on descendants not only of victims but also of perpetrators while being mindful not to equate one group with the other. In her study, Schwab maintains that in order for the next generation, i.e., descendants of perpetrators, to move forward, mentally and socially, they need to begin the work of delayed mourning and address the effects of transgenerational trauma.[16] According to her line of argument they have to find their own voice – and autobiographical or literary writings play a significant role in this process. While Schwab focuses primarily on trauma connected to the discovery of the parent generation's responsibility for the Holocaust and war crimes, she also mentions violence and trauma experienced by this same parent generation during intense air raids at the end of the war. The task of the postwar generations to pursue "delayed mourning" (i.e., mourning that was often aborted

14 See also Beer's critical review of the 2010 book by Hahn and Hahn (Beer 2011).

15 For a helpful discussion of 'translation' as an analytical term for the transfer and transformation of approaches (and associated terminology) across genres, contexts, and disciplines, see Bachmann-Medick 2016, 256–260. She also mentions critical responses to a broad use of 'translation' (268–269).

16 For related arguments regarding the connection between memory work, social change, and human rights activism, see Rothberg 2009, 21–29 and Hirsch 2012, 6.

by the war generation, Schwab 2010, 103) would then extend to both groups of victims – to victims of Nazi Germany and to "German" victims of air raids and expulsions. This constellation is complicated by the fact that the latter group may have condoned Nazi ideology or participated in the victimization of the first group. But precisely these complicating factors can be addressed in works of fiction: literature provides the space to explore complex character constellations that do not neatly fit into victim-perpetrator binaries.

Schwab's study marks a significant shift in theorizing postmemory. Based on the analysis of literary and autobiographical texts, she refines our understanding of the dynamics of transgenerational trauma and delayed mourning as it applies not only to descendants of victims (Holocaust survivors) but also to those of perpetrators or those who occupied more complex positions on the victim-perpetrator spectrum. But unlike Schwab, who sees literature primarily through a psychoanalytical lens and an interest in trauma, my own emphasis is on the relationship between individual and collective memory discourses and on the creative and multi-voiced dimension of literature. In short, the novels discussed in the subsequent sections of this contribution are relevant for but not limited to working through transgenerational hauntings.[17] Other important dimensions that intersect with these psychological processes include intertextual references, the role of re/mediation, and the use of irony, parody, and other distancing devices. Furthermore, as Anne Fuchs and Mary Cosgrove have pointed out, the imaginative dimension of postmemories may also result in the misrepresentation or sentimentalization of history, potentially sidestepping questions of power relations, guilt, and responsibility (Fuchs and Cosgrove 2006, 11–12). This critical perspective assumes special significance when it comes to literary representations of German wartime suffering as this focus inherently raises the question of historical contextualization.

To illustrate how the analysis of literary representations of flight and expulsion benefits from theories of postmemory and how, conversely, the genre of the novel enhances our understanding of the functioning of post/memory, I now turn to two novels. A close look at Walter Kempowski's *Mark und Bein* (*Bone and Marrow*) from 1992 is followed by a brief discussion of Ulrike Draesner's 2014 novel *Sieben Sprünge vom Rand der Welt* (*Seven Leaps From the Edge of the World*): two distinct literary responses to the historical events of forced migration and their effects on subsequent generations. Taken together, these novels span more than two decades and show how authors of different generations address

---

17 For a critical discussion of the privileging of trauma in the study of memory, see Huyssen 2003, 9–10.

memories and postmemories of flight and expulsion. Walter Kempowski, born 1929, is one of very few authors of the war generation who addressed the long-term effects of flight and expulsion in his creative work. He did so at a time when these historical events were still largely associated with the political right. More than 20 years later, Ulrike Draesner, an author of the subsequent generation (born in 1962) explores the changing responses to and memories of these events across four different generations and two different nationalities, Germans and Poles. In contrast to Kempowski, whose novel predates the "obsessive" public and scholarly attention to memory (Huyssen 2003, 3), Draesner draws on insights of memory studies, including the very concepts of 'postmemory' and 'transgenerational trauma.' In short, these novels throw into relief not only the changing memory scapes from which they emerged but also the academic discourses they either predate (Kempowski) or incorporate (Draesner).

2 Literature and Post/Memory Within a National Paradigm

Large parts of Walter Kempowski's novel *Mark und Bein* read like a satire of West Germany's attempts to come to terms with the 'German past.' The novel was published in 1992, a decade earlier than *Im Krebsgang* (*Crabwalk*). Both novels examine the lingering effects of the past, including National Socialism, the Holocaust, and flight and expulsion, on subsequent generations. But while *Im Krebsgang* takes place at the turn of the millennium, Kempowski's novel portrays German society in the 1980s with special attention to Germans born after the war. Perhaps the best example of Kempowski's satirical take on dominant public attitudes towards the past is chapter 13, which portrays tourists at the Marienburg/Majbork. This major historical landmark and medieval remnant of the German presence in the East is located close to Gdansk in Poland; the use of both the German and the Polish name points to the complexity of this "lieu de mêmoire" (Nora) or site of memory. The chapter illustrates how this multi-layered site of memory is instrumentalized by two groups of tourists: students from the leftist *Rosa Luxemburg High School* and members of a *Landsmannschaft*, an expellee association representing the political right. By portraying the groups' diverging responses to the Majbork tour guide, the novel highlights engrained ideological positions: those who are determined to see Germans merely as perpetrators and those who continue to believe in German cultural superiority *vis-a-vis* the East. Both perspectives come across as highly problematic.

This chapter stands out in its parody of West German memory contests, but throughout the novel, lingering stereotypes about Poles and the East are juxtaposed with official pronouncements regarding Germany's responsibility for the War – a contrast that calls into question the sincerity of these assertions. Furthermore, many characters recognize German guilt but at the same time lack interest in those who were affected by the violent past. This attitude extends to both Germany's victims and German victims. Regarding the latter, the novel underscores that an engagement with German wartime suffering was deemed politically inappropriate. For instance, the main protagonist (Jonathan Fabrizius) refrains from reading a book documenting the expulsions[18] on the subway because he fears to be perceived as a "Kalter Krieger," a Cold-War warrior who challenges the postwar borders (Kempowski 1992, 57).

Jonathan Fabrizius, who lacks regular employment but leads a comfortable life in Hamburg, might be seen as a caricature of a particular segment of the educated West German middle class. Yet he is also the most fully developed character whose generational position is central to the plot and the novel's larger significance. Jonathan's birth during the flight in 1945 coincides with the end of the war but his life is marked by traumatic events of the war,[19] specifically his mother's death during childbirth and his father's death at the collapsing Eastern front. This constellation marks the protagonist as prototypical member of the "second" generation or, as Hirsch has termed it, the "postgeneration": growing up in postwar West Germany, Jonathan's familial origin combines both guilt (via his soldier father)[20] and suffering (via his mother's death during the flight). While the protagonist initially shares the dominant dispassionate attitude towards the past, the novel chronicles his increasing personal involvement with his parents' fate, a shift that is tied to his travel to Poland.

An example of Jonathan's initial attitude towards his past is the seemingly casual manner in which he comments on the loss of his parents at the end of the war: "'Meine Eltern hab ich nicht gekannt', sagte er oft gleichmütig, meinen Vater

18 This is most likely a reference to the federally funded *Dokumentation der Vertreibung der Deutschen* (*Documentation of the Expulsion of Germans*) that appeared in multiple volumes between 1951 and 1963.

19 This constellation resembles the one in *Im Krebsgang* where the narrator's birth coincides with the sinking of the Gustloff.

20 For reasons that are not entirely clear, Jonathan extends a general association of the war generation with guilt – or potential guilt – to his uncle (who serves as father figure). Jonathan repeatedly draws attention to the fact that he resembles Julius Streicher (a prominent Nazi and founder of the propaganda paper *Der Stürmer*) while also emphasizing that the uncle was a "nice man" (Kempowski 1992, 112, 238).

hat es auf der Frischen Nehrung erwischt, und meine Mutter ist bei meiner Geburt draufgegangen, in Ostpreussen, 1945"[21] (Kempowski 1992, 21). The protagonist's indifference ("Gleichmut") informs his word choice: for instance, he uses "draufgegangen," a colloquial term for dying (to kick the bucket) to refer to his mother's death during childbirth – i.e., the birth of the protagonist – on her flight from East Prussia. In this context, the notion of "Leidensvorsprung" is mentioned, a sarcastic reference to the "edge in suffering" that Jonathan gains from the tragic death of his parents. This term, mentioned several times in the novel (Kempowski 1992, 21, 32, 56), underscores not only the character's apparent lack of affect regarding the fate of his parents but also Jonathan's instrumentalization of his victim status.

Despite this professed indifference, the narrative indicates that Jonathan suffers from the traumatic events that preceded his life. The most obvious sign is that he has a recurring vision of his dead mother whose body is placed in front of a church by her brother (Jonathan's uncle who later raised his orphaned nephew Jonathan) during the flight. This image recurs multiple times early on in the narrative, mirroring the persistent intrusion of traumatic experiences. Recalling Hirsch's definition of "postmemory," Jonathan "'remembers' only by means of the stories, images, and behaviors" among which he grew up; the haunting vision of his mother suggests that his uncle's experiences were transmitted to him "so deeply and affectively as to seem to constitute memories in their own right" (Hirsch 2012, 5).

At one point, the narrative comes close to providing a definition of "postmemory" *avant la lettre*. When Jonathan accepts a journalistic assignment that involves extended travel through the former East Prussian region of contemporary Poland, he decides against seeking advice from his uncle: "Ihm brauchte der Onkel nichts von Ostpreußen zu erzählen, er wußte alles, auch das, was er ihm nicht erzählt hatte"[22] (Kempowski 1992, 59). This comment suggests that the protagonist is shaped not only by the uncle's memories and anecdotes but also by what was communicated to him implicitly and through other media of collective memory, including the documentation mentioned earlier. Both, explicit and implicit dimensions are central to the concept of 'postmemory' and contribute to its powerful effects on the postgeneration.[23]

---

21 "I never knew my parents, he often remarked with indifference, my father was killed on the Vistula Spit, and my mother perished during my birth, in East Prussia, 1945" (my translation).
22 "There was no need for the uncle to tell him anything about East Prussia; he knew everything, even those things that he [the uncle] had never told him" (my translation).
23 These implicit or partial transmissions of memories – through gaps, nonverbal cues, silences – make up an important dimension of postmemories (Hirsch 1997, 22–23).

The novel highlights the extent to which postmemories emerge from multiple forms of mediation and remediation. Jonathan's anticipation of the trip and his observations during the trip are a case in point. These passages are saturated with iconic images, clichés, anecdotes, and snippets of historical information, crowding out his interest in the (Polish) presence. Put differently, Jonathan is so preoccupied with his postmemories that his travel to the East does not result in any serious engagement with contemporary Poland. Instead the visit to specific places conjures up images and stories of the past, acquired via multiple sources and media.[24] From the vantage point of Kempowski's novel, postmemories can thus be conceptualized as a particular kind of remediation of familial memories mixed with collective memories of German-Polish history.

Against this backdrop, the encounter with his parents' sites of burial or death takes on special significance in the narrative. During his travel through Poland, Jonathan suddenly and inexplicably realizes that he has reached Rosenau (the novel only references the German name), the place of his mother's burial. This encounter with a traumatic place of the past is marked by a shift in narrative style. He intuitively identifies the unmarked gravesite even though he only knows about it through his uncle's memories: "[E]r starrte auf eine Stelle an der Mauer, und er wusste: Dort liegt *sie*"[25] (Kempowski 1992, 202). In stark contrast to the indifference with which the protagonist had repeatedly commented on his parents' death, the visit of his mother's place of burial overwhelms and temporarily immobilizes him. Significantly, this is one of the rare passages narrated without distancing devices;[26] instead the immediacy of the experience – and Jonathan's inability to communicate his state of mind – is highlighted (Kempowski 1992, 202–203). When he is about to leave the cemetery, a fellow traveler's question about his father finally triggers a visible emotional response in Jonathan, as well as vivid memories of his father and the overall desperate situation at the end of the war and, beyond that, of humankind.

Unlike the remainder of the novel, which underscores the highly mediated nature of the protagonist's travel to the East, the encounter with his mother's gravesite and, shortly thereafter, with the presumed place of his father's death,

---

24 This applies even to Jonathan's encounter with a Polish woman and her sick daughter in Gdansk; his visit to their home triggers images of belonging (*Heimat*) and associations with his own mother (whom he never met).

25 "He stared at the spot close to the wall and he knew: There *she* lies" (my translation).

26 Berger reads the novel's distancing devices as a way of portraying German suffering without "uncritical empathy or sentimentalized victimhood" (Berger 2014, 125).

are marked by immediacy.[27] These central passages also help to explain the novel's title *Mark und Bein* (*Marrow and Bone*). Readers are more likely to relate Jonathan's psychosomatic response to the colloquial saying "es geht durch Mark und Bein" ("to be shaken to the core") than to the biblical passage from which the saying originates.[28] In sum, both content and style point to the unprocessed nature of Jonathan's postmemories as well as to the role of place as a powerful trigger for the raw emotions attached to them.

Overall, the 1992 novel illustrates a particular dynamic between memory and space that has since been explored in the study of memory and postmemory. As Hirsch points out in her reading of stories of return after dispossession and displacement: "Return to place literally loosens the defensive walls against the sorrow of loss that refugees build up over decades and that they pass on to their children" (2012, 207). Kempowski's novel is a case in point: Jonathan's encounter with the presumed places of his parents' death turns the distant past temporarily into the present.

Building on Hirsch, in their introduction to *Germany, Poland and Postmemorial Relations* (2012) Kristin Kopp and Joanna Niżyńska have coined the term "post-memorial space." The concept captures the powerful effect particular locations may have on members of the postgeneration who only know about them in a highly mediated fashion, that is, via stories, images, and memories that, strictly speaking, are not their own.

> 'Space' is understood here in the literal terms of material surroundings, but these surroundings are, in turn, understood as the source of a sense of inhabiting history; post-memorial space thus carries the potential of exerting a powerful effect on its inhabitants. Their experiences may range from living in a house left behind by German expellees, to touring through a village recounted in one's grandfather's memoirs, to passing through Kraków's train station on a pilgrimage to Auschwitz. The affective charge of encountering a historical legacy in its sheer materiality may, however, not always be as direct as one expects; the imaginary qualities of postmemory enter our experience of space as well. (Kopp and Niżyńska 2012, 19)

---

27 Later in the novel the protagonist also visits the presumed location of his father's death on a peninsular stretch at the Baltic Sea (in German, "Frische Nehrung"). Part of this passage is similarly marked by immediacy and sudden insight ("hier war es," Kempowski 1992, 231) as well as the appearance of the specter of his father (Kempowski 1992, 232).

28 I would like to acknowledge Emily Sieg and Willi Barthold for this insight into the relevance of the novel's title. In a joint paper (for my graduate seminar "Fluchtgeschichten," spring 2016), they analyzed the protagonist's visit of the cemetery and his father's place of death with reference to Benjamin as an "aura of postmemorial space that cannot be mediated."

This definition of postmemorial space goes beyond Hirsch's account on the role of place[29] in two important ways: it explicitly includes examples of postmemorial sites related to flight and expulsion, thereby further expanding the methodological reach of the notion of 'postmemory' beyond Holocaust studies. At the same time, Kopp and Niżyńska also introduce an element of caution regarding postmemory's "imaginary" qualities, recalling Fuchs's critique of the concept. As they explain with reference to the contribution by Erica Lehrer, "Holocaust tourism" to former concentration camps has resulted in the association of the Holocaust not with Germany but with Poland, the location of Auschwitz, among some visitors. Lehrer thus points to the "necessity of resisting pure affect" in the context of postmemorial constellations (Kopp and Niżyńska 2012, 20). In sum, the volume *Germany, Poland and Postmemorial Relations* in general and the notion of "postmemorial space" in particular are examples of how approaches and theoretical terms developed primarily in Holocaust studies are productively employed in broader historical and cultural contexts, and how these concepts in turn have been shaped by these new contexts.

To return for a moment to Kempowski's novel, the personal and socio-political situation that constitutes its backdrop does not provide any corrective to the 'pure affect' triggered by Jonathan's visit of postmemorial places. His trip neither results in a more sustained exploration of the (family) past nor of his own life. Instead, the novel emphasizes on the one hand Jonathan's overpowering sense of past suffering and guilt[30] and, on the other, his friends' lack of empathy and interest. When nobody wants to hear his 'story' upon his return to Hamburg, Jonathan muses: "Wem sollte er nun von seinen ostpreußischen Tagen erzählen? Von seinen Vergangenheitserlebnissen, dass es nicht ungefährlich ist, sich mit Sachen zu beschäftigen, die man besser ad acta legt"[31] (Kempowski 1992, 237). The wording of this passage underscores the dynamics of postmemo-

29 While space and place have distinct meanings in spatial theories they are used interchangeably (or at least without clear definitions) in the sources discussed here.

30 After visiting his mother's gravesite, Jonathan's despair is described in the following way: "Es ist alles umsonst! dachte er immer und immer wieder. Und: Wer hat die Schuld?" ("It is all for nothing! he thought over and over again. And: Who bears the guilt?") (Kempowski 1992, 204). Just a few pages earlier, the protagonist makes an explicit connection between German guilt (and responsibility for the war) and the expulsion of Germans (Kempowski 1992, 197). – *Alles umsonst* (*All for Nothing*) is the title of Kempowski's last novel (2006) that also focuses on flight and expulsion [It appeared in English translation].

31 "Whom should he tell now about his East Prussian days? From his experiences in the past; that it was not without danger to preoccupy oneself with matters that one should rather put aside" (my translation).

ries within the larger socio-political situation in the (West) Germany of the 1980s: the fact that the protagonist speaks "von seinen ostpreußischen Tagen" ("from his East Prussian days") and "Vergangenheitserlebnissen" ("experiences in the past") confirms that his trip was not one to contemporary Poland but to postmemorial experiences and places located in a past that was not his own. "Dass es nicht ungefährlich ist" ("That it is not without danger") references the traumatic character of these postmemories, exemplified in the recurring image of his dead mother and his strong response at her gravesite. Furthermore, the comment "die man besser ad acta legt" ("that one should rather put aside") suggests that he intends to discontinue the 'dangerous' engagement with his family past, a plan that seems in direct response to the lack of interested interlocutors ("Wem sollte er nun [...] erzählen?" ["Whom should he tell now?"]).

Such a reading is further supported by Schwab's study *Haunting Legacies*, discussed earlier. Intratextually, the novel narrates the main protagonist's faltering attempt to engage with the effects of a painful past. If he fails to find "a voice,"[32] it is in no small part the result of a situation in which nobody is willing to listen. The novel thus ends with the prospect that the 'dangerous' issues related to the protagonist's and, by extension, to Germany's past will be put aside ("ad acta legen") – curtailing not only a better understanding of his family history, but also pointing more generally to a troubled identity formation of the post-war generation.[33] Read in the context of West Germany in the 1980s, the pervasive disinterest in German wartime suffering that the novel foregrounds suggests that at a collective level victims of flight and expulsion were not considered "grievable lives" (Butler).[34]

From today's perspective, Kempowski's novel throws into relief the dramatic changes in public discourse and in scholarship on flight and expulsion over the past two decades. Through content and narrative style, *Mark und Bein* presents dominant memory discourses prior to unification in a highly critical manner by foregrounding a pervasive sense of numbness. West German society is portrayed

---

32 According to Schwab, "finding a voice" is a first step toward a successful engagement with the familial and collective past (Schwab 2010, 81).

33 According to this psychoanalytically informed approach, short of finding such a voice, trauma will continue to speak through the body or through repetition of the same image or fantasy (Schwab 2010, 83–84).

34 Schwab references Judith Butler's *Precarious Life: The Powers of Mourning and Violence* (2004). Butler argues along similar lines, albeit in the markedly different context of the US military response to 9/11, against the "violence of derealization" (Butler 2004, 33), that is, against making the enemy's victims invisible. Instead, she makes a plea for the important process of grieving as precondition for working towards social change and social justice.

as displaying a general awareness of but no real engagement with the Nazi past, and as lacking any interest in the fate of expellees and the lingering effects on the next generation. Instead, the novel suggests that public recognition of German wartime suffering was considered politically inappropriate, especially among the educated middle-class.[35] While this latter critique was traditionally associated with the political right, Kempowski places the lack of engagement with German wartime suffering in the larger socio-political context of German guilt and coming to terms with the Nazi past. The novel maintains that, contrary to official pronouncements, this process was also still at its beginning.

Furthermore, *Mark und Bein* is one of the first novels to explore transgenerational hauntings of flight and expulsion including the powerful impact of postmemorial places.

It is no small irony that the political and social constellation portrayed within the novel was mirrored in the novel's reception when it was published in 1992. From today's perspective it is astounding that critics paid hardly any attention to the postmemorial constellation the novel captures so aptly. Instead, several reviewers adopt a condescending tone and reduce the plot to a "nostalgic pilgrimage into to the past" (Ross 1992)[36] while others mimic the author's sarcastic style in their comments on the novel.[37] Overall, the response in the media and the scant scholarship on the novel illustrates that the conceptual and socio-political context for recognizing and critically commenting on these dynamics was largely missing in the early 1990s.[38]

As my reading of the novel has shown, drawing on memory studies, and in particular on the concept of 'postmemory' that was initially developed in the context of Holocaust studies, enables us to make sense of the novel in new and

---

35 The novel's critical perspective on both memory discourses is significant and explains why Berger lauds the novel for achieving a balanced representation of German guilt and victimhood (125).

36 In his review Werner Ross also points to the novel's lack of engagement with contemporary Poland, one of the few substantive critiques.

37 Reviewers are critical of Kempowski's consistent use of satire, irony, and caricature and his mixing of autobiographical and fictional elements, among other aspects (see Schacht 1992 – *Welt am Sonntag*; Goertz 1992 – *Tagesspiegel*; Kramberg 1992 – *SZ*; Ross 1992 – *FAZ*, among others). The condescending tone of many reviews reflects the consensus of the 1980s and early 1990s that Kempowski was an author of popular literature and that his political stance did not comply with the dominant liberal stance among intellectuals (see Hempel 2004, 208–209).

38 By contrast, the novel was well received when it was reissued in 2006, at a time when Kempowski had gained broad success with the publication of his *Echolot* series and when public and academic discourses on forced migration had changed considerably (for an insightful account of this shift, see Berger 2014, 109–152).

productive ways. The troubled origin of the protagonist – symptomatic of post-war German society – contributes to a complex postmemorial constellation: the transgenerational effects of German wartime suffering are linked with the question of (German) guilt. At the same time, the multi-voiced literary dimension of the novel – combined with the specific historical constellations it references – challenges us to critically reflect on the very notion of 'postmemory.' In the novel, postmemories are inseparable from multiple layers of mediation and remediation all of which contribute to a highly subjective notion of the past with repercussions for the present. By highlighting the continued influence of national stereotypes and a general lack of engagement with contemporary Poland and its people, the novel points to the dangers when such a preoccupation with the past remains closed off from empathy and from scrutiny.[39] From a meta-critical perspective the novel also reminds us of the cultural biases and blinders that postmemories might unwittingly transport. *Mark und Bein* is a first invitation for this kind of critical engagement.

3 Literature and Post/Memory: Toward a Transnational Paradigm

In the post-war period and through the early 1990s, literary accounts of flight and expulsion, Kempowski's 1992 novel chiefly among them, were largely ahead of public discourse and of scholarship. By contrast, contemporary literary and scholarly discourses are frequently intertwined, drawing on one another and yielding new insights. Memory studies now serves as one of the main conceptual frameworks for literature *and* scholarship on forced migration.

Put differently, authors of the second and third generations who did not experience the war and forced migration first-hand, have continued the creative memory work that Kempowski – via his main character in *Mark und Bein* – breaks off at the end of the novel.[40] Significantly, some of these younger authors, including Tanja Dückers, Olaf Müller, Jörg Bernig, Sabrina Janesch, and Ulrike Draesner,

---

39 On the role of stereotypes of Poles in the novel, see Jaroszewski 1998, 246–247. The novel's consistent use of irony and sarcasm makes it difficult to determine the narrator's position. In my reading, the novel highlights and thus implicitly challenges these stereotypes.

40 In his last novel *Alles Umsonst* (Kempowski 2006), Kempowski does not continue to investigate the legacy of these events, instead he returns to the historical events themselves and narrates them by drawing on multiple sources of remediation (see Berger 2014, 136–152).

address post/memories of lost homes, trauma, and guilt no longer in an exclusively German context. Instead, they also engage with aspects of Eastern European history, most frequently with Polish history.[41]

If Kempowski's 1992 novel represents an early literary account of the transgenerational effects of forced migration, Draesner's 2014 novel *Sieben Sprünge vom Rand der Welt* (*Seven Leaps from the Edge of the World*) illustrates how similar issues are addressed more than two decades later by an author of the post-war generation. Two interrelated aspects of Draesner's novel exemplify current literary and scholarly trends: *Sieben Sprünge vom Rand der Welt* challenges an exclusively national approach to the histories of forced migration. And it draws, at times explicitly, on memory studies, especially on the notion of 'postmemory' and the role of re/mediation.

In her comprehensive 550-page novel, Draesner reworks aspects of her own family history and combines this 'German' tale of the flight from Silesia with the (fictional) story of a Polish family that, due to the westward expansion of the Soviet Union, also lost their home and was forced to resettle in what became Polish Silesia. The novel is organized around first person accounts of four generations of Polish and German characters: the two generations who experienced the war and forced migration first-hand (parents and children), and two post-war generations: those who were born in the 1960s and their children, born around 1990. The title, *Sieben Sprünge vom Rand der Welt,* references members of the two older generations – some German, some Polish – who were forced to "leap" from their respective homeland. Structured as a multi-voiced narrative, the novel frequently presents the reader with different perspectives on the same time period or event.

In stark contrast to Kempowski's self-absorbed protagonist who barely acknowledges contemporary Poland during his travels, some of Draesner's protagonists engage with one another across different national backgrounds. Regarding the representation of forced migration, the novel incorporates the histories and memories of Poles and of Germans not in a competitive but rather in a multidirectional manner as discussed by Rothberg. He developed his approach in the context of the globalization of Holocaust discourses; yet his notion of memory as "multidirectional," that is, "as subject to ongoing negotiation, cross-referencing, and borrowing, as productive and not privative" is useful for other contexts as well (Rothberg 2009, 3, 28). In regard to Draesner's novel, the notion of 'multidirectional memory' is helpful for a consideration of its multi-voiced organiza-

41 As mentioned, Polish authors started to engage with similar issues much earlier (Eigler 2014, 151–154; Sywenky 2013).

tion and overall impact. The focus on multi-dimensionality pertains to the plot, e.g., characters 'borrow' or 'cross-reference' each other's memories. But it also involves the narrative structure, that is, the placement of German and Polish voices next to one another. This organization has the effect that histories and memories which most readers are used to viewing from a national perspective gain new significance within a transnational context. At the most basic level this means that German readers are likely to relate the memories of Poles who were forced to leave the Eastern territories of Poland (due to the expansion of the Soviet Union) to those of German expellees. Collective 'German' memories are refracted via 'Polish' memories.[42]

Sieben Sprünge also includes convergences across national lines regarding traumatic flight experiences. An example is when the Polish character Halka, who is forced to resettle in Silesia, witnesses the rape of several German women by Soviet soldiers. Halka describes herself in hindsight as a person who turned away and told herself "es sind nur Deutsche, und das stimmte, es waren nur Deutsche, doch Frauen waren sie auch, wie ich" (Draesner 2014, 474).[43] Through internal focalization, intensified by a shift to the first person, this episode highlights Halka's ambivalent response: her distancing from the enemy ("nur Deutsche;" "only Germans") while empathizing with the women who are subjected to sexualized violence. At the level of literary representation, this passage can serve as an example of multidirectional memory. The narrative breaks open a purely national perspective on this act of violence without glossing over differences that result from opposing positions in the war ("sie waren nur Deutsche, doch Frauen waren sie auch, wie ich;" "they were only Germans, but they were also women, just like me;" my translation).

Yet there are limitations to the novel's transnational approach and, by implication, to the role of multi-directional memories. In an effort to highlight similar human conditions of Germans and Poles during the experience of forced migration, *Sieben Sprünge* hardly mentions significant differences in the historical circumstances and events that preceded the respective resettlements, specifically Nazi Germany's brutal occupation of parts of Poland (which included forced population movements) and the systematic crimes committed by the German army against the Polish population during the war. Of course, a novel cannot be held

42 Janesch's 2010 novel *Katzenberge* goes even further than Draesner by focusing primarily on the flight of Poles and the difficult process of establishing a new home in the formerly German region of Silesia (Eigler 2014, 155–176).

43 "These are only Germans, and that was correct, they were only Germans, but they were also women, just like me" (my translation).

to the same standards as a historiographical study. *Sieben Sprünge* is in many other ways quite successful in challenging a single national perspective, but I mention these aspects because they relate to some larger concerns: the kind of shortcomings that Traba and Zurek identify in German public discourses as well as in some scholarship on forced migration (Traba and Zurek 2011); the difficulties of developing substantive transnational approaches; finally, they raise questions about the very concept of 'multi-directional memory.' Even if one agrees with Rothberg that memories of violent histories often draw on one another in a productive manner, at the level of political discourse and critical analysis it might sometimes be necessary to insist on the competitive nature of particular memories. In the case of Draesner's novel this means keeping in mind the vastly diverging historical contexts that remain beyond the purview of the narrative but that constitute the backdrop to 'shared' experiences.

Finally, a few comments on the role of postmemories in *Sieben Sprünge*. At the plot level, the novel's transnational dimension extends to postmemorial constellations. For instance, one of the Polish characters (Boris) is an expert on transgenerational effects of the war, including forced migration. The German character Simone, like Boris a member of the postgeneration, is attracted to him because they share generational experiences and sensibilities across national lines. Beyond these explicit references, and a somewhat contrived German-Polish romance that emerges from this encounter, some of the most compelling parts of the novel revolve around the memories of one particular character with the name Emil. Physically and mentally handicapped, Emil barely survives Nazi racial policies but then disappears during the flight from Silesia at the end of the war. The novel's narrative composition and language highlight the unclear circumstances of his death. For instance, the reader learns about Emil up to the time of his disappearance primarily through the diverging memories of several other characters. Only at the very end do we hear Emil's own voice. In a fragmented yet lyrical style he speaks from a liminal space in the past, anticipating the loss of his home and his life and ending in a farewell ("ich wünsch euch Glück;" "I wish you good luck"). The fragmented language of the epilogue defies any attempt to turn his account into a coherent narrative. By not providing closure to the story of Emil, the novel alludes to the historical events, that is, the fate of countless expellees who disappeared and perished during the flight.

In the context of this fictional retelling of a traumatic past, the memories of Emil's parents and his brother, characters who survived the flight, approximate the work of mourning for Emil. From an extradiegetic perspective, the entire novel and especially the poetic epilogue, illustrate the author's own attempt of working through postmemories of flight and expulsion.[44] *Sieben Sprünge* can thus be considered an example of "delayed mourning" (Schwab 2010, 103) that Draesner performs in lieu of the generations that experienced the war and the flight first-hand. This contrasts sharply with Kempowski's novel where an exploration of the past and the process of mourning are cut short. Draesner incorporates insights into post/memory not only into the novel's plot but also into multiple paratexts (essays, interviews, introductions, among others), some of which are accessible on a carefully designed website accompanying the book publication. *Der-siebte-sprung.de* serves multiple functions: it provides access to multiple documents and sources that contributed to the writing of the novel, including recordings of the author's trips to Wroclaw and her interviews with Poles;[45] under the rubric "Selber Erzählen" ("Narrate yourself"), it invites readers' responses and contributions; and the website provides space for metaliterary reflections. For instance, a "Lexikon der reisenden Wörter" (Lexicon of traveling words) includes historical and theoretical terms as well as some words in Silesian dialect and in Polish, all of which played a role in writing the novel.[46] Entries that draw on memory studies and often include quotations and references to pertinent scholarship include "postmemory," "Kriegskind" ("child of war"), "Schweigen" ("silence"), among many others. Overall, the *der-siebte-sprung.de* functions not only as the introduction to the novel but also as its continuation. (In fact, Draesner has described the website as the novel's last chapter.) The website builds a bridge between the fictional realm of the novel and a virtual space that allows for commentary, reflection, and dialogue. Taken together, the novel and the website counteract the closed-off space within which Jonathan – the protagonist in Kempowski's novel – grapples with post/memories of flight and expulsion in relative isola-

44 See essays 1 and 2 under the rubric "Romanwege" at <https://der-siebte-sprung.de/rendez-vous-7tersprung/> [accessed: 03 August 2016], where the author comments on her family background and the motivation for writing the novel.

45 Although the website includes Polish sources and some documents in Polish translation *der-siebte-sprung.de* is designed primarily for a German audience, underscoring once more the challenges of a transnational approach.

46 In her introduction (in German and in the Polish translation), Draesner explains that the words included in the lexicon were part of the writing process: "Das Lexikon war Teil meines Schreibprozesses. Es wanderte in den Text und wieder aus ihm heraus. Im *Tractatus logico-philosophicus* spricht Ludwig Wittgenstein von Leitern: die man anlegt, anlegen muss, um weiterzukommen. Doch ist man den Weg einmal gegangen, kann man sie zur Seite stoßen. Die reisenden Worte bewegen sich zwischen Sprachen, zwischen Erleben und Nachdenken, zwischen damals und heute. [...] Sichtbar werden Gedankengänge und Gefühlswege, die sich vielleicht nicht begrifflich, wohl aber atmosphärisch im Roman niederschlugen." [accessed: 08 August 2016].

tion.[47] By including voices of different generational and national backgrounds, Draesner's novel and website cut open this insolated memory space. And even if one recognizes that a lot more remains to be done to achieve a truly transnational perspective, the novel and website begin to transcend a narrow focus on national memories and traumas by considering Polish perspectives on forced migration.

To conclude, the vastly different literary approaches of Kempowski and Draesner throw into relief significant changes in literary and scholarly discourses on forced migration over the past two and a half decades. The main character of Kempowski's novel is stuck between collective memories of German guilt on the one hand and unprocessed postmemories regarding his parent's death at the end of the war on the other. Predating developments in memory studies, *Mark und Bein* addresses the roles of transgenerational memories and of postmemorial spaces. Draesner's novel, by contrast, clearly draws on aspects of memory studies and works creatively with and through postmemories. Furthermore, the author uses remediation – via the website *der-siebte-sprung.de* – to open up the generic space of the novel.[48] She does so by incorporating interactive features and by providing multiple entry points to the subject matter of the novel. Considering the integration of Polish characters and aspects of Polish history, *Sieben Sprünge* indicates a shift towards a growing transnational orientation in both German literature and scholarship on forced migration.

References

Aly, Götz. "Auschwitz und die Politik der Vertreibung." *Zwangsmigration in Europa. Zur wissenschaftlichen und politischen Auseinandersetzung um die Vertreibung der Deutschen aus dem Osten.* Eds. Bernd Faulenbach and Andreas Helle. Essen: Klartext, 2005. 35–44.

Assmann, Aleida. *Der lange Schatten der Vergangenheit. Erinnerungskultur und Geschichtspolitik.* Munich: Beck, 2006.

Bachmann-Medick, Doris. "The Translational Turn." *Cultural Turns: New Orientations in the Study of Culture.* Trans. Adam Blauhaut. Berlin/Boston: De Gruyter, 2016. 175–210.

Beer, Mathias. *Flucht und Vertreibung der Deutschen. Voraussetzungen, Verlauf, Folgen.* Munich: Beck, 2011a.

47 In Draesner's novel, this enclosed space is represented by the grandparents' living room where the older generation shares its memories of the flight.

48 The website thus also undercuts clear boundaries between primary and secondary literature, and between literary text and (scholarly) commentary.

Beer, Mathias. "Review of Hahn, Eva, and Hahn, Hans Henning: *Die Vertreibung im deutschen Erinnern. Legenden, Mythos, Geschichte.*" *H-Soz-Kult*, 03 June 2011b. <https://www.hsozkult.de/publicationreview/id/rezbuecher-15114> [accessed: 06 March 2016].
Berger, Karina. *Heimat, Loss and Identity: Flight and Expulsion in German Literature from the 1950s to the Present.* Oxford/Bern: Lang, 2014.
Bernig, Jörg. *Niemandszeit. Roman.* Munich: DVA, 2002.
Breuer, Lars. *Kommunikative Erinnerung in Deutschland und Polen. Täter- und Opferbilder in Gesprächen über den Zweiten Weltkrieg.* Wiesbaden: Springer VS, 2015.
Butler, Judith. *Precarious Life: The Powers of Mourning and Violence.* London/New York: Verso, 2004.
Draesner, Ulrike. *Sieben Sprünge vom Rand der Welt.* 2013– , <https://der-siebte-sprung.de/rendez-vous-7tersprung/> [accessed: 08 March 2016].
Draesner, Ulrike. *Sieben Sprünge vom Rand der Welt. Roman.* Munich: Luchterhand, 2014.
Dückers, Tanja. *Himmelskörper. Roman.* Berlin: Aufbau, 2003.
Eigler, Friederike. *Heimat, Space, Narrative: Toward a Transnational Approach to Flight and Expulsion.* Rochester, NY: Camden House, 2014.
Erll, Astrid, and Ann Rigney, eds. *Mediation, Remediation, and the Dynamics of Cultural Memory.* Berlin/Boston: De Gruyter, 2009.
Fuchs, Anne, and Mary Cosgrove. "Germany's Memory Contests and the Management of the Past." *German Memory Contests: The Quest for Identity in Literature, Film, and Discourse Since 1990.* Eds. Anne Fuchs, Mary Cosgrove, and Georg Grote. Rochester, NY: Camden House, 2006. 1–21.
Goertz, Heinrich. "Polenfahrt in Luxuslimousine." *Tagesspiegel* 23 February 1992.
Grass, Günter. *Im Krebsgang. Roman.* Göttingen: Steidl, 2002.
Hahn, Eva, and Hans Henning Hahn. "Die 'Holocaustisierung des Flucht- und Vertreibungsdiskurses.' Historischer Revisionismus oder alter Wein in neuen Schläuchen?" *Deutsch-Tschechische Nachrichten.* Dossier Nr. 8. May 2008.
Hahn, Eva, and Hans Henning Hahn. *Die Vertreibung im deutschen Erinnern. Legenden, Mythos, Geschichte.* Paderborn: Schöningh, 2010.
Halicka, Beata. "Erinnerungskultur." *Die Erinnerung an Flucht und Vertreibung. Ein Handbuch der Medien und Praktiken.* Eds. Stephan Scholz, Maren Röger, and Bill Niven. Paderborn: Schöningh, 2015. 89–99.
Hempel, Dirk. *Walter Kempowski. Eine bürgerliche Biographie.* Munich: Random House, 2004.
Hirsch, Marianne. "Family Pictures: *Maus*, Mourning, and Post-Memory." *Discourse* 15.2 (Winter 1992–93): 3–29.
Hirsch, Marianne. *Family Frames: Photography, Narrative, and Postmemory.* Cambridge, MA: Harvard University Press, 1997.
Hirsch, Marianne. *The Generation of Postmemory: Writing and Visual Culture after the Holocaust.* New York: Columbia University Press, 2012.
Huyssen, Andreas. *Present Pasts: Urban Palimpsests and the Politics of Memory.* Palo Alto, CA: Stanford University Press, 2003.
Janesch, Sabrina. *Katzenberge. Roman.* Berlin: Aufbau Verlag, 2010.
Jaroszewski, Marek. "Danzig und Ostpreußen in Walter Kempowskis 'Mark und Bein.'" *1000 Jahre Danzig in der deutschen Literatur.* Gdańsk: Institut Filologii Germańskiej, 1998. 233–247.
Kelletat, Andreas. "Von der Täter- zur Opfernation? Die Rückkehr des Themas 'Flucht und Vertreibung' in den deutschen Vergangenheitsdiskursen bei Grass und anderen."

Triangulum. Germanistisches Jahrbuch für Estland, Lettland und Litauen 2003/2004. Eds. Silvija Pavidis and Thomas Taterka. Vilnius: Vilnius Academy of Fine Arts Press, 2005. 132–147.

Kempowski, Walter. *Mark und Bein. Roman.* Munich: Albrecht Knaus Verlag, 1992.

Kempowski, Walter. *Alles Umsonst. Roman.* Munich: Knaus, 2006.

Kopp, Kristin, and Joanna Niżyńska. "Introduction: Between Entitlement and Reconciliation: Germany and Poland's Postmemory after 1989." *Germany, Poland, and Postmemorial Relations: In Search of a Livable Past.* Eds. Kristin Kopp and Joanna Niżyńska. New York: Palgrave Macmillan, 2012. 1–24.

Kramberg, K.H. "Ein Narr am Rande der Tragödie. Mit Walter Kempowski auf Ostpreußen-Fahrt." *Süddeutsche Zeitung* 02/03 May 1992.

Lange, Simon. *Der Erinnerungsdiskurs um Flucht und Vertreibung in Deutschland seit 1989/90. Vertriebenenverbände, Öffentlichkeit und die Suche nach einer 'normalen' Identität für die 'Berliner Republik.'* Dissertation, University of Heidelberg, 2015.

Möller, Kirsten. *Geschlechterbilder im Vertreibungsdiskurs: Auseinandersetzungen in Literatur, Film und Theater nach 1945 in Deutschland und Polen.* Frankfurt a.M.: Peter Lang, 2016.

Müller, Olaf. *Schlesisches Wetter. Roman.* Berlin: Berlin Verlag, 2003.

Naimark, Norman M. *Fires of Hatred: Ethnic Cleansing in Twentieth-Century Europe.* Cambridge, MA: Harvard University Press, 2001.

Niven, Bill. *Representations of Flight and Expulsion in East German Prose Works.* Woodbridge: Camden House, 2014.

Nora, Pierre. "Between Memory and History: Les lieux de mémoire." *Representations* (1989): 7–24.

Piskorski, Jan M. *Die Verjagten. Flucht und Vertreibung im Europa des 20. Jahrhunderts.* Trans. Peter Oliver Loew. Munich: Siedler, 2013.

Röger, Maren. *Flucht, Vertreibung und Umsiedlung. Mediale Erinnerungen und Debatten in Deutschland und Polen seit 1989.* Marburg: Verlag Herder-Institut, 2011.

Röger, Maren. „Ereignis- und Erinnerungsgeschichte von 'Flucht und Vertreibung:' Ein Literaturbericht." *Zeitschrift für Geschichtswissenschaft* 62.1 (2014): 49–64.

Ross, Werner. "Umsonst. Walter Kempowski in Polen." *Frankfurter Allgemeine Zeitung* 25 April 1992.

Rothberg, Michael. *Multidirectional Memory.* Palo Alto, CA: Stanford University Press, 2009.

Schacht, Ulrich. "Eine Episode aus deutsch-deutscher Vergangenheit um einen melancholischen Anti-Helden. Am Ende ein Windstoß verblaßter Bilder." *Welt am Sonntag.* 02 February 1992.

Scholz, Stephan, Maren Röger, and Bill Niven, eds. *Die Erinnerung an Flucht und Vertreibung. Ein Handbuch der Medien und Praktiken.* Paderborn: Schöningh, 2015.

Schwab, Gabriele. *Haunting Legacies: Violent Histories and Transgenerational Trauma.* New York: Columbia University Press, 2010.

Schwartz, Michael. *Ethnische 'Säuberungen' in der Moderne. Globale Wechselwirkungen nationalistischer und rassistischer Gewaltpolitik im 19. und 20. Jahrhundert.* Munich: Oldenbourg, 2013.

Spiegelman, Art. *Maus: A Survivor's Tale.* New York: Pantheon Books, 1986.

Stickler, Matthias. *Jenseits von Aufrechnung und Verdrängung. Neue Forschungen zu Flucht, Vertreibung und Vertriebenenintegration.* Stuttgart: Franz Steiner, 2014.

Sywenky, Irene. "Representations of German-Polish Border Regions in Contemporary Polish Fiction: Space, Memory, Identity." *German Politics and Society* 31.4 (2013): 59–84.

Traba, Robert, and Robert Zurek. "'Expulsion' or 'Forced Resettlement'? The Polish-German Dispute about Notions and Memory." *Breakthrough and Challenges: 20 Years of the Polish-German Treaty on Good Neighbourliness and Friendly Relations*. Ed. Witold M. Góralski. Warsaw: Elipsa, 2011. 363–401.
Traba, Robert, and Hans Henning Hahn. *Deutsch-Polnische Erinnerungsorte. Vol. 4: Reflexionen*. Paderborn: Schöningh, 2013.
Treichel, Hans-Ulrich. *Der Verlorene. Roman*. Frankfurt a.M.: Suhrkamp, 1998.
Warley, Linda. "*Flucht und Vertreibung* and the Difficult Work of Memory." *Life Writing* 10.3 (2013): 329–350.

Encarnación Gutiérrez Rodríguez

Conceptualizing the Coloniality of Migration

On European Settler Colonialism-Migration, Racism, and Migration Policies

The months of August and September 2015 were marked by media images of refugees making their way from Syria via Turkey, Greece, Macedonia, Slovenia, or Hungary to Austria and Germany, or even further afield to Denmark, Sweden, Finland, and Norway. It was moving to see adults, children, and the elderly walking with determination, fleeing war, misery, and despair. As they faced the destruction of their homes, the dismantling of their neighborhoods, and the deaths of their loved ones, the feeling of impotence that might otherwise have overcome them was converted into a driving force of collective hope. They shared meager resources and supported one another in a fraught journey filled with dehumanizing conditions and acts. Nonetheless, they continued their journey, sometimes with the support of local volunteers gathering food, clothing, blankets, and other necessary items. Cruelty was met with solidarity; the brutality of violence was confronted by a heartening love for life. However, as this wave of solidarity unfolded, politicians purportedly representing the 'ordinary citizen' publicly expressed vastly different sentiments about the situation.

As I will discuss using the trope of 'refugee crisis,' migration is a metonym for the coloniality-modernity juncture. It expresses one of the current social formations evolving within Aníbal Quijano's framework of the coloniality of power (Quijano 2000, 2008). Following Quijano's observation, I will develop an analysis of migration within this framework by proposing the 'coloniality of migration' as an analytical framework. As Quijano explains in his analysis of coloniality, European colonialism is characterized by the implementation and development of a global system of racialization, on which basis social hierarchies were established locally in distinct ways. Racism is at the center of the coloniality of power. The analysis of contemporary forms of racism requires that we attend to new forms of population differentiation, categorization, and classification as developed in migration and asylum policies. These policies contribute to the degradation and dehumanization of people by subjecting them to an objectifying logic of racial differentiation through contemporary migration discourses and governance practices of asylum and migration. This article will examine how racism works through migration discourses and policies in contemporary European societies by engaging with a historical perspective on how migration policies in former European colonial territories introduced in the late 19th century, operated within

https://doi.org/9783110600483-011

a colonial logic of racial differentiation. Within this context, I will examine the question of whiteness in transatlantic European settler colonialism-migration and overseas migration policies, elaborating upon how these contribute to conceptualizing the coloniality of migration. On this basis, I will subsequently address the relationship between racism and the contemporary migration-asylum nexus, and will conclude with remarks on how racial capitalism operates within the coloniality of migration.

1 Whiteness and Transatlantic European Settler Colonialism-Migration

Despite the transculturation (Gutiérrez Rodríguez 2010b) and creolization (Gutiérrez Rodríguez and Tate 2015b) of European societies, the 'ordinary citizen' is largely represented through a white racial ethnic-national lens. Myths about cultural homogeneity, racial and ethnic endogamy, immobility, and territorial rootedness prevail and are re-actualized in European nationalisms in the rise of right-wing national populisms. Tropes invoking the Ur-myths of racial purity, cultural endogamic authenticity, and ethnic encapsulation within accounts of European nationalism disregard the global historical entanglements in which Europe has evolved. Contemporary Europe has been marked by these issues: by trade, by political expansion by different regional powers, and by movements of people due to poverty, social deprivation, and religious and political persecution. Thus, it is remarkable that in contemporary popular media and political representations alluding to European national identity and belonging, Europe is imagined as unsettled by its own global entanglements, resulting from its history of colonialism, enslavement, imperialism, settler colonialism, indentured labor, global labor migration, exile, capitalist ventures, global governance, and wars.

Considering Europe's entangled global history, it is surprising that political and media discourses often perceive contemporary migratory movements as singular phenomena, and as external to Europe's history. In fact, transatlantic European migration has been foundational in the nation-states of the American continent[1] and the nations of Australia (Jupp 2002), New Zealand, and South Africa, which are marked by histories of European colonialism, settler colonialism, and transatlantic migration (see Lowe 2015). Defining themselves in the 18th

1 With the term 'American continent' I refer to North, Central and South America as well as the Caribbean.

and 19th centuries as countries of settlers and immigrants, these nation-states' public discourses on national, cultural, and linguistic representation oscillated in the 19th century between negation and partial acknowledgement of the transcultural fabric of their societies (see Martinez-Echazabal 1998). Through this day, these national narratives have masked the exploitation and dehumanization of the original inhabitants of these territories, ending in some instances with their genocide (see Lowe 2015). Also, the African presence in these territories due to transatlantic slave trade during the 16th and 19th centuries is omitted from the official national representation (Hall 2015), despite that approximately 13 million people from West and East Africa were enslaved and shipped to Europe and the American continent during this time (see Eltis and Richardson 2008, xiv–xv).

In the 21st century, the indigenous populations of these territories are still treated as non-citizens or second-class citizens (see Coulthard 2014; Simpson 2014). They are erased from national historiographic, cultural, and political representation; when they are remembered, they appear as objects of past times, displayed in a folkloristic manner in museums. Nonetheless, the presence of indigenous intellectuals, artists, and activists challenging the white supremacist discourse of the Eurocentric narratives of 'discovery' and 'country of immigration' is more public than ever (see Tuck and Young 2012; Tzul Tzul 2015).

This same narrative discloses the continuity between European colonialism and European transatlantic migration. Europe's colonialism defined world immigration patterns from 1500 to 1800: while Europe was establishing its colonial rule in Africa and Asia, between 1800 and 1925 approximately 48 million emigrants left Europe for the American continent, Australia, and New Zealand (Massey 2000, 62). Settlers arriving in the American continent from Britain, Ireland, Italy, Norway, Portugal, Spain, and Sweden in the 18th and 19th century formed part of settler colonialism. This transatlantic migration, coupled with the ongoing settler colonialism in Oceania, forms part of the modern European overseas colonial settlement project. Driven by the annexation of land, the appropriation of raw materials, and the subjugation of the indigenous population to pure exploitable labor, this project was also propelled by the economic boost produced by transatlantic slave trade, enabling industrialization in England and other parts of Europe as well as in the Americas. Within this context, many impoverished, religiously and politically persecuted Europeans were recruited to work in the rapidly expanding overseas plantation industry during the late 19th to early 20th centuries.

Though migration due to religious persecution, poverty, and epidemic menaces represents a constant feature of European history, it was not until the late 19th century that migration came to center stage in the regulation and control of the nation, differentiating between national citizens and the nation's Other, 'alien' or 'migrant.' As a biopolitical tool of governance, migration policies were engineered and implemented first in countries transitioning from colonial rule to sovereign national power. The first modern migration policies were developed in the late 19th century in North, Central, and South America, and in parts of the Caribbean (Lesser 2013; FitzGerald and Cook-Martín 2014). Guaranteeing the political, economic, and cultural influence of former colonial powers, migration policies established a set of instruments prioritizing the recruitment of white European migrants. This process took place in Canada, the United States, the Spanish-speaking Caribbean, and Latin America, or in territories kept in political dependency to the British Crown until the second half of the 20th century, such as Australia[2] and New Zealand.[3]

Due to the expansion of various modes of transportation and the need for workers in rapidly expanding industries, 19th-century white European transatlantic migration signaled the advent of racially structured capitalist progress, technological advancement, and urbanization. The system of racial capitalism (Robinson 2005) constituted the nation-states' rationale for the racial coding in migrant labor recruitment policies, introducing a racial division in the workforce in former European colonies. The recruitment of white European migrants was also determined by a cultural and educational project of nation-building in these former colonies, and the colonial discursive mantra of Europe as the cradle of civilization, modernity, culture, and progress underpinned the nation-state project.

The newly constituted sovereign nation-states in the American continent reacted to increasing immigration by establishing policies banning certain social, national, religious, and racial groups from entry. For example, after Britain introduced its first immigration service to promote the emigration of Irish and poor people to Canada in 1827, the Nova Scotia Assembly reacted in 1828 by establishing a bond system for immigrants entering the country. The bond system set a £ 10 tax on the master of any migrant vessel aiming to land on Canadian shores, if that could guarantee that the migrant workers arriving would not become a "burden" for the Canadian state, they would be refunded the bond within a year. At this time, the Canadian state refuse elderly, mentally and physically sick and

2 While Australia gained colonial independence from the British Crown in 1931, Britain's power remained as the Queen was responsible for appointing state governors and giving assent to state bills until the 1980s. This changed through the introduction of the Australian Act in 1986, which stated that the British government was no longer responsible for the government of any state and that the Westminster parliament could no longer legislate for Australia.

3 The 1852 New Zealand Constitution Act granted the colony's settlers the right to self-government, yet New Zealand did not become a sovereign nation-state until 1919. Like Australia, political ties to the Queen remained significant.

poor migrants to enter the country. This was later changed to a head tax system, which was applied to all migrants. In the 1870s, the United States followed the lead of the Canadian head tax system by passing legislation prohibiting certain groups of migrants from entering the country. In 1875, Congress prohibited the entry of prostitutes, convicts, and persons with mental health issues or physical incapacities; in 1891, the ban was expanded to persons suffering from contagious diseases, and allowed the deportation of migrants not complying with entry requirements. Some years later, this biopolitical screening began to include the categories of race and nationality in its selection criteria. A group of Black Bermudians entering Canada as British colonial subjects caused an uproar in the Nova Scotia Assembly in 1815. The Assembly complained to the British Crown about sending black people to Canada, asking the Crown to repatriate this group of people and prevent further migration of black people (Plender 1988). Similar reactions to non-European migration also took place in the United States, the Caribbean and Latin America.

In the United States, as Gerald Neuman (1993) notes, similar laws were passed in the 18th century, uncovering the myth that the United States of America was a country of free borders until the introduction of migration laws in 1875. In the 18th century movement between states was already regulated by the British Crown, and after independence, by the United States itself, in the form of incipient migration regulations. Also David Scott FitzGerald and David A. Cook-Martín (2014) concede that the United States has been one of the first nations to initiate racially coded naturalization and migration policies in the 18th and 19th centuries. The introduction of the Naturalization Law in 1790 reserve eligibility to naturalize to "free whites," excluding the indigenous and enslaved population from citizenship. Further, at this time the first federal migrations laws are past. For example in Massachusetts a law passed in 1794 penalizes "any person who knowingly brought a pauper or indigent person into any town in the Commonwealth" or the masters of vessels bringing "unauthorized" colonial settlers, banning in particular poor people from entering the state (Neuman 1993, 1849). In 1803, the first federal laws coded by race were passed. In the same year, in the southern states of the US an "enactment of a federal statute prohibiting the importation of foreign blacks into states whose laws forbade their entry" was passed (Neuman 1993, 1869). This federal law articulated the racist attitude of this state towards the movement of black people fleeing from slavery in the Southern States of the United States and seeking political asylum in states that had abolished slavery. Further, in the aftermath of the anti-colonial and anti-slavery rebellion in Saint Domingue in the early 19th century, black people coming from abroad to the United States were not only already considered suspicious as "free blacks," but their revolutionary engagement was feared as they were perceived as potential rebels that could organize resistance against racism (Neuman 1993, 1849). It was not until the beginning of the 19th century that a difference between 'aliens' and 'colonial settlers' was made. In 1831, laws passed in Massachusetts penalized the entry of 'aliens' into US territory.

In the second half of the 19th century, migration regulations were explicitly guided by racial differentiation. For example, the United States government reacted to Chinese migration by passing the Chinese Exclusion Act in 1882 (Plender 1988). The Chinese Exclusion Act established a system of registration whereby all Chinese workers were obliged to register or face deportation. Though lawyers partially challenged this act, restrictions on Chinese immigration were nonetheless tightened throughout the next decades. Canada also passed a Chinese Immigration Act in 1885 introducing a head tax of fifty dollars on Chinese migrants. Eleven years later, Australia passed an Immigration Restriction Bill to prevent access by Southeast Asian immigrants, followed by the White Australian Policy in 1904, which banned immigration from South Asia, particularly from India, and from Africa (Plender 1988). This policy continued into the second half of the 20th century.

In Latin America, as Tanya Ketarí Hernández (2013, 23) notes, "debates over immigration policies in Spanish America were often couched in racial language." At the beginning of the 19th century, the Congress of Gran Colombia (constituting what is now Colombia, Ecuador, Panama, and Venezuela) promoted settler colonialism by granting land to European migrants. Brazil and Argentina followed suit and prioritized European migration in their constitutions, arguing that this would be beneficial for technological and economic progress. After the constitutional emancipation of the enslaved population in 1853, Argentina actively promoted and sought European immigration. Between 1869 and 1895, the European population in Argentina increased from 1.8 million to 4 million people, and in 1914 it was 7.9 million, which made up 30 % of the population (Ketarí Hernández 2013, 23). This increase was not coincidental, but resulted from a concerted effort by the Argentine government, lobbying abroad for European workers, gifting them land, and, in the early years, partially covering white European migrants' transportation costs. Similar developments took place in Brazil, Cuba, and Uruguay, and 90 % of the 10–11 million European migrants that arrived between 1880 and 1930 settled in these countries or in Argentina. In Venezuela, after the constitutional emancipation of the enslaved population in 1854, the government's interest in white European migration was confirmed in public intellectual debates around the *blanqueamiento* (whitening) of the nation. Through the biological metaphor of a "transfusion of blood," the Venezuelan government recruited migrants from Ireland, Gran Canaria, Germany, and Italy, with the aim of keeping the nation white. In 1891, legislation was passed that prevented non-white migrants entering the country. This policy was integrated into the 1906 constitution, which also explicitly prohibited immigration by anyone of African descent (Ketarí Hernández 2013, 23). In a similar vein, in 1890, Brazil instituted Decree No. 528, which excluded all migrants from Africa and Asia from entering the country. This Decree instituted the primacy of whiteness and dispossessed the inhabitants of *Abya Yala*[4] from their entitlement to land they had inhabited for centuries as the migrants arriving mainly from Portugal, Italy, Spain, and Germany were thought to legally owned and settled in individual allotments of this territory. Brazil states in its 1853 constitution: "The federal government shall foster European immigration, and may not restrict, limit or burden with any tax whatsoever, the entry" (Ketarí Hernández 2013, 24). In 1921 Brazil's Federal Law prohibited the entry of "undesirables" (Lesser 1999). This legal regulation was factually executed, when Brazil rejected the settlement of a group of African-Americans that were planning to create a settler colony in Mato Grosso (FitzGerald and Cook-Martín 2014). Legislation was also passed in other parts of Latin America and the Caribbean that prevented Chinese and non-European immigration in particular. In Haiti, legislation forbidding the entry of Syrian immigrants was introduced in 1903 (Plender 1988, 69). Similar laws banning Arabs, Armenians were also passed in Costa Rica (1914), Panama (1909–1917), and Venezuela (1919) (Plender 1988, 69). Although Europe expanded its economic, political, legal, and cultural control over colonized territories overseas through settler colonialism-migration until the mid-20th century, in Europe itself, migration was not problematized until the second half of the 20th century.

2 The Coloniality of Migration and Racial Capitalism

Caribbean and South-East Asian immigration to the United Kingdom in the late 1940s and 1950s, and North African immigration to France in the 1950s, challenged the public myth that the European nation-state was cut off from the circuits of transatlantic settler colonialism-migration. In the 1960s and 1970s, immigration as constitutive of the nation-state was unavoidable, if we are to

4 *Abya Yala* is the term that the World Council of the Indigenous Nations meeting since every four years since 1977 has opted to use for this continent. *Yala* in the Kuna language means "land, territory," *Abya* means "whole of blood," "maternal maturity," "virgin maturity" and "land in its full maturity."

understand Fordism in Germany, France, and Britain as characterized by the recruitment of a labor force from the disfranchised territories of southern Europe, Turkey, Morocco, and the (post-)colonial territories. In the case of the migration from still or former colonial territories, this was happening within a new organization of territorial imperial power. Thus, the person arriving in France or the United Kingdom from the still or former colonies were subjected to imperial membership and in the case of Britain after independence those countries form part of the Commonwealth or in France, some became the Overseas Countries of France. This form of migration thus did not take place between nations, but within the territory of one imperial power and was negotiated between the citizens of the Empire and their colonial subjects, excluded from citizenship or semi-included as 'second citizen.' Yet, all these migratory movements did not push to the fore Europe's memory about its own history of colonialism, slavery, and imperialism, and neither of its own migratory movements due to settler-colonialism, transatlantic migration and exile to Africa, Asia, Oceania and the American continent. The 1980s made the global entanglement of Europe with its former colonies evident, the movements of people due to political persecution, poverty, war, austerity, social constraints, cultural restrictions, lack of employment, study, leisure, or a simple wish for change constitutes the fabric of current societies. It is within this context of transnational migration that migration and border control measures, technologies, devices, and tropes have been engineered in the last three decades. Set within this context, analysis of the connection between transatlantic European migration and racism in the 19th century, and of current migratory movements occurring within the framework of the migration-asylum nexus, requires an analytical framework that reconsiders migration as a metonym of the modernity-coloniality juncture. It is in this regard that I propose the analytical framework of the coloniality of migration.

The coloniality of migration operates within the matrix of a nomenclature of social classification based on racial hierarchies reminiscent of colonial differentiation. Colonial difference departs from the idea that the colonized population is fundamentally different and inherently inferior to the colonizer (Chatterjee 1993). It conceives the Other as a projection of the colonizer as radical and unassimilable other. In both cases, the Other is defined from a hegemonic position of the self as oscillating between the position of strangeness or similarity (Hall, Evans, and Nixon 1997). In discourses on migration and migrants, a similar dynamic of differentiation is played out, which is reminiscent of the logic of coloniality. Migration regulation remains a fundamental societal field in which the Other of the nation, of Europe, and of the Occident is reconfigured in racial terms. The logic generated in this context constructs and produces objects to be governed through restrictions, management devices, and administrative categories such as 'refugee' and 'asylum seeker,' or through a variety of migrant statuses. The entry, mobility, and settlement of migrants are strictly regulated and administered. Migration policies reiterate a matrix of objectification reminiscent of colonial times, and as such, operate within the nomenclature of racialization.

As W.E.B. Du Bois (2007) noted, modern societies are constituted through processes of racialization.[5] Exported from the 16th century to Europe's colonized and occupied territories, and developed further by European philosophical and scientific discourses in the 18th and 19th centuries (Bernasconi and Coole 2003), racism has become the fundamental matrix through which a world order is constituted and the world's population is divided (Bethencourt 2014). Within this system of racial classification, social categories of ethnicity, indigeneity, race, and religion emerged, classifying the population and developing a system of power through which relationships of governance, labor, economy, and culture were shaped. This matrix of power has been coined the "coloniality of power" by Aníbal Quijano (2000, 2008). The coloniality of power defines a matrix of knowledge deriving from European colonialism, which represents the epistemological grounds on which rational discourses, technologies, and practices of governing the global population were drafted. Coloniality defines the endurance of this mental script governing our present times; the link to power further refers to Foucault's understanding of power as ubiquitous and potentially productive, but also as a repressive apparatus. The coloniality of power refers to the racial matrix within which occurs the contemporary organization of labor, the configuration of relations of production and social reproduction, the production of cultural and political representation, and the circulation of knowledge and educational endeavors.

Though not spelled out through the term coloniality, this understanding of the colonial condition and its social, political, and cultural endurance has been outlined by anti-colonial thinkers such as Cyril L.R. James (1989), Eric Williams (1994), Claudia Jones (see Davies 2011), and Kwame Nkrumah (2006) in their analyses of European colonialism, pointing at racisms as the primordial axis of modernity. As they demonstrate, the differentiation between citizen and non-citizen (alien and others), which regulates access to the labor market, education, political participation, the health system, media, and cultural representation, was not only established in the colonies but also ruled the metropoles.

5 Du Bois's analysis relates to industrial societies, but processes of racialization already took place before and during colonialization, as shown in the introduction of laws for the persecution of the Jewish, Muslim, and Roma populations in the Spanish kingdom in the 14th and 15th century (see Eliav-Felden et al. 2009).

Thus, racism was not just exported to the colonies, but existed within the fabric of European societies before colonization (Eliav-Felden et al. 2009). Racism is not an exception to European modernity, but is at its very foundation.[6] Thus, following Eric Williams's observation (1996), the transatlantic slave trade foregrounds the entanglement between European modernity and the colonial plantation economy. It is in this entanglement that migration emerges in the 19th century as a modern nation-state colonial tool of governing the population in racial, ethnic, national, religious, and cultural terms.

Prior to this date, regions in Europe differentiated between citizens and aliens. For example, in 4 B.C. in Athens a differentiation between citizen, *metic*[7], and strangers/travelers existed, which regulated the entry of these persons to the city. In ancient Rome and the Roman Empire, the control of aliens determined access to imperial territory. Legislation regulating the relationship between citizens and aliens was crucial for the creation of the nation's Other in European migration policies. Historically, the 'alien' has been defined in Europe by two lines of membership. In early medieval times, large parts of Europe were ruled by a tribal kinship and a feudal system. While the tribal system departed from a biological kinship model, the feudal system established a relationship to the territory through property or serfdom. The tribal kinship model relies on the notion of sanguinity – blood ties – while the feudal system model depends upon the relationship to property (Plender 1988). The first constitutes the *jus sanguinis* principle, the second the *jus solis* principle.

The perception of the nation's Other in association with migration in Europe is related to these systems of imagining national communities according to (a) a tribal kinship model or (b) as property-related communities. In both cases, migration is perceived as exterior to the nation. In the first, migration seems unassimilable by the nation, and in the second, favorable political and economic conjunctures might promote forms of regulated immigration. However, in both cases the matrix of racialization is played out in the definition of the nation's Other. As we have seen, in the course of the introduction of nation-state migration policies in the 19th century, the status of 'alien' was developed into an administrative category occupying various regional and national imaginaries related to colonial difference. Within this setting, who would be considered an alien depends on

6 These authors deployed here develop a similar argument to Horkheimer and Adorno (1988) in regard to the analysis of antisemitism. As they argue, antisemitism represents the underside of modern Enlightenment.

7 In ancient Greece, the *metic* defined the status of a resident *alien* who did not have citizen rights and who paid a tax for the right to live in the *polis*.

specific historical, national, and local genealogies of Othering, the establishment of racial formations configured on legal and political terms. As I will argue next through the example of the migration-asylum nexus, migration policies still develop within the entanglement of thinking about the White national citizen and the racially different Other as two opposing sides mirroring the entrenchment of modernity and coloniality.

3 The Migration-Asylum Nexus and the 'Refugee Crisis'

In the 1970s, Chileans, Argentinians, and Uruguayans were recognized as exiles in the UK, Germany, France, and Spain (Gutiérrez Rodríguez 2010a). Nowadays, the term 'exile' has almost disappeared from public discourse, and has been replaced by asylum policies and discourses on 'asylum seekers.' These policies and discourses are characterized by a perspective on 'asylum' that undermines the meaning of this word, which means entitlement to sanctuary or shelter for persons fleeing from violence and persecution. The current debate about the 'refugee crisis' has mirrored this trend towards the regulation of asylum based on political and economic interests. As such, the right to asylum has been eroded by increasingly coupling it with national migrant labor recruitment strategies.

In this context, I borrow Stephen Castles's (2006) term "migration-asylum nexus." When Castles (2006) introduced this term in his essay "Global Perspectives on Forced Migration," he intervened in a political and academic debate discussing the distinction between economic and forced migration. This distinction assumes that economic migration is a voluntary option, and as such, is not the result of conditions that force people to migrate. Yet, as Castles argues, behind the idea of "economic migration" there is a societal analysis that disregards how migration can be an outcome of complex political, economic, and social constraints that force people's movement. When people migrate because of poverty, unemployment, and deprivation, these societal conditions are connected to political constellations, very often tied to the exercise of power by authoritarian regimes. Thus, political or religious persecution might interact with economic deprivation, and economic constraints might be connected to political repression. The differentiation between asylum as forced migration, and labor migration as voluntary migration, approaches complex societal constellations in a schematic way, disregarding the interplay between the social, the economic, and the political. Castles challenges this divide by suggesting that we examine the relationship between migration and asylum. Castles's observations are relevant for the analysis of the contemporary articulation of the migration-asylum nexus as articulated through the European 'refugee crisis.' The interdependency between migration and asylum has increased due to three elements: (a) the increasing erosion of the humanitarian aspect of the right to asylum, (b) the tightening of restrictions in regard to migration, and (c) the economic demand for labor migration. In this regard, migration is regulated politically through asylum as the field of asylum is increasingly regulated by labor migration demands. These two fields are further coupled through the deployment of securitization measures.

Since September 11, 2001, asylum and migration policies revolve around the theme of securitization. Antonio Negri's analysis of war as an integrative principle in the formation of the social order (*"guerra ordinativa"*)[8] is more relevant than ever (see Negri 2003, 23; Gutiérrez Rodríguez 2010a). In this sense, war has become integrated in the everyday social order due to its virtual omnipresence, through the development of a rhetoric of war, and by war tactics outside of physical war zones. As such, the rhetoric of war is not just "the continuation of politics by other means; it becomes the fundamental aspect of politics and legitimation" (Atzert and Müller 2003, 136). The migration-asylum nexus serves this logic on three levels. First, it represents a field of management and administration of collateral damage and victims of global war and conflicts. Second, it functions as a tool for guarding and securing borders, such that asylum seekers are increasingly treated as invaders. Third, policies differentiating causes, patterns, and trajectories of persecution and escape undermine the ethical legitimation of asylum as a humanitarian resource. Within this context, the definition of countries as 'safe countries' or 'countries of persecution' depends increasingly on global political conjunctures and national or European interests. Yet, the migration-asylum nexus is not only characterized by policies of securitization, but also in that the categorization of refugees into different statuses attached to the process of application and recognition of asylum produces a hierarchical order, a nomenclature drawing on an imaginary reminiscent of Orientalist and racialized practices of European colonialism and imperialism. As such, asylum policies are coupled to the logic of racialization inherent to Quijano's coloniality of power and represent a contemporary expression of the coloniality of migration.

8 With the term *"guerra ordinativa,"* Negri attempts to define the principle organizing the social order globally. This principle preconditions the formation of new power elites, state institutions, political organizations, and the emergence of new security measures and actors organizing the field of war prevention and war logistics. Within this context, Negri discusses the formation of new subjectivities and the potential of resistance against this order organized by the multitude (Negri 2003, 74).

While current EU migration and asylum policies do not explicitly operate within a matrix of racial or ethnic difference, by coupling nationality and the right to asylum they have the effect of formulating hierarchies in the recognition or rejection of asylum in terms of nationalities. This places people in zones of recognition or rejection of the human right to livability (Pannett 2011). This coupling follows from the foundation of racialized notions of the Other. For instance: in autumn 2015, while Syrian refugees were being accepted into Germany, people from Kosovo, Albania, and Montenegro were being deported. While these latter countries were declared 'safe countries of origin' on October 24, 2015, those affected by the deportations were primarily Roma families who had fled these areas because of racist violence. The perception and categorization of this group was determined in Germany not only by their national origin, but also because through Western European racism against Roma and Sinti, deeply rooted since the Middle Ages and re-articulated in the new political constellations, this group of people became objects of securitization measures (Jonuz 2009; Randjelović and Schuch 2014; Ivasiuc 2015).

Further, the discourses on the 'refugee crisis' operate within the duality of the Self and the Other, determined either by a humanitarian perspective or a regulatory approach. The humanitarian perspective, appealing to Christian and humanitarian traditions of charity and empathy, emphasizes the duty of wealthy nations to provide support to people fleeing from wars and conflict zones. The regulatory approach argues instead for the prioritization of securing local wealth. This debate is taking place in countries with strong welfare regimes, such as in Western Europe and Scandinavia, and is present across all political party ideologies. In Germany, for example, very different political actors debate the argument about the limits of the welfare state in providing support for refugees. While there are, of course, ideological and policy-related differences between the political camps, it is surprising to see how they converge in the use of the figure of the refugee as the 'Other' of the nation, or, in Enrique Dussel's (1995) terms, as Europe's "exteriority." This Other is evoked in the logic of what Stuart Hall (1997) has defined as "the spectacle of the 'Other.'" Produced on the identitarian dichotomy of Self and Other, the Other becomes a negative template reflecting the "hegemonic Self's" anxieties, worries, contempt, and fear (Popal 2011).

However, the migration-asylum nexus not only follows the logic of the production of a racialized exteriority to the nationally imagined and proclaimed norm of European whiteness. Instead, this nexus operates within the dynamics of exploitation that have functioned for the last five centuries within the colonial-modern world system, and particularly within the context of nation-state migration policies since the 19th century.

4 Conclusion: Racial Capitalism

As I have shown through the discussion of the coloniality of migration, as developed from Quijanos's "coloniality of power," the discussion of migration as an articulation of a modern world-system needs to be connected to the cultural, political, economic, and social legacies of European colonialism. The development of transnational migratory movements and migration policies unfold within this context. This observation leads us to consider the coloniality of migration as a conceptual perspective for examining current articulations of racial capitalism as discussed through the trope of the 'refugee crisis.' A differential logic of racialization emerges in this context, which finds expression in nation-state institutional and organizational parameters of managing and controlling migration and asylum.

As we have seen above, the joining of productivity, migration, and racism marked the rise of migration policies in the Americas and Oceania (Reinhardt and Reinhartz 2006; Oltmer 2016). The recruitment of migrant workers took place within European racial notions of sameness and strangeness. Until the mid-20th century, as we have seen, countries like the US, Argentina, Brazil, Australia, New Zealand, and South Africa explicitly recruited white Europeans. These policies resulted in a settler colonialism-migration that constructed these nation-states as extensions of a white Christian Europe. The recruitment of white European migrants was officially legitimized as a means for national industrial achievement, technical progress, and urban industrialization. Yet despite the restrictions applied to migration movements from non-European territories, people from the Middle East, North Africa, China, and the Caribbean still immigrated to these areas, despite regular attempts to stop them.

For migration policies in Western Europe, a racialized logic of exploitation was decisive as well. We are speaking here, for example, about the "guest-worker" programs in Germany in the 1950s and 1960s (see Türkmen 2011, 2010), which were originally intended to temporarily recruit workers from Southern Europe, Turkey, Tunisia, and Morocco. Another example may be found in the *Empire Windrush*, a ship from the Caribbean that arrived at Tilbury Dock in London in 1949, carrying a highly skilled workforce intended for work as nurses, bus drivers, and teachers. In the current migration-asylum nexus, economic interests also shine through. One of the measures agreed upon across all party lines in the German parliament in June 2016 was the introduction of "One-Euro jobs for refugees" (see Öchsner 2016). This measure was intended to initially create jobs for 100,000 people who had arrived in Germany as refugees. The German newspaper *Süddeutsche Zeitung* reported that these people only received 80 cents for every Euro, because the costs for travel and work clothes were deducted. Here too, recommendations from the

International Labor Organization for working standards were ignored for those given asylum. These forms of utilization recall Enrique Dussel's "objectification" of the indigenous and Afro-descendent populations during the Portuguese and Spanish colonization of the Americas (1995). Migration policies tend to follow up this logic by reducing human beings to exploitable labor. Asylum policies seem to be turning into a new way of regulating and controlling racialized labor migration. The coloniality of migration not only draws attention to the racism in contemporary migration regulation, policies, and official national discourses, but it also addresses the link between racism and capitalism.

References

Atzert, Thomas, and Jost Müller, eds. *Antonio Negri. Kritik der Weltordnung: Globalisierung, Imperialismus, Empire*. Berlin: ID, 2003.

Bernasconi, Robert, and Syboll Coole, eds. *Race and Racism in Continental Philosophy*. Bloomington, IN: Indiana University Press, 2003.

Bethencourt, Francisco. *Racism: From the Crusades to the Twentieth Century*. Princeton: Princeton University Press, 2014.

Castles, Stephen. "Global Perspectives on Forced Migration." *Asian and Pacific Migration Journal* 15 (2006): 17–28.

Chatterjee, Partha. *The Nation and Its Fragments*. Princeton: Princeton University Press, 1993.

Coulthard, Glen Sean. *Red Skin, White Masks: Rejecting the Colonial Politics of Recognition*. Minneapolis: University of Minnesota Press, 2014.

Davies, Carole Boyce, ed. *Claudia Jones: Beyond Containment*. Oxfordshire: Ayebia Clarke Publishing, 2011.

Du Bois, W. E. B. *Black Reconstruction in America* (1939). New York: Oxford University Press, 2007.

Dussel, Enrique. *The Invention of the Americas: Eclipse of "the Other" and the Myth of Modernity*. New York: Continuum, 1995.

Eliav-Felden, Miriam, Benjamin Isaac, and Joseph Ziegler, eds. *The Origins of Racism in the West*. Cambridge: Cambridge University Press, 2009.

Eltis, David, and David Richardson, eds. *Extending the Frontiers: Essays on the New Transatlantic Slave Trade Database*. New Haven: Yale University Press, 2008.

Fanon, Frantz. *Les damnes de la terre*. Paris: Points, 1971.

Fiddian-Qasmiyeh, Elena. "Embracing Transculturalism and Footnoting Islam in Accounts of Arab Migration to Cuba," *Interventions: International Journal of Postcolonial Studies* 18.1 (2016): 19–42.

FitzGerald, David S., and David A. Cook-Martín. *Culling the Masses: The Democratic Origins of Racist Immigration Policies in the Americas*. Cambridge, MA: Harvard University Press, 2014.

Gutiérrez Rodríguez, Encarnación. *Migration, Domestic Work and Affect*. New York: Routledge, 2010a.

Gutiérrez Rodríguez, Encarnación. “Transculturation in German and Spanish Migrant and Diasporic Cinema: On Constrained Spaces and Minor Intimacies in ‘Princesses’ and ‘A Little Bit of Freedom.’” *European Cinema in Motion: Migrant and Diasporic Film in Contemporary Europe*. Eds. Daniela Berghahn and Claudia Sternberg. New York/London: Palgrave Macmillan, 2010b. 114–131.
Gutiérrez Rodríguez, Encarnación. “Archipelago Europe: On Creolizing Conviviality.” *Creolizing Europe: Legacies and Transformations*. Eds. Encarnación Gutiérrez Rodríguez and Shirley Tate. Liverpool: Liverpool University Press, 2015a. 80–99.
Gutiérrez Rodríguez, Encarnación, and Shirley Tate, eds. *Creolizing Europe: Legacies and Transformations*. Liverpool: Liverpool University Press, 2015b.
Hall, Stuart. “The Spectacle of the ‘Other.’” *Representation: Cultural Representation and Signifying Practices*. Eds. Stuart Hall, Jessica Evans, and Sean Nixon, eds. London: Sage, 1997. 223–290.
Hall, Stuart. “Creolité and the Process of Creolization.” *Creolizing Europe: Legacies and Transformations* (2003). Eds. Encarnación Gutiérrez Rodríguez and Shirley Tate. Liverpool: Liverpool University Press, 2015. 12–25.
Horkheimer, Max, and Theodor W. Adorno. *Dialektik der Aufklärung*. Frankfurt a.M.: Fischer, 1988.
Ivasiuc, Ana Nichita. *Provincializing Citizenship: Critical Anthropological Notes on the Uses and Usefulness of “Citizenship” in the Context of the Roma Political Subject.* <https://www.academia.edu/11754903/Provincialising_Citizenship._Critical_anthropological_notes_on_the_uses_and_usefulness_of_citizenship_in_the_context_of_the_Roma_political_subject> [accessed: 15 February 2017].
James, Cyril L.R. *The Black Jacobins: Toussaint L’Ouverture and the San Domingo Revolution.* New York: Vintage Books, 1989.
Jonuz, Elizabeta: *Stigma Ethnizität. Wie zugewanderte Romafamilien der Ethnisierungsfalle begegnen*. Opladen: Budrich UniPress, 2009.
Jupp, James. *From White Australia to Woomera: The Story of Australian Immigration.* Cambridge: Cambridge University Press, 2002.
Ketarí Hernández, Tanya. *Racial Subordination in Latin America: The Role of the State, Customary Law, and the New Civil Rights Response*. Cambridge: Cambridge University Press, 2013.
Lesser, Jeffrey. *Negotiating National Identity: Immigrants, Minorities and the Struggle for Ethnicity in Brazil*. Durham, NC: Duke University Press, 1999.
Lesser, Jeffrey. *Immigration, Ethnicity, and National Identity in Brazil, 1808 to the Present.* Cambridge: Cambridge University Press, 2013.
Lowe, Lisa. *The Intimacies of Four Continents*. Durham, NC: Duke University Press, 2015.
Martinez-Echazabal, Lourdes. “Discourse of National/Cultural Identity in Latin America, 1845–1959.” *Latin American Perspectives* 25.3 (1998): 21–42.
Massey, Douglas S. “The Social and Economic Origins of Immigration.” *Annals of the American Academy of Political and Social Sciences* 510 (2000): 60–72.
Negri, Antonio. *Cinque lezioni di metodo su Multitudine e Impero.* Soveria Mannelli: Rubbettino, 2003.
Neuman, Gerald L. “The Lost Century of American Immigration Law (1776–1875).” *Columbia Law Review Association* 93.8 (1993): 1833–1901.
Nkrumah, Kwame. *Class Struggle in Africa*. New York: International Publishers, 2006.

Öchsner, Thomas. "Ein-Euro-Jobs für Flüchtlinge sind nur 80-Cents-Jobs." *Süddeutsche Zeitung* (10 June 2016).
Oltmer, Jochen. *Migration vom 19. bis zum 20. Jahrhundert. Enzyklopädie Deutscher Geschichte.* Berlin/Boston: De Gruyter, 2016.
Pannett, Loraine. *Making a Livable Life in Manchester: Doing Justice to People Seeking Asylum.* (Unpublished doctoral dissertation). Manchester: University of Manchester, 2011.
Plender, Richard. *International Migration Law.* Dordrecht: Martinus Nijhoff Publisher, 1988.
Popal, Mariam. "Zivilisiert/wild [Civilized/wild]." *Wie Rassismus aus Wörtern spricht. (K)Erben des Kolonialismus im Wissensarchiv deutsche Sprache. Ein kritisches Nachschlagewerk.* Eds. Susan Arndt and Nadja Ofuatey-Alazard. Münster: Unrast, 2011. 278–287.
Quijano, Aníbal. "Colonialidad del Poder, Eurocentrismo y América Latina." *International Sociology* 15.2 (2000): 201–246.
Quijano, Aníbal. "Coloniality of Power, Eurocentrism, and Social Classification." *Coloniality at Large: Latin America and the Postcolonial Debate.* Eds. Mabel Moraña, Enrique Dussel, and Carlos A. Jáuregui. Durham, NC: Duke University Press, 2008. 181–224.
Randjelović, Isidora, and Jane Schuch. *Dossier Perspektiven und Analysen von Sinti und Roma in Deutschland.* Heinrich-Böll Stiftung: Heimatkunde/Migrationspolitisches Portal. 2014. <https://heimatkunde.boell.de/dossier-sinti-und-roma> [accessed: 12 February 2018].
Reinhardt, Steven G., and Dennis Reinhartz. *Transatlantic History.* College Station: Texas A&M University Press, 2006.
Robinson, Cedric J. *Black Marxism: The Making of the Black Radical Tradition.* Chapel Hill: The University of North Carolina Press, 2005.
Simpson, Audre. *Mohawk Interruptus: Political Life across the Border of Settler States.* Durham, NC: Duke University Press, 2014.
Steinmetz, George. "'The Devil's Handwriting:' Precolonial Discourse, Ethnographic Acuity, and Cross-Identification in German Colonialism," *Comparative Studies in Society and History* 45. 1 (2003): 41–95.
Tuck, Even, and K. Wayne Young. "Decolonization is not a Metaphor." *Decolonization: Indigeneity, Education & Society* 1.1 (2012): 1–40.
Tzul Tzul, Gladys. "Mujere indígenas: Historias de la reproducción de la vida en Guatemala." *Bajo el Volcán* 15.2 (2015): 91–99.
Türkmen, Ceren. "Vom Klassenkampf zum Kampf ohne Klassen? Ein Kommentar zu Rassismus und Klassenanalyse." *Perspektiven der Demokratie.* Ed. Promotionskolleg "Demokratie und Kapitalismus." Siegen: Selbstverlag, 2010. 168–177.
Türkmen, Ceren. "Diskontinuität und Kohärenz. Gastarbeitsmigration und die Organisierung der Arbeitsteilung in Deutschland." *Geschlecht – Migration – Integration.* Eds. Jane Angerjärv and Hella Hertzfeldt. Berlin: Karl Dietz, 2010. 51–65.
Williams, Eustace Eric. *Capitalism and Slavery.* Chapel Hill: The University of North Carolina Press, 1994.

Kader Konuk

What Does Exile Have to Do with Us?

Academic Freedom in Turkey

In light of today's surge of authoritarian regimes and the overwhelming number of exiles and refugees that these create, we must recall April 1933, when civil servants and academics from German institutions of higher education were expelled on the basis of race and political dissent. Eliminating the intellectual elite is often the first step toward suppressing criticism and bringing citizens in line with their governments, yet history has taught us that disabling the critical elite creates a specter for other parts of society and prepares the ground for further atrocities. The arrest of Armenian intellectuals in Istanbul in April 1915, for example, was the precondition that enabled the mass deportation of Anatolian Armenians to Ottoman Syria, as well as the ensuing Armenian genocide. The goal of these measures was to silence the Armenian elite in order to facilitate governmental control and to consolidate its power.

In today's various populist environments, 'elite' has become a morally charged and tainted term, wherein members of 'the elite' are set against 'the people' as if they counter the interests of 'ordinary citizens.' In a recent article, Rogers Brubaker identifies a common "discursive, rhetorical and stylistic" repertoire among various forms of populism (Brubaker 2017, 360). According to Brubaker, one of the markers of populist discourse is the construction of a "vertical opposition between people and elite and/or the horizontal opposition between inside and outside"; the 'people' are simultaneously construed as outside, and at the bottom (Brubaker 2017, 364). Critique of 'the establishment' and 'academic elitism' is a common and defining feature of populist leaders in the 21st century.[1] Populism translates into rejection of pluralism and resentment toward the establishment in favor of the 'will of the people.'[2] Arguably, a populist democracy represents a degraded form of democracy. And as can be observed in Turkey in

1 The 2016 report compiled by Scholars at Risk demonstrates that apart from populist, authoritarian leaders, "Extremists and militants target universities because they see a free, open university space as a threat to their quest for power. Such extremists and militants have committed mass attacks on universities in Pakistan and Afghanistan, as well as targeted killings of individual scholars in Bangladesh, India, Iraq, and Syria." <https://www.scholarsatrisk.org/wp-content/uploads/2016/11/Free_to_Think_2016.pdf> [accessed: 15 August 2017].

2 Jan Werner Müller points out that we "draw on a set of assumptions derived from modernization theory that had its heyday in the 1950s and 1960s." In the 1950s, political theorists and

https://doi.org/9783110600483-012

recent years, when populism combines with Islamism, the resultant Islamist populism relegates secular communities to the periphery, deems secularists an elitist minority, and forces dissident academics into exile.

In many countries around the world, critics of authoritarian regimes have their civil liberties curtailed: they are subject to arbitrary and often indefinite detention, solitary confinement, torture, and other acts of violence. Speaking truth to power poses an intolerable threat to despotic regimes that regulate information flow to maintain control. We have grown almost too accustomed to the vulnerability of journalists, activists, and writers who challenge such regimes. What perhaps we don't expect, however, is that academics working on seemingly innocuous topics in ostensibly liberal countries may meet similar fates, through state-directed and state-sponsored intimidation, extradition, incarceration, and state-directed and state-sponsored violence. Whether openly critical of authoritarian governments or not, the higher-education community is a new target of repression; globally, academics are increasingly under attack.

The US-based Scholars at Risk Network is one of a few major organizations monitoring the growing threat to scholars around the world. This international network of higher education institutions, associations, and individuals works to protect scholars and to promote academic freedom. Scholars at Risk, the Institute of International Education's Scholar Rescue Fund, and their British counterpart CARA (the Council for At-Risk Academics) act as advocates for threatened scholars and seek host universities for these scholars in countries that protect their civil liberties. Yet even in the so-called free world, academic communities are challenged as elitist, and democratically elected governments seek to interfere with research agendas. This is partly due to the fact that state universities are vibrant interlocutors for the democratic process, and those who work in public institutions are particularly accountable to the public. As vehicles for democracy and healthy public discourse, universities have historically played a key role. In this era of post-truth politics and the active dispersal of misinformation, we are called upon to protect educational institutions that foster inquiry and produce reliable knowledge.

In the past two years, most applications to Scholars at Risk have originated in Turkey, suggesting that Turkish academics are currently one of the most threatened groups of scholars worldwide. As a member of the United Nations, Turkey is bound to uphold rights protected by the Universal Declaration of Human Rights and the International Covenant on Civil and Political Rights. Recent actions of

---

social scientists described populism as a "helpless articulation of anxieties and anger by those longing for a simpler, 'premodern' life" (Müller 2017, 17).

the Turkish state, however, do not comply with its "human rights obligations, including those relating to freedom of association, due process, and academic freedom."[3] During the past two years especially, thousands of regime critics have been criminalized, imprisoned, and banned from their professions in Turkey. Without the possibility of employment, hundreds of scholars have left the country in an attempt to continue their critical work and live in safety abroad.

The erosion of academic freedom is not merely symptomatic of the rise of political Islam, and is far from being isolated in Turkey. Examples from other countries are perhaps less comprehensive in their verbal assaults on academic freedom, but are nonetheless equally characteristic of the current conjuncture. In 2015 for example, the Russian government accused the Center for Independent Social Research in St. Petersburg of acting as a foreign agent, and in 2016 gravely endangered the future of the European University at St. Petersburg, a distinguished private graduate school founded at the time of the Soviet Union's collapse in 1991, by revoking its teaching license. Also under threat in 2017 was the Central European University in Hungary, an American-Hungarian institution founded with the aim of promoting liberal values by the philanthropist George Soros. Other initiatives funded by Soros in Eastern Europe have likewise become targets, with the former Macedonian prime minister even calling for a "de-Sorosization" of society.[4] In Poland, on the other hand, the government has taken specific measures in 2018 to penalize suggestions of any complicity by the Polish state or people in the Nazi Holocaust. As historian Jan Grabowski reminds us, these measures constitute a "threat to the liberty of public and scholarly discussions," and represent "a dramatic departure from the democratic principles and standards which govern the laws of other members of the European Union."[5]

It is commonly agreed that such backlash in Russia and Eastern Europe is connected to developments in the US, where, among other draconian measures,

---

3 <http://monitoring.academicfreedom.info/reports/2017-07-06-konya-selçuk-university-necmettin-erbakan-university> [accessed: 20 August 2017].

4 In Romania, Soros was named a "financial evil": <https://www.nytimes.com/2017/03/01/world/europe/after-trump-win-anti-soros-forces-are-emboldened-in-eastern-europe.html> [accessed: 05 April 2017].

5 Jan Grabowski likens Poland's legislative measures to Turkey's laws that control the debate about the responsibility for the Armenian genocide: <http://www.macleans.ca/news/world/as-poland-re-writes-its-holocaust-history-historians-face-prison/> [accessed: 01 February 2018]. For further repercussions against historical scholarship and the representation of Polish antisemitism in the Museum of the Second World War in Gdańsk: <http://www.sueddeutsche.de/kultur/kulturpolitik-der-polnische-standpunkt-1.3791929> [accessed: 01 February 2018]. I thank Egemen Özbek for drawing my attention to this discussion in public history.

President Trump has deemed climate change a hoax, slashed funds for the National Institute of Health, threatened to eliminate the financial backing of the National Endowment for the Humanities, and issued a warning that he would cut funding for recalcitrant public institutions like the University of California, Berkeley. These threats and measures point to the fact that academic freedom, established as *Lehr- und Lernfreiheit* in the 19th century at Humboldt University, can no longer be taken for granted.

Not only in Turkey but all over the world, private lives and academic careers are being ruptured by war and revolution, as well as by a populist backlash against academic freedom. Academic freedom is also under threat because of the rise of authoritarian regimes worldwide. Of primary responsibility for the growing numbers of scholars seeking refuge abroad are the ongoing conflict in Syria and its spillover effect on neighboring nation-states, the unstable and repressive governments in countries like Egypt and Iraq, and the outlawing of dissent by democratically elected governments. Once understood as a system that protects the rights of the weakest members of society, democracy has now become a tool for demagogues to secure the rule of the majority. In governments that repeatedly invoke the 'will of the people,' academic freedom is construed as the concern of an elite part of society. It poses a threat to demagogues because of its capacity to dismantle propaganda through the practice of critique in public institutions.

The Turkish government has gone so far as to outlaw and persecute dissent in all forms. President Erdoğan closed entire universities after the coup attempt in July 2016 to minimize the influence of his former ally Fetullah Gülen, a leading Islamic cleric based in the US.[6] Erdoğan's AKP (Justice and Development Party) regime is particularly threatened by secular, leftist, and Kurdish academics, and by journalists, artists, and writers who claim the right to the freedom of speech and research. Because of the government's anti-intellectual stance, numerous intellectuals have been imprisoned, silenced, or forced into exile. Here, Turkey is taking a leaf from the pages of history. Parallels to this crackdown can be found in the US during the McCarthy era, when scholars were subject to political tests and loyalty oaths, universities were pressured to purge communists, and those charged with communism were dismissed and imprisoned. The fear of communism could be felt on the other side of the Atlantic, too. The 1972 *Radikalenerlass* (Anti-Radical Decree) in West Germany was designed to permanently ban communist academics from universities. This constraint on employment meant

6 For an article investigating the reasons for the conflict between the Gülenist movement and the AKP see Demiralp 2016. The article identifies the AKP's interest in initiating a 'peace process' with Kurds in 2011 as the main reason for the conflict between the rival groups (Demiralp 2016, 4).

the reinstatement of an earlier law, the *Berufsbeamtengesetz* (Professional Civil Service Law) of 1933, which allowed Hitler to discharge academics on either political or racial grounds. The *Gleichschaltung* (elimination of opposition) ensured that German academia was purged of Jews and socialists and stood in the service of National Socialist objectives, to the extent that German philologists who served the Nazis developed volumes in the name of the *Kriegseinsatz der Geisteswissenschaften* (war service of the humanities) in which they put literature and even literary criticism directly in the service of the war and the idea of a *neue geistige Ordnung Europas* (a new intellectual foundation for Europe).[7]

While April 1933 constitutes a moment of unparalleled horror and tragedy, it also marks the beginning of a process of intellectual emigration that had a tremendous impact on universities in the United States, the United Kingdom, Turkey, and elsewhere. The academic refugee organizations of the 1930s shaped conditions for knowledge production across disciplines. The Society for the Protection of Science and Learning in the UK, the *Notgemeinschaft deutscher Wissenschaftler im Ausland* (Emergency Association of German Scientists) initiated by Philipp Schwartz in Switzerland, and the concept of the University in Exile that was incorporated into The New School in the US are prime examples of the impact that refugee academics have had on the way contemporary scholars have come to think and work. The majority of scholars in the humanities and social sciences owe their critical training to the continuing engagement with the exilic work of the German Jews who were expelled. In Turkey, the secularization and modernization of universities, museums, libraries, and other educational institutions would have been unthinkable without the impact of hundreds of exiled German Jewish and socialist scholars, artists, musicians, and scientists.

1 Historical Legacy

The founding of secular institutes of higher education in Turkey was based on the 1924 law for the Unification of Education, which centralized education throughout the nation. To implement the centralization process, the US philosopher and educational reformer John Dewey was invited to Turkey in 1924 for a three-month visit to advise the Turkish government. For Dewey, freedom of inquiry and freedom of education were inextricably linked to social change. Dewey consid-

7 Frank-Rutger Hausmann's study of the "Aktion Ritterbusch" provides comprehensive insights into the politics of philology during National Socialism (Hausmann 2007).

ered academic freedom essentially a social issue, since the freedom of teaching and learning was intimately bound up with the success of democracy. Indeed, he coined the phrase: "Since freedom of mind and freedom of expression are the root of all freedom, to deny freedom in education is a crime against democracy" (Dewey 1987, 378). Even if his visit to Turkey did not result in the desired outcome, Dewey's views may have helped shape the Turkish government's decisions regarding the modernization of education (Büyükdüvenc 1994).

The establishment of Istanbul University, which replaced the Ottoman Darülfünun in 1933, was a major accomplishment in the reform process. Yet it is impossible to imagine the secularization of Turkish universities and the establishment of new disciplines, from psychology and philology to entire branches of the sciences, without the flight of mostly Jewish academics from Germany to Turkey in 1933. Émigré scholars indeed played a central role in the formation of the intellectual, secular elite that dominated Turkish universities for decades – the same elite that has now come under attack (Konuk 2010).[8]

Exile, critical thinking, and academic freedom are deeply connected within the historical context of German and Turkish universities. There is irony in the fact that scholars now flee Turkey when historically Turkey has been a host to scholars fleeing Europe. Turkey is being Islamized and scholars are leaving the country in great numbers. By the same token, Western Europe is concerned about its own secular status. If the historical example is anything to go by, Western Europe's ability to host the current exiles from Islamization might hinge on a renewed commitment to secularism, or at least to religious pluralism.

In the modernization of Istanbul University, the reforms were part of a national agenda that linked its success to its capacity to overcome cultural differences between East and West. Humanist worldviews were preserved and transformed as the core of European culture in Turkey, at the same time that these views were simultaneously under attack by fascism in Europe. During the postwar period, however, humanists lost the kind of influence that they once enjoyed. If there had ever been a window of opportunity for the democratization of cultural politics, this was now effectively closed by autocratic leaders, anticommunist campaigns, attempts to subvert secular education, and a series of disastrous military coups.[9] During the transition from a single-party regime to a democratic multiparty

8 After the hire of 40 German professors in 1933, almost half of Istanbul University's faculty was German (see Hirsch 1950, 460). On the reforming role of these emigrants from a translational perspective, see the article by Doris Bachmann-Medick in this volume.

9 For a detailed study of educational politics after the military coup in 1980, see Kaplan 2006.

system in 1946, Turkey's humanist project ground to a halt.[10] The minister of education, Hasan Ali Yücel, who spearheaded the humanist reforms, came under increasing criticism. The system of village institutes that he had introduced for the training of primary school teachers and promoting a rural intellectual elite was branded a communist breeding ground (Sakaoğlu 1993, 99). Yücel himself was accused of passing off communism as humanism. Conservatives blamed him for using the cultural reforms to create a Greco-Roman basis for Turkish culture (Çıkar 1994, 79). In response to this smear campaign, Yücel mounted a lawsuit on the grounds of slander. He eventually won, but he nonetheless retired from his official duties in 1947. The translation project that was an essential part of the humanist culture reform and had facilitated and coordinated the translation of hundreds of Western classics into Turkish ended in 1950 (Sakaoğlu 1993, 106; Çıkar 1994, 59). Before he retired from his position, the minister of education ensured universities the basis for academic freedom by granting them autonomy.

The concept of autonomous universities and intellectual freedom that Yücel introduced did not survive for long, however. It was first interrupted by the military coup of 1960; in the aftermath of this coup, the movement of 1968 further affected Turkish universities. The reason for Turkish student protests of this era was not an effort to de-Nazify universities, as was the case in Germany, nor to protest the Vietnam War, as in the United States: it was dissatisfaction with overcrowded universities and the curtailment of the freedom of speech. In 1968, only eight public universities provided higher education for a population of 35 million Turks. Deniz Gezmiş, a law student at Istanbul University who became a revolutionary political activist, embodied the spirit of the late 1960s. Because of his involvement in an armed struggle against the state that involved robbery and the taking of hostages, Gezmiş was brought to trial and received a death sentence. His struggle against imperialism and his execution in 1972 turned him into a martyr.

In the 1970s, the young theater actress Emine Sevgi Özdamar left Turkey for Germany in self-imposed exile. She would later write short stories and novels that elaborated in varying ways the contamination of the Turkish language by military and state violence. Her literary texts echo with the philologist and Holocaust survivor Victor Klemperer's reflections on the Nazi appropriation of the German language. It is once again an irony of history that Özdamar turned away from the Turkish language after emigrating to Germany to embrace the language of exiles and Jewish writers such as Brecht, Kafka, and Else Lasker-Schüler.[11]

---

10 For an overview of this period, see Zürcher 1998, 209–214.

11 For the context of transnational literatures, see Konuk 2007, 2017.

In Turkey, the student protest movement continued throughout the 1970s and became increasingly threatening to the Turkish state. Despite Deniz Gezmiş's execution, campuses continued to be a breeding ground for the leftist protest movement until the military coup in 1980, which had even graver consequences for the pursuit of knowledge and free speech that questioned the very foundation of the nation-state. As a result of this military intervention, Turkish universities lost the autonomy that they had been granted in 1946. In 1982, the Council of Higher Education (YÖK) was founded in an effort to centralize and better control Turkish universities. Thousands of activists, students, and intellectuals were imprisoned, disappeared, or driven into exile in the 1980s, many of them to Germany. When in the 1990s the process of democratization began once more, writers and scholars sought to raise the veil that had been cast over Turkey's religiously, ethnically, and linguistically diverse past. Once this Pandora's box was opened, momentum increased in academic pressure to lift the ban on the Kurdish language and to face the atrocities of the Armenian genocide. Sociologist Fatma Müge Göçek considers the "pockets of public space not controlled by the Turkish state" at this time as signs for the advent of the new era toward a "post-nationalist period" (Gocek 2006, 98).[12] Hopes for a lasting post-nationalist historiography and the diversification of knowledge production were soon challenged by the government's reaction to the Gezi protest of 2013, which challenged the plans of the AKP government to resurrect Ottoman barracks in one of Istanbul's rare parks close to the iconic Taksim square. The Gezi protest was arguably the single most important act of civil disobedience in Turkey's recent history, and was a turning point for Erdoğan's successful establishment of the AKP's neo-Ottomanist vision. Since then, various forms of state decrees have fostered a climate of self-censorship among writers and journalists, contributing to the increasingly conservative and religious tenor of cultural production.

2 Purge and Exodus: Academics for Peace

For many years, Turkey represented a bulwark in the struggle for liberal humanism in the Middle East, but during the past two decades, new forms of Islamism have emerged in Turkey as a reaction to this secularism. As a result, Turkey's powerful secular tradition has been gradually dismantled. The nationalist under-

12 For an analysis of the transformation of society after the introduction of neoliberal measures in 1982, see Gocek 2006.

pinnings of secularism in Turkey and the interpretation of French *laïcité* into the Turkish constitution are undoubtedly problematic. Yet, however critical one might be of the interpretation of secularism in Turkey, it is important to recognize that the recent rise of religious fundamentalism and the politicization of Islam has put a stay on the secularization of Turkish society and has profoundly impacted academia.

The erosion of secular education could be observed as early as 2009, when the Turkish Scientific and Technological Research Council (TÜBİTAK) censored an issue of the science journal *Bilim ve Teknik* that discussed Darwin's theory of evolution. The council fired the journal's editor, arguing that the Darwin issue was "too controversial" for the country's political climate.[13] Since then, TÜBİTAK and other academic and educational institutions have come under even tighter control by the AKP government.[14] In 2017, the government announced that the theory of evolution was to be dropped from school curricula, after the chair of the Board of Education – a theologian appointed after the coup attempt in 2016 – insisted that the topic is both inappropriate and too controversial to be taught to Turkish students. In its stead, the Board decided to expand Islamic education in schools,[15] and, with this decree, suddenly and dramatically curtailed the pursuit of scientific knowledge in Turkey. The government continues to systematically eradicate the secular foundation of Turkey's democracy and along with it, academic freedom – a cornerstone of any dynamic society open to the advancement of knowledge.

One group presently targeted for persecution in Turkey is a collection of 1,128 scholars known as Academics for Peace. These scholars appealed to the Turkish government to stop the atrocities against Kurdish civilians in the winter of 2015 and signed a peace petition in January 2016 entitled "We will not be party to this crime." The petition asked the government to find ways of ensuring lasting peace and demanded to stop the massacre and exodus of civilians. President Erdoğan denounced these signatories of the petition as "pseudo-intellectuals"

---

13 <http://www.spiegel.de/international/world/0,1518,613768,00.html> [accessed: 20 August 2017].

14 For an article on further repercussions and the changes regarding the funding agency TÜBİTAK and the Turkish Academy of Sciences (TÜBA), see N.N. "A Very Turkish Coup." *Nature* 477.131 (08 September 2011). <http://www.nature.com/articles/477131a> [accessed: 01 July 2017].

15 Other examples of new taboos in higher education are discussed in Genç 2015. The article refers to the removal of displays of nudity in sculptures at Gazi University, paintings at Çukurova University, and of an art history conference poster at Istanbul University. Genç connects these acts of censorship on moral grounds to others based on political grounds, for example, the refusal to host a conference on the Armenian genocide at Bilgi University in 2015.

and charged them with supporting terrorism. Criminalized by the Turkish state, 70 of these academics were placed in police custody, and three were imprisoned. To date, 498 scholars have not only been dismissed from their positions, but also banned outright from practicing their professions at any institute of higher education in Turkey.[16] All signatories of the 2016 peace petition were immediately cut off from state funding and since then have had either limited or no means of pursuing their research. Self-censorship and the threat of the professional ban also affect scholarship and teaching at Turkish universities more generally. By forcing all deans to withdraw from their positions in July 2016, issuing a travel ban, and canceling passports of scholars and their spouses, Turkey is in violation of both academic freedom and the freedom of movement and travel guaranteed by international human rights law. Since academic freedom is not a right protected under the Turkish constitution, its enforcement is contingent upon each university's pledge, as a member of the European Universities Association, to uphold the right to teach and research. To date, 62 universities in Turkey have become members of the EUA and as such committed to the values outlined in the *Magna Charta Universitatum*.[17] Yet since the commitment is not legally binding, violations of the principles have no legal ramifications. Instead, hundreds of scholars who were dismissed by government decrees are reported to have applied to the European Court of Human Rights.[18]

Since early 2016, numerous Academics for Peace signatories have left Turkey. To date, approximately 150 have sought refuge in Germany. Exiled scholars have been threatened with withdrawal of their Turkish citizenship and, consequently, statelessness. Some scholars who have been banned from their professions and remain in Turkey are likely to undergo a process of 'inner emigration,' a contested concept historically applied to German intellectuals under fascism, which tries to capture the ambiguity between open resistance and the need for self-preservation. The current flight of academics begs the question of how to respond to the crisis other than in humanitarian ways. In the concluding section, I investigate what Edward Said once termed the "executive value of exile." What, we might ask,

---

16 With the decree laws, by January 2018, 498 Academics for Peace signatories were reported to have been either removed or banned from public service, dismissed, forced to resign, or retire. <https://barisicinakademisyenler.com/node/314> [accessed: 27 January 2018].

17 The *Magna Charta Universitatum* states the values of European universities and was signed in Bologna in 1988. By becoming a member of the European University Association, each university commits to the principles of the freedom of teaching and research: <http://www.magna-charta.org/resources/files/the-magna-charta/english> [accessed: 01 February 2018].

18 For a broad overview of the statistics of applications to the ECHR see <http://www.echr.coe.int/Documents/Stats_analysis_2017_ENG.pdf> [accessed: 01 February 2018].

is the redeeming feature of exile today? Is there something we can learn from the past? How can we respond in ways that help curtail the rise of anti-academicism and populism in the world today?

3 The Figure of Exile: We Refugees

In 1943, Hannah Arendt warned that outlawing the Jewish people in Europe was "closely followed by the outlawing of most European nations. Refugees driven from country to country," she wrote, "represent the vanguard of their peoples—if they keep their identity." In Arendt's essay, the assimilated Jew who constantly looks for approval is the counterpart to the "conscious pariah," who was at the vanguard of his people (Arendt 1994, 119). Traditionally, the figure of the 'refugee' is not necessarily associated with that of the 'exile.' It is as if, following a class division, we imagine the refugee to be less educated, and the exile (or the émigré, as they were also commonly termed) to have valuable cultural and intellectual capital. Arendt's essay begins by stating: "In the first place, we don't like to be called 'refugees.' We ourselves call each other 'newcomers' or 'immigrants.'" She points out that the refugee has commonly been understood as "a person driven to seek refuge because of some act committed or some political opinion held." According to Arendt, the Jewish community rejected identification with the refugee because "we committed no acts and most of us never dreamt of having any radical opinion" (110). While criticizing this apolitical stance, her essay acknowledges the changed meaning of the term 'refugee' and tries to restore the identity of the Jewish refugee by connecting it to the image of the "conscious pariah."

In 1993, during the immediate aftermath of the Cold War, Edward Said took up the question of the intellectual in exile, and presented his well-known Reith Lectures, which explored the concept of displacement and the condition of marginality. Said's interest lay less in his present than in displacements caused by the revolutions, fascism, deportation, and genocides of the first half of the 20th century, and in the great masterpieces that Adorno, Auerbach, and Naipaul generated in exile. In Said's view, exilic displacement allowed the intellectual to be liberated from his or her usual career or prescribed path. While Said did not deny the challenges and hardship of exile, in a vein similar to Arendt's he emphasized the condition of marginality as a potential asset to the intellectual. To Said, "the *exilic* intellectual does not respond to the logic of the conventional but to the audacity of daring, and to representing change, to moving on, not standing still" (Said 1994, 64).

For many contemporary critics, exile represents a state of critical detachment and superior insight that is supposed to arise when intellectuals are expelled from their homes and forced to take up residence elsewhere. This line of thought, however, too readily reduces 'exile' to be merely a metaphor for 'uprootedness': disconnected from his or her social and political context, the exile is granted the potential for cultural transfer and transnational exchange. Too easily does the exilic condition acquire almost utopian possibilities. Suddenly unencumbered by his or her background, the exile emerges instead as a new mediator between systems, a perspicuous commentator on both the endogenous and exogenous. To me, this view of exile distorts the historical record, diminishing the existential plight of those who are expelled even as it elevates the individual case to a general paradigm. Against this view of exile qua detachment, I propose a condition of multiple attachments. The task, then, is to investigate these attachments and tease out their implications both for the individual and for the respective societies at large. Rather than merely salvaging the positive in the exilic condition, we need to ask what it means to go into exile and what arises therefrom.

In 1995, around the same time that Said published his Reith Lectures, Giorgio Agamben also revisited Arendt's essay, writing an article that used Arendt's original title, "We Refugees." Here, Agamben shows no interest in the question of the intellectual in exile but discusses the status of the refugee. He sees the refugee as "nothing less than a border concept that radically calls into question the principles of the nation-state and, at the same time, helps clear the field for a no-longer-delayable renewal of categories" (Agamben 2018, 117). Agamben's reevaluation of the status of the refugee is informed by the collapse of the Soviet Union and the outbreak of ethnic violence in former Yugoslavia. Rather than idealizing the potential of the refugee – Arendt's "conscious pariah" – Agamben calls for reconstructing "our political philosophy beginning with this unique figure" (114).

Perhaps our task today is to reconstruct our idea of academic freedom and the role of universities via the figure of the refugee academic. By inquiring into the ways in which we have conceptualized exile and the exilic intellectual, throughout the 20th century and into the present, we create opportunities both to understand the conditions that have brought us to this juncture, and to reevaluate the premises of our profession. Exile studies has long been associated with the exodus of Jewish writers, scholars, and artists from Nazi Germany. Investigations into the conditions of exile and its far-reaching consequences for the arts and humanities continue to engage scholars who are interested in displacement, diasporic communities, memory studies, or the Holocaust. Although comparative approaches are encouraged by longstanding initiatives in both Germany and Europe at large, the main aim is to shed light on the unique conditions of mass expulsion by Nazi persecution. Building upon the research and methodologies developed in Europe in the field of exile studies, it would be timely to expand the field beyond 1945 and acknowledge, for example, postwar Germany as a country of exile.

As part of restorative justice for the crimes committed during WWII, the West German constitution introduced the right to asylum in 1949. Towards the end of the Cold War Era, West Germany granted asylum in increasing numbers, among them writers, journalists, and scholars who fled the 1979 Iranian Revolution and the 1980 coup in Turkey. Those who found refuge in Germany and elsewhere in Europe created diasporic networks to support political resistance in their respective home countries. The establishment of the Institut Kurde de Paris, the recognition of the Armenian Genocide in some European countries or the flourishing of a transnational Kurdish literature is a case in point. So far, exile studies has on the one hand neglected to recognize the refuge of Middle Eastern literati, scholars, and journalists in Europe and on the other not connected to refugee studies or the larger field of migration studies in the 21st century.

Following Vanessa Agnew, a transnational approach that connects historical experiences of forced migration to the present is one that promises to build knowledge and link experiences of contemporary refugees from the Middle East and North Africa to those of Europeans displaced in large numbers during the 20th century. Agnew stresses the significance of "fostering awareness of historical continuities and a commemorative culture around forced migration" (Agnew 2017, 9). Exile studies would benefit from historicizing the experience of forced migration and differentiating between the refugee as recognized under the 1951 Refugee Convention, the asylum seeker, internally displaced person (IDP), stateless person, and returned refugee. If exile is to remain a useful concept to characterize critical distance to imperialism and authoritarian nation-states, each case of exile needs to be contextualized within the framework of asylum legislation, immigration policies of the respective host countries, and the reasons for forced migration. Historicizing the figure of the 'refugee academic' that has been called exile, émigré, refugee, expatriate, conscious pariah, or, as is the trend now, simply a person 'at risk,' might shed light on the uneasy marriage between nation-states and universities.

The current 'brain drain' from Turkey and the Middle East demands that we rethink the paradigms for exile studies. Connecting exile studies to the burgeoning field of refugee studies promises to overcome the conventional class divide that these assume, and to produce further insight into the ranges and reasons of forced migration. Since the study of intellectual exile in the 21st century necessitates the wider study of refugees and other forms of forced migration, joint networks to connect exile studies with the field of refugee studies are called for. Such an approach encourages inquiry into the forced migration of the intellectual and the political activist, as well as the politically disenfranchised and illiterate. Questions about issues such as the transfer of knowledge, the formation of diasporas, the definition of exile, and the building of transnational political networks are imperative to responding to increasingly Islamist, populist, or other authoritarian states.

References

Agamben, Giorgio. “We Refugees.” *Symposium* 49.2 (2018): 114–119.

Agnew, Vanessa. *Refugee Routes: Commemorating Forced Migration in Australia and Europe.* Research Proposal, Australia Research Council, 2017.

Arendt, Hannah. “We Refugees.” *Altogether Elsewhere: Writers on Exile*. Ed. Marc Robinson. Boston/London: Faber and Faber, 1994. 110–119.

Brubaker, Rogers. “Why Populism?” *Theory and Society* 46.5 (2017): 357–385.

Büyükdüvenc, Sabri. “John Dewey’s Impact on Turkish Education.” *Studies in Philosophy and Education* 13.3/4 (1994): 393–400.

Çıkar, Mustafa. *Hasan-Âli Yücel und die türkische Kulturreform.* Bonn: Pontes, 1994.

Demiralp, Seda. “The Breaking up of Turkey’s Islamic Alliance: The AKP-Gulen Conflict and Implications for Middle East Studies.” *Middle East Review of International Affairs* 20.1 (2016): 1–7.

Dewey, John. *The Later Works 1925–1953, Vol. 11: 1935–1937.* Ed. Jo Ann Boydston. Carbondale/ Edwardsville: Southern Illinois University Press, 1987 (original article was published in 1936).

Genç, Kaya. “Silence on Campus.” *Index on Censorship* 44.2 (2015): 10–13.

Gocek, Fatma Muge. “Defining the Parameters of a Post-Nationalist Turkish Historiography Through the Case of the Anatolian Armenians.” *Turkey Beyond Nationalism: Towards Post-Nationalist Identities.* Ed. Hans-Lukas Kieser. London: I.B. Tauris, 2006. 80–99.

Hausmann, Frank-Rutger. *Deutsche Geisteswissenschaft im Zweiten Weltkrieg. Die Aktion Ritterbusch (1940–1945).* 3rd ed. Heidelberg: Synchron, 2007.

Hirsch, Ernst. *Dünya Üniversiteleri ve Türkiyede Üniversitelerin Gelişmesi.* I. Istanbul: Ankara Üniversitesi Yayımları, 1950.

Kaplan, Sam. *The Pedagogical State: Education and the Politics of National Culture in Post-1980 Turkey.* Stanford: Stanford University Press, 2006.

Konuk, Kader. “Taking on German and Turkish History: Emine Sevgi Özdamar’s *Seltsame Sterne.*” *Gegenwartsliteratur: German Studies Yearbook* 6 (2007): 232–256.

Konuk, Kader. *East West Mimesis: Auerbach in Turkey*. Stanford: Stanford University Press, 2010.

Konuk, Kader. “Genozid als transnationales historisches Erbe. Literatur im Kontext türkischer und deutscher Geschichte.” *Gegenwart schreiben. Zur deutschsprachigen Literatur.* Eds. Corina Caduff and Ulrike Vedder. Paderborn: Fink, 2017. 165–176.

Müller, Jan Werner. *What is Populism?* Pennsylvania: University of Pennsylvania Press, 2017.

N.N. “A Very Turkish Coup.” *Nature* 477. 131 (08 September 2011). <http://www.nature.com/articles/477131a> [accessed: 19 March 2018].

Said, Edward W. *Representations of the Intellectual: The 1993 Reith Lectures*. New York: Vintage Books, 1994. 47–64.
Sakaoğlu, Necdet. *Cumhuriyet Dönemi Eğitim Tarihi*. Istanbul: İletişim Yayınları, 1993.
Zürcher, Erik J. *Turkey: A Modern History*. London: I.B. Tauris, 1998.

IV: New Contexts – Changing Concepts

Evangelos Karagiannis and Shalini Randeria

Exclusion as a Liberal Imperative

Culture, Gender, and the Orientalization of Migration

1 Introduction

In the summer of 2016, the leader of the Austrian right-populist party FPÖ in Graz uploaded a video to Facebook that led to considerable public controversy. It depicted the then 36-year-old local politician lecturing male refugees about appropriate norms of behavior toward European women. While his addressees for these rules of conduct were meant to be male asylum seekers ("Sehr geehrte Herren Asylanten"), the German language video would, of course, never reach them. In reality, its populist rhetoric was aimed at Austrians, especially at his own party members with their strong anti-immigration views. Its message was that refugees did not know how to behave around European women, and that the FPÖ would uncompromisingly oppose the harassment of women by refugees. The text of the message is as follows:

> Dear (male) Asylum Seekers, In light of recent unpleasant events that we have heard of, and in light of the beginning of the bathing season, and since in our culture, thank God, women are permitted to dress more liberally, it is important to keep some rules in mind: (*a blond female manikin, in summer attire appears to the left of the picture*) Here, one does not look at women provocatively (*a sign with the word "No" in German and misspelt in Arabic appears in the upper right corner of the picture*), whistle at women provocatively, or act in any way so that they might be made to feel uncomfortable. And one certainly does not harass women in a club (*he comes closer to the puppet*), or grab them by the butt or the breasts (*he does both*). Anyone who is guilty of these offenses must know that there is a party (*an FPÖ sign appears to the left*) that will ensure that those who abuse the right of asylum will be quickly returned to where they originate (*a sign with an image of an airplane is seen*). So, dear (male) asylum-seekers: behave yourselves. And most importantly: keep your hands off our women!!! (*a sign with this demand appears on screen*).[1]

Though purportedly aimed at protecting Austrian women from sexual harassment at the hands of a surging refugee population, the politician's misogynist

1 FPÖ Video "Sehr geehrte Herren Asylanten." *Spiegel Online* (1 June 2016) <http://www.spiegel.de/politik/ausland/fpoe-video-fuer-fluechtlinge-sorgt-fuer-spott-a-1095360.html> [accessed: 14 March 2018].

https://doi.org/9783110600483-013

and sexist performance was so brazen and crass that he was forced to 'temporarily' remove the video the very next day from his website (perhaps under pressure from his party).[2] His response to the storm of protests was to make light of the video, suggesting that its provocative humor was intended to attract attention. But he also justified his position by arguing: "I am interested in having a discussion on the rates of assault and rape, which have risen sharply since last summer, and that we take all necessary steps to prevent a repetition of the events of Cologne and Darmstadt, as well as of numerous isolated cases."[3]

The explicitly and embarrassingly sexist language used in the video may lead one to believe that it could only be aimed at a small segment of die-hard FPÖ supporters on the political fringe. However, the bizarre video reflects a specific grammar of exclusion, which has gained ground lately even within mainstream European discourses on migration. A core characteristic of this register of exclusion is its justification in the name of liberal values. It posits a liberal gender and sexual regime common to Austrian society as a whole that (Muslim male) migrants alone presumably do not share. The use of a language of gender and sexual equality as a form of border control (Ticktin 2008) characterizes the Orientalization of the European migration discourse by thus positing a radical, cultural alterity between European host societies and (Muslim) migrants.

The argument of this chapter unfolds in two steps. We first scrutinize this recent discursive turn using an anthropological perspective, which is *not* anchored in any expertise in 'foreign cultures' that the discipline is often presumed to possess. What defines a specific socio-cultural anthropological *approach* to societal phenomena in our view is its sensitivity to context, its perspectivity and reflexivity. The first section delineates a few basic principles underlying our understanding of migration and culture. It begins with the definition of the category of a 'migrant.' Although the question of *who* is a migrant, or *what* constitutes migration, is rarely posed, we argue that the answer to these questions is neither self-evident nor inconsequential. Because it shapes the discourse on migration in academia, as it does in politics, the very idea of who is a migrant needs careful scrutiny. We then address the concept of culture, which is central to both anthropology as a discipline and to the current discourse on migration. We point out that the use of the concept of culture can have radically different and even con-

2 Colette M. Schmidt. "Die kurze Filmkarriere des Armin Sippel." *derStandard.at, Blog: Colette M. Schmidt* (3 June 2016) <https://derstandard.at/2000038145880/Die-kurze-Filmkarriere-des-Armin-Sippel> [accessed: 14 March 2018].

3 Armin Sippel FPÖ. "Sehr geehrte Herren Asylanten ..." *Youtube*, Videobeschreibung nach 5. Bearbeitung. <https://www.youtube.com/watch?v=smwsVwplyDc> [accessed: 14 March 2018].

tradictory political implications. Finally, the first section addresses the specific pitfalls of a culturalist interpretation of migration, while exploring parallels with other discourses in which culture assumes a pivotal position.

The discussion of cultural fundamentalism, which postulates a radical cultural alterity between the (European) host society and incoming (Muslim) migrants, bridges the first and the second sections of the chapter. The latter then examines the shifts in the discourse on migration against the backdrop of European integration. We go on to analyze the increasing equation of the category of migrants with Muslims and explore the resultant Orientalism that pervades current discussions of migration and (absence of) integration. Some dilemmas of the discursive dichotomy between the European Self, posited as 'liberal,' and the Muslim Other, constructed as 'illiberal' that is used to stigmatize primarily migrants and asylum seekers from North Africa, the Middle East, and South Asia are also delineated in this context. In conclusion, we outline what in our view would constitute an adequate response to this highly problematic binary opposition.

2 Conceptual Remarks

2.1 Problematizing Migration

The very problematization of migration is predicated on its becoming a political challenge for host states. In Central Europe, the earliest theoretical work on migration was first conducted at a time when the phenomenon had already become politically problematic (Hoffmann-Nowotny 1973; Esser 1980). As long as workers from abroad were needed and thus welcome, neither politics nor academia recognized the need to understand various facets and implications of migration for the host society or the migrants themselves. What lent urgency to an academic understanding of migration was the need for policy prescriptions as migrant workers stayed on in host countries instead of returning home. Migrants as permanent residents came to be seen as a *problem for the state* (due to the financial burden placed on the welfare state and the presumed divided loyalties of migrants) rather than a *solution to the problem* of temporary labor shortages after 1945. That the first comprehensive study on migration in German-speaking countries was titled 'Sociology of the Problem of Foreign Workers' (*Soziologie des Fremdarbeiterproblems*) speaks volumes (see Sökefeld 2004, 11–15). One recognizes here how the category of the 'migrant' is less a neutral tool of sociological analysis, but rather belongs to the realm of socio-political semantics – what anthropologists term an 'emic' category. Such a contextualization of an 'analytical category' like 'migrants' is useful in tracing semantic shifts over time in allegedly timeless categories. It also reveals how the seemingly neutral descriptive or analytical term is situated in a concrete socio-political discourse.

A migrant is not simply someone who moves and relocates her place of residence, even if such naive definitions sometimes appear to be self-evident. In the global North, mobility of citizens *within* a nation-state is not only regarded as unproblematic; it is, on the contrary, welcomed and promoted as characteristic of modern economic flexibility. The term 'internal migration' has largely disappeared from use in the global North, where one is encouraged to be mobile, not to migrate. The difference goes far beyond terminology. Mobility reflects the desired norm – migration, an undesirable anomaly. Mobility of comparatively affluent pensioners even within Europe does not make them migrants. Migration is thus not a special form of mobility, but rather a particular interpretation of mobility, which is linked to the political-normative discourse of nation-states and the global political hierarchy in which the respective states are embedded (Karagiannis and Glick Schiller 2006, 163). 'Migration' can thus be unrelated to the movement of people or the crossing of nation-state borders. For example, while Germans settled in Mallorca or Londoners owning homes in the south of France are not considered migrants, the so-called 'second-generation migrants' are subsumed under the category of migrants although they are born in the countries of their residence.

Citizens of powerful states in Europe or America, who have moved all over the world, never regarded themselves as migrants and were never perceived as such in their homelands either. Their relationship to their new place of residence was described by using other categories, namely traders, entrepreneurs, investors, missionaries, teachers, civil servants, bringers of civilization, development aid workers, expatriates, estate-owners, or colonizers, and explorers. In short, they were anything but migrants. Citizens of powerful Western states, irrespective of where they have found themselves, could always rely on the energetic support of their states of origin whenever their rights or interests were seen to be violated. The difference to the fundamental vulnerability and the feelings of helplessness of most migrants could not be greater. In short, a German in Uganda is by definition not a migrant, while a Ugandan in Germany is very likely to be classified as one.

The ultraconservative former Bavarian Minister of the Interior (and for a short time Bavarian Chief Minister), Günther Beckstein, notorious for his advocacy of a highly restrictive migration policy, coined a slogan in the 1990s, which epitomizes the prevalent understanding of who migrants are: "We need people who are useful for us; not those who use us" (*Wir brauchen Menschen, die uns nutzen und nicht die uns ausnutzen*). Indeed, migrants are first and foremost *undesirable people*; people who are not welcome where they want to live. However, the demarcation between welcome/desirable and unwelcome/undesirable people is anything but a neat one, which leads to ambiguities in the public perception of who a 'migrant' is. For example, the Indian IT expert, who is sought after by *Siemens* but not necessarily by German society, is often not classified as a 'migrant.' What sets him apart from unskilled migrants are the expectations directed at him (for instance, unlike 'migrants' he may not even be expected to learn German). The term 'highly qualified migrants,' which is sometimes used for this category of foreigners, suggests their grudging acceptance as a 'necessary evil.' The controversial slogan coined in the 1990s by the German CDU politician Jürgen Rüttgers – "(We need) Children rather than Indians" (*Kinder statt Inder*) attests to ethno-nationalist preferences. In short, in Europe migration continues to be a question of how to deal with foreigners considered to be undesirable. Foreigners who are welcome are not subsumed under the category of migrants. In public discourse migrants are an unwelcome burden on the exchequer whose entry should be prevented at best or, if they are already there, whose integration, assimilation, incorporation, is by definition a problem. The centrality of these concepts within migration research reflects the traffic between academic and demotic discourses, revealing not only the political ends that much academic research on migration serves but also how politics saturate the conceptual foundations of academic discourse.

2.2 Uses of 'Culture' in Anthropology and Political Discourses

A key term in the description of the so-called 'migration problem' is 'culture.' What is striking about the use of the concept in academic and political discourse is its enduring ambiguity. It is not only reflected in a multiplicity of understandings, differing considerably in their semantic structure, and resulting in widely divergent, even contradictory, analyses and conclusions. It is also clearly borne out in a comparison of the anthropological concept of culture with those deployed in current mainstream descriptions of migration.

Although the term "has undergone a career of multilinear development" (Brightman 1995, 527) and embodies anything but a substantial consensus within anthropology, it encompassed two fundamental innovations. First, it introduced a decentralized view of the world and, thereby, a *new understanding of the world*. Whereas academic racism and evolutionism ordered human diversity hierarchically and placed white West (Europe and USA) at the pinnacle of civilization, the anthropological concept of culture initiated a kind of Copernican shift. The concept, which affirmed both the psychic unity and diversity of mankind, radically questioned the postulated hierarchy. It also allowed Western culture to appear as one culture among many, at par with the others and not as a model for all these. The anthropological concept of culture was thus an anti-hegemonic and emancipatory tool against the superiority and civilizing mission of the Western world. Moreover, such a concept of culture also brought with it a *new understanding of man* that challenged the classical Enlightenment conception shared by liberalism (Boggs 2004). In a now classic article, Clifford Geertz noted that humans without culture would not be rational beings, but rather lacking in orientation. He concluded that "there is no such thing as a human nature independent of culture" (Geertz 1973, 49).

Mainstream anthropology meanwhile is reserved towards the concept of culture at best, and, at worst, is completely dismissive. Ironically, just as 'culture' became increasingly salient in the humanities and social sciences (the 'cultural turn'), a critical stance towards the concept attained hegemonic status in anthropology (see Brightman 1995). Serious distortions in the assessment of societal reality were seen as a result of the concept. Criticism was directed at the idea and treatment of culture as territorially bounded, spatially isolated, or as an immutable, functionally integrated system. Further, the concept was seen to contribute to the Othering of the anthropological research subjects (Abu-Lughod 1991; Gupta and Ferguson 1992; see Karagiannis and Randeria 2016).[4] The increasing prominence of the concept in various political discourses was also a cause for concern among anthropologists, for culture has become the central concept of political practice and the 'key semantic terrain' of our time (Benthall and Knight 1993, 2). For many in the discipline, however, who often feel more comfortable on the political margins, the popularization of culture caused unease. In sum: increasing scepticism towards the analytical fruitfulness of the concept of culture, on the one hand, and its parallel ascent in political discourses, on the other, led anthropologists to treat 'culture' not as an analytical term but as an emic category – that is, as one belonging to the semantics of the societies under study.[5]

---

4 Due to its central conceptual importance for the discipline, the criticism of the concept of culture was always also part of a reckoning with the discipline's past. Thus the concept was often construed by its critics as the embodiment of all the errors and weaknesses of the discipline (Brightman 1995, 510). Arguably, evolutionism, structural functionalism, and area studies in the post-war era bear far more responsibility for the notorious tendency in anthropology to treat communities as territorially bounded and isolated wholes.

5 Criticisms of the anthropological concept of culture notwithstanding, the aspects it encompasses – pluralism, historicism, holism, relativism, and behavioral determinism (see Stocking 1968) – remain topical for social-scientific thinking, as is evident in recent discussions of plu-

Edward Said rightly pointed out that "the notion of a distinct culture [...] always get(s) involved either in self-congratulation (when one discusses one's own) or hostility and aggression (when one discusses the 'Other')" (Said 1978, 325). Thus, political discourses in which 'culture' plays a central role can be divided roughly into two groups, depending on who speaks and whose culture is referred to. In the first case, culture constitutes a central category of collective self-representation to which political demands are linked. Here, culture is considered as something positive, valuable and in need of protection. In the second set of discourses, 'culture' is employed to depict the Otherness of certain groups, especially Western representations of non-Western collectivities. The concept of culture, which carries negative connotations, is viewed here as a problem. While nationalism and multiculturalism can be assigned to the first category of discourses, to the second belong, for instance, the developmental discourse, the UN discourse on human rights and, last but not least, the discourse on migration. European discourses on multiculturalism and migration constitute a special case. The discourse on multiculturalism is primarily sustained not by migrants themselves but by the elites of host societies, who view the cultures of migrant minorities as contributing to cultural diversity. Yet this recognition implies a self-congratulatory liberalism on the part of host society. The ethnic majority celebrates itself, here, as an open and tolerant society welcoming of cultural diversity. Since multiculturalism has remained marginal to the mainstream discourse on migration, especially on the continent, and has thus had little political impact, it is not examined here.

2.3 Culturalist Representations of Migration

The status of the concept of culture within the discourse on migration has changed over time. While the culture of migrants was considered in early conceptions to be only *one* aspect of the larger problem of integration, later interpretations increasingly elevated culture to be the central issue. Despite repeated reminders of the inadequacy of culturalist interpretations of migration and/or the academic treatment of migrants as strangers/Others (Meillassoux 1980; Çağlar 1990; Sökefeld 2004), the view that the culture of migrants is the key to understanding their attitudes, behavior, and degree of integration into host societies has now attained hegemonic status. Culturalist interpretations of the migration problem

ral modernities over the last couple of decades (Eisenstadt 2000; Gaonkar 2001; Therborn 1995, 2003; see Karagiannis and Randeria 2016, 80).

are rooted in a concept of culture that locates it in spatial terms, tends to interpret it as static and assigns it the power to determine behavior. Leaving individual migrants little agency, it reduces them to their culture and assumes them to be identical with it. It is typical for this cultural-behavioral determinism to eclipse competing interpretations based on class affiliation, education level, rural or urban background, gender etc., factors which would be used, as a rule, to explain the individual behavior of members of the host society. Migrants on the contrary are treated as a unified and homogeneous community whose members are imagined to be at the mercy of the overwhelming force of their respective 'culture.' It is not only assumed to completely shape their attitudes and conduct but they are also seen as powerless and unwilling to act against the dictates of their culture. Thus, while Euro-Americans are considered to *have* culture or cultures, migrants are imagined to be automatically and fully determined by their culture/s (Brown 2006, 150–151).

In such a culturalist understanding of migrant behavior as inevitably rooted in local customs, all explanations focus on their home societies rather than on the receiving society, regardless of how long migrant families have been living in the latter (see Schiffauer 1983). The territorial boundedness of culture suggests that migration also constitutes an *unavoidable* problem for the host society, and above all, a problem for the unity (homogeneity) and cultural integrity of the host society (Stolcke 1995, 8). Migration is perceived as a threat that could have a disintegrating effect on the host society as the sheer number of foreigners in their midst 'overwhelms the public.'[6]

As a discourse that *culturalizes* political and social problems, the discourse on migration exhibits striking similarities to culturalist interpretations in the human rights discourse supported by UN organizations, as well as within development discourse, which hold the culture of various non-western societies responsible for their development deficits or poor human rights records (see Merry 2003; Sökefeld 2004). Common to all discourses that use culture as an explanation for an issue is the attribution of undesirable practices, institutional weaknesses, or political problems to cultural norms. Interestingly, migration discourse is unique in doubly defining something as both a violation of norms of the host society

6 The 'overwhelming of host populations' is an important rhetorical figure in political discourse on migration that assigns the vulnerable victim the blame. In the early 1990s, leading politicians of the ruling Christian Democratic Union in Germany explained away the racist riots against immigrants in German cities like Hoyerswerda and Rostock by arguing that the massive abuse of the right of asylum had "overwhelmed the local population" that felt threatened by the numbers of the newcomers (see Deutscher Bundestag. *Drucksache* 12/3162 (13 August 1992): 1–2; DISS (1992): 53, 65).

and as behavior in conformity to the norms of the migrants' own cultures. The simultaneous construction of a phenomenon as a norm violation in Europe but as standard practice in other societies is predicated on defining it as behavior *out of place*, which parallels an understanding of migrants as people *out of place*. The term 'culture crime' (*Kulturdelikt*), coined by a conservative Austrian minister of the interior about ten years ago, makes this normative contradiction explicit: it is a violation of (our) norms but acting according to (their) norms.[7]

Interestingly, the same politicians who preach the importance of individual choice and responsibility resort to cultural, collectivist arguments when migrants are involved in offences (Fernando 2013, 154). Here, culture then points to the perceived insolubility or impossibility of dealing with a problem that is deeply ingrained. Thus, the resort to culturalist interpretations increases with the degree of difficulty in addressing a problem, the responsibility for which is simply attributed to the Other. As in the case of the UN human rights discourse and the discourse on development or corruption, recourse to the culture of the Others in the migration discourse attempts to explain away one's own responsibility by shifting the blame squarely on to the Other. All these discourses relieve the global North of its share of responsibility for the emergence of, and solution to, a problem (see Merry 2003, 64; Fernando 2013, 161). Perhaps the persuasiveness of culturalist explanations is due to their self-exculpatory function. Culturalist interpretations thus not only appear highly plausible, but it is also remarkably difficult to articulate objections against them.

A radical version of this culturalism in the domain of migration is cultural fundamentalism, which has gained considerable ground since the early 1990s. It can be understood as a discourse of political exclusion, which overstates cultural difference by overemphasizing the territorial boundedness of culture, its invariability over time, and its capacity to determine behavior. Among its central premises is the fundamental incompatibility of cultures (Stolcke 1995, 4). Cultural fundamentalism, thus, subjects culture to a dual naturalization. For one, the underlying ahistorical concept of culture erases the historical processes of cultural homogenization within the space of the nation-state. The newly formed national monoculture appears instead as *natural* and normal, reflecting an eternal and desirable state of affairs. Moreover, it claims that human beings *naturally* desire to live among those with whom they share a common culture. Interactions with cultural Others are said to constitute a challenge that could prove overwhelming. Claims that cultural similarity produces primordial ties, or that

7 "Ausländerkriminalität: Fekter will 'Kulturdelikt' einführen." *diepresse.com* (7 August 2008) <http://diepresse.com/home/innenpolitik/404624/> [accessed: 14 March 2018].

cultural difference leads to primordial mistrust and hostility, allows xenophobia to appear natural, and can be used to justify the presumed disloyalty of migrants (Stolcke 1995, 5–8). The cultural fundamentalism of recent decades in Europe frames migrants as a principal threat to the host society due to an alleged fundamental and insoluble incompatibility of their cultures with 'ours.' The imputation of incommensurability of cultures and its very naturalness reminds one how close the attribution of such fundamental 'cultural difference' is to the old language of 'racial differences.' Verena Stolcke goes so far as to suggest that the concept of culture enables the revival of a racist discourse that does not need to resort to the now highly discredited biological race concept (Stolcke 1995, 12). Cultural fundamentalism turns the underlying premises of the anthropological culture concept on their head. Whereas within anthropology the culture concept was conceived as an emancipatory and anti-hegemonic tool, in cultural fundamentalist discourse it justifies demands for unequal political treatment and political exclusion. If the anthropological culture concept was developed to overcome academic and political racism, cultural fundamentalism promotes racism in thinly veiled form.

3 The Exclusion of Migrants as a Liberal Imperative

3.1 European Integration, 'Fortress Europe' and the Orientalization of Migration

Cultural fundamentalist descriptions of the migration problem are intertwined with changes in the imagination of the 'Self' (the 'familiar') and the 'Other' (the 'foreign'), which are accompanied by processes of inclusion and exclusion. In our view, the central categories of the discourse on culture and migration in Europe have shifted considerably as a result of the process of European integration (especially an entitlement to freedom of movement within EU member states). Within the EU, national demarcations have increasingly lost their salience, though admittedly it is not always clear who belongs to Europe. But citizens of European states, who used to be classified as foreigners and therefore as migrants, have increasingly come to be included in the category of the Self. As a result of these changes, the differences between those who were considered only a few decades ago to be problematic migrants from southern Europe and their affluent western European host societies are no longer foregrounded, but instead their similarities to the receiving societies are being emphasized. The erstwhile southern European migrant community considered to be a problem, for instance in respect of criminality and lack of education, has now completely disappeared from the migration discourse. Nothing of significance has changed in these communities of southern European migrants except for their politico-legal status as fellow EU citizens, which has altered how they are perceived and treated. The politics of desired mobility that is now promoted within the EU has normalized these groups.[8]

The erosion of the borders of the nation-states and of national communities in Europe along with enhanced opportunities for mobility for citizens of the global North since the 1990s have fostered a view of globalization as a flattening of the world (Friedman 2005). Yet to those in the global South facing all kinds of barriers to mobility it was always evident that globalization implies a highly selective mobility, or rather a reorganization of opportunities for crossing nation-state borders, for it results in a dismantling of only some borders, in the erection of others as well as in the reordering of some demarcations (Ferguson 1999, 234–254). Indeed, the dismantling of the intra-European borders went hand in hand with the construction of Fortress Europe, with attempts to make external borders of Europe as impermeable as possible to keep away those from the global South. Europe has never been more difficult to reach for those outside EU borders. The same is true of the USA, well before the election of Trump.[9]

The formation of a new collective European Self forged a new image of migrants. Following Frank-Olaf Radtke's caustic differentiation between 'foreign' and 'all too foreign' (Radtke 1996), one could say that, as a result of the inclusion of former 'foreigners' (i.e., those from the European periphery), the focus in the migration discourse has shifted to the 'all too foreign' (i.e., to those from the global South and, above all, Muslims). This semantic shift in the category of

8 This is, however, neither a linear nor an irreversible process. Brexit, whose advocates sought to bring about an end to free movement, could turn EU citizens in the United Kingdom once again into migrants. The financial crisis, which was quickly represented as a crisis of the European South, allowed for a revival of exclusionary discourses (one thinks of the stereotypes of 'lazy and fraudulent Greeks,' who live at the expense of 'hardworking Germans') and thus revealed the fragility and contingency of the European project. The hundreds of thousands of Greeks, who have found work in European countries (mainly in Germany) since the beginning of the financial crisis, are not classified as 'migrants' as of yet. But that too could change.

9 Thus there is something hypocritical about the European outcry against Trump's plans to erect a wall along the border with Mexico, given that the European Union has for years now been spending billions for security and surveillance designed to prevent migration from outside the EU. The outsourcing of this task to non-European states is unlikely to prove any more humane in its consequences than a wall.

migrants parallels an Orientalization of the migration discourse. This in turn has far-reaching ramifications for the cultural coding of migration, which is increasingly structured by narratives of the incommensurability of Western values with those of the rest of the world. One consequence is the radicalism of cultural fundamentalism with its premise of the incompatibility of cultures along with the Orientalization of the discourse on migration.

Today, Muslim migrants bear the brunt of the hostility to immigration in Europe. Muslims are seen to embody a dual threat for the host society. The first is clearly evinced in the saturation of migration discourse by the language of securitization. The Latvian actor and director Alvis Hermanis succinctly expressed this controversial view: "Perhaps not all refugees are terrorists, but all terrorists are refugees or their children." Hermanis decided to cancel his contract with the *Thalia* Theatre in Hamburg because the latter is committed to the support of refugees in Germany. In his view, "a simultaneous support for terrorists and for the victims of Paris is out of the question. [...] The Paris attacks show that we are at war. In every 'war' one must decide which side to support. [...] The era of political correctness is over."[10]

Muslim migrants are also perceived as a threat to European (Western) culture and civilization. The exclusionary character of references to Europe in their association with Islamophobia are made explicit in the German reactionary, populist *PEGIDA* movement ('Patriotic Europeans against the Islamization of the Occident'). Ironically, some of its leaders are radical nationalists who have taken upon themselves to protect European culture, despite their full-throated criticism of, or even outright hostility to, the strengthening of European integration. Europe appears to be a mere fig leaf that thinly veils their exclusionary agenda. Just as cultural arguments replaced discredited racist arguments justifying discrimination, the reference to Europe sidesteps the German nation by providing an alibi against the charge of 'old-fashioned' nationalism. This is especially important in Germany, where nationalist language is often discredited politically on historical grounds. In short, resorting to concepts of 'culture' and 'Europe' allows such discredited and marginalized positions to gain legitimacy within public discourse, bestowing exclusionary politics with a potential for mass mobilization.

---

10 "Ein Volksfeind," *Spiegel Online* (12 December 2015) <http://www.spiegel.de/spiegel/print/d-140390066.html> [accessed: 14 March 2018].

3.2 The Construction of Muslims as Illiberal Others

A critical analysis of the construction of Muslims as 'undesirable Others' would have to begin not with the Muslims themselves, but first and foremost with the protagonists, who exclude. In other words, in order to deconstruct the construction of Islam in dominant contemporary discourse, it is necessary to consider how the exclusionary Self is invented. Moreover, current fundamentalist depictions of the migration problem differ in one important respect from earlier culturalist descriptions in the 1980s. Not unlike the representation of culture in the UN discourse on human rights and the discourse on development aid, within the latter culture was always portrayed as a (problematic) characteristic of the Other. Culture was thus an explanation for the behaviour of foreign Others but not one's own. The argument had two aspects: a) the culture of migrant groups hinders them from adopting the universal, liberal values that prevail in western societies; b) the failure to assimilate to these ideal values has far-reaching negative consequences for the migrants themselves. In cultural fundamentalist discourses, by contrast, some liberal Western values, earlier understood to be universal, are increasingly particularized. These are now framed as fundamental elements of European culture (Uitermark, Mepschen, and Duyvendak 2014, 235). It is for this European Self that the culture of the (Muslim) Other poses a problem.

The "human right to homeland" – that is, the right to cultural familiarity and homogeneity within a demarcated space free of foreigners – formulated by Jörg Haider, the leader of the Austrian far right in the early 1990s, exemplifies this trend. One's own culture is imagined here not only as worthy of protection but also represented as threatened, even under siege. Keeping migrants away now becomes a legitimate demand in defence of one's own culture. This form of cultural protectionism, also referred to as neo-culturalism (Uitermark, Mepschen, and Duyvendak 2014; see Mepschen, Duyvendak, and Uitermark 2013), allows for another inversion of the anthropological culture concept that we have discussed briefly above. Whereas the anthropologists who coined and advocated the concept of culture imagined Western economic and cultural imperialism as a threat to the cultures of the rest of the world, the cultural fundamentalist discourse on migration reverses the relationship by placing 'Western' culture as under threat from the rest of the world.

But what constitutes this vulnerable Western culture? The following anecdote from a panel discussion in a European capital in early 2016 provides us with some clues. The panel addressed the highly-charged topic of state welfare and benefits for refugees. The intention of the organizers was to question the argument often put forward by critics of a liberal refugee policy that refugees constitute a financial burden on the state's social welfare budget. But in the absence of critics of a liberal refugee policy on the panel or in the audience, the critical discussion failed to materialize. The consensual panel discussion was followed by assenting statements from the audience in favor of taking in refugees. The panel chair, however, felt compelled to play devil's advocate, so she began by questioning the compatibility of the values of the refugees with those of 'our own' liberal order. She posed the provocative question as to whether the refugees share 'our' opposition to anti-Semitism and to homophobia.

The focus on the treatment of Jews and homosexuals as the epitome of a particular liberal order of values is reminiscent of the argument of the anthropologist Matti Bunzl, who has argued that in Europe the treatment of both groups can be taken as indicators of entire eras and their political projects (Bunzl 2004). Bunzl reads the shift in the current hegemonic imaginations of Jewish and homosexual alterity as major signs of the postmodern construction of Europe. In his view, European integration undermines the prerequisites for the construction of national Others (such as Jews and homosexuals), or even makes them obsolete as objects of Othering by unsettling the integrity of the nation-state. Drawing especially on the rhetoric of extreme right-wing parties, whose traditional images of the Jewish enemy had been replaced by corresponding Muslim ones, Bunzl posits that in Europe today Muslims have taken the place of Jews or homosexuals as the quintessential Others (Bunzl 2005).[11]

The chairwoman's aforementioned remark, however, goes beyond the new valorization and inclusion of former Others into the postmodern, transnational European Self that Bunzl points to. What is striking here is that the valorization of the previously despised and discriminated Other is employed today for the demarcation and exclusion of Muslims. Abandoning longstanding images of foreigners and enemies in 'post-national' Europe thus seems to reinforce the earlier colonial Orientalist schism between East and West. Though the features that characterize the West and the East may have undergone a change, the dichotomy remains intact and is confirmed anew.

The chairwoman's remark reveals a crucial aspect of contemporary constructions of European-ness and its Muslim alterity. Regardless of its opposition to Islam, the core of the emergent European Self is neither Christian nor religious, but rather *secular*, *liberal*, *enlightened*, and *tolerant of differences*. So the educational courses that are envisaged as key components of new integration plans for migrants address not Christian but liberal values, such as freedom of speech,

---

11 It should be noted here, however, that Bunzl's argument is based on rhetorical shifts in Austria, which perhaps have parallels in some Western European countries. Extreme right-wing parties in the European periphery are most definitely anti-Semitic and homophobic.

freedom from violence and abuse, equality of men and women, tolerance of diversity, and coexistence, etc. The increasing prominence of 'tolerance' in recent hegemonic migration discourse is anything but accidental. Wendy Brown comments on the relationship between 'tolerance' and culturalist interpretations of conflict:

> When political or civil conflict is explained as a cultural clash, whether in international or domestic politics, tolerance emerges as a key term for two reasons. The first is that some cultures are depicted as tolerant while others are not: that is, tolerance itself is culturalized insofar as it is understood to be available only to certain cultures. The second is that the culturalization of conflict makes cultural difference itself into a (if not *the*) salient site for the practice of tolerance or intolerance. The border between cultures is taken to be inherently volatile *if* those cultures are not subdued by liberalism. So tolerance, rather than, say, equality, emancipation, or power sharing, becomes a basic term in the vocabulary describing and prescribing for conflicts rendered as cultural. (Brown 2006, 150)

European nations celebrate themselves as open and free of prejudice, as societies that function in accordance with the principle 'live and let live' (laissez-faire liberalism). In these societies of allegedly free self-fulfilment of the individual, Muslims are *out of place* because Islam is construed as irreconcilable with liberal principles. The making of the liberal Self thus emerges in tandem with the construction of an illiberal Other, whom entry into Europe should be rightly denied. These discursive figures reveal the function of liberal values as a new register of exclusion. Exclusion here is not related to one's own illiberal and intolerant attitudes, but is instead held up in support of a liberal, tolerant, and progressive world-view. As noticed by Fernando, this paradoxical liberal ethic of tolerance – namely, intolerance of intolerance in the name of tolerance – reveals the inherent contradiction of liberalism, which results from the simultaneous commitment to individual autonomy on the one hand, and particular moral norms on the other (Fernando 2014a, 224–238).

3.3 Gender and Sexuality as Markers of Cultural Difference

The strategic character of the polarization between the liberal, enlightened, and tolerant culture of European-ness and the authoritarian, reactionary, and intolerant culture of the Muslim Other is especially visible in the selectivity with which the dichotomy is deployed. Although 'culture' comprises a broad spectrum of values, norms, and forms of conduct, the opposition as framed above is reduced by and large to gender, and is focused on sexuality. Whereas sexual equality and tolerance towards the sexual self-determination of the individual are construed as core elements of the liberal Self, the rigid control of female sexuality, and the rejection or persecution of homosexuality are considered essential to the nature of the Other (see, among others, Ticktin 2008; Butler 2008; Fassin 2010; Scheibelhofer 2013; Fernando 2014; Uitermark, Mepschen, and Duyvendak 2014). The notorious interview script (also known as the *Gesinnungstest* or the Muslim questionnaire)[12] designed by the state of Baden-Württemberg to interrogate the attitudes of Muslims toward the German liberal order is a good example of this tendency. Twenty of the thirty questions included in the interview form concern gender and sexuality.

Questions of gender lend themselves to the drawing of strict boundaries and have, during the course of history, repeatedly proven to be potent instruments of polarization. The position of women in family life and society has long been at the center of modernization and development policies in the (semi-)periphery, and it has frequently been used as an index of a society's level of development or civilization in earlier narratives. Even critical references to culture in the global South (and the associated criticism of cultural relativism) in the UN discourse on human rights are deeply interwoven with, and at times dominated by, questions related to gender. Both the *focus on gender* and the *culturalization of gender relations* are thus characteristic of numerous discourses that offer the West the opportunity to set itself apart from the rest and to occupy the moral high ground (Strobel 1993). In short, gender has unsurprisingly developed into a key cultural terrain, on which the superiority of the West – and correspondingly the 'lamentable' state of non-western societies – is often staged. Honor killing, widow immolation, forced marriage, veiling of women, and clitoridectomy are key elements of this dramatic language, which implies the call on white men (and women) to save brown women from brown men (Spivak 1999, 284).

Over the last decade, questions of sexuality have moved sharply to the center of the public debate on gender. The fundamental opposition between the European (or Western) Self and the Muslim Other has increasingly been articulated as a *sexual clash of civilizations* (Fassin 2010; Fernando 2013, 2014a, 2014b). Eric Fassin observes, regarding the sexualization of the French Republic with respect to migration:

> More and more, [...] the French republican motto has been redefined as *sexual* liberty, but also *sexual* equality, while the third term, *fraternity*, has generally been replaced by *laïcité*. This is manifest in particular in all the documents concerning immigrants, such as the 'integration contract' they have been required by law to sign and observe since 2006. In particu-

12 "30 Fragen für den Pass." *Zeit Online* (11 January 2010) <http://www.zeit.de/online/2006/02/gesinnungstest> [accessed: 14 March 2018].

> lar, equality is now defined exclusively in terms of gender, thus leaving out race or class. In the same way, *laïcité* is primarily understood as *sexual* secularism – insofar as it pertains to women and sexuality [...]. (Fassin 2010, 513–514; see Fernando 2014b, 694)

If deviation from the liberal sexual norms now championed ironically by reactionary right wing parties establishes one's cultural alterity and is the basis for calls for exclusion, sexuality has developed into a discourse of border control and social closure (Ticktin 2008; Fassin 2010). Complete integration thereby implies the sexual normalization of migrants (Fernando 2014a, 2014b), which has been expressed succinctly by Dutch Prime Minister Mark Rutte in his warning to migrants "to be normal or leave."[13] Interestingly, the remarkably persuasive force of this liberal register of exclusion has led to shifts in politically conservative discourse as well. Since migrants are increasingly being identified with conservative values, European conservatives, who earlier were neither advocates of women's emancipation or gay rights, are beginning to adopt progressive values, or at least a progressive rhetoric on these issues (see Uitermark, Mepschen, and Duyvendak 2014, 242). These developments have lead to a shift in some of the long-standing fault lines within European host societies. Coupling the demand for exclusion to a commitment to liberal values enables committed Europeans to deploy the same arguments as xenophobic nationalists. Similarly, the critique of the veil enables an unholy alliance between blatant sexists and women's rights advocates.

Secular women from Muslim migrant families and high-profile gay men often figure prominently in exclusionary discourses that stage a confrontation between the sexual progressiveness of the West and the sexual backwardness of Islam, since women and gay men are often regarded as particularly vulnerable to Islamic, patriarchal aggression.[14] The British journalist, former editor of *Breitbart News*, and founder of the *Gays for Trump* movement, Milo Yiannopoulos, points to the danger Islam poses for the achievements of the feminist and gay movement:

---

13 "Dutch prime minister warned migrants to 'be normal or be gone,' as he fends off populist Geert Wilders in a bitter election fight," *The Telegraph* (23 January 2017) <https://www.telegraph.co.uk/news/2017/01/23/dutch-prime-minister-warns-migrants-normal-gone-fends-populist/> [accessed: 14 April 2018].

14 The Maghreb-French feminist organization *Ni Putes Ni Soumises*, which enjoys remarkable political support in France, constitutes a prominent example of the depiction of the Islamic threat from the perspective of women who are at risk and thus require protection (Fernando 2013). The Dutch populist politician Pim Fortuyn, by contrast, was one of the first openly gay politicians to successfully combine traditional left-wing themes like secularism, gender equality, and the emancipation of gays with a neoliberal, populist, and anti-migration agenda (Mepschen, Duyvendak, and Uitermark 2013).

> I'm not talking about Islamists. I'm not talking about terrorists. I'm not talking about radical Islam. I'm talking about mainstream Muslim culture. [...] There are eleven Muslim countries in which I could be killed for being a homosexual. The state penalty is death. One hundred million people live in countries where the penalty for homosexuality is death. This is not radical Islam. This is mainstream Muslim society. Look what's happening in Sweden. Look what's happening anywhere in Germany, anywhere there are large influxes of a Muslim population. Things don't end well for women and gays. The left has got to make a decision. Either they want female emancipation and it wants gay rights or it wants Islam. It's got to pick [...].[15]

The stereotypical juxtaposition of orders of gender and sexuality is enabled by the interpretation of politically dominant positions that are historically contingent, and yet rendered as universally applicable and unalterable values of entire religious collectives. Such an Orientalist view ignores, for one, the diversity of ideas and practices among migrants, as well as among members of the host society,[16] and conceals, for another, the similarities and overlaps between the ideas and practices within the receiving society and the migrants. While, for example, the criticisms of sexual freedom and same-sex marriage in conservative Western and religious circles are interpreted as political issues, similar positions, when advocated by Muslims, are framed as cultural problems (Fernando 2014a, 254). The notion of the fundamental incompatibility of two opposing cultures can only be established by virtue of double standards. Furthermore, the dramatization of cultural differences is exaggerated by means of the juxtaposition of a few 'significant' symbols, which function as the symbolic embodiment of entire cultural orders. The concentration on the veiling of women in Islam is characteristic for this discursive strategy. It reveals a parallel with the preoccupation with clitoridectomy (or harmful 'traditional' practices) in the discourse on human rights as well as criticisms of multiculturalism and cultural relativism, whereby reference is repeatedly made to such practices as a way to stigmatize entire cultures.

That the practice in question is often controversial within the societies where it is practiced is concealed by its elevation to the status of a definitive cultural

15 "Milo: 'The Left Has Got to Choose between Gay Rights or Islam.'" *Breitbart Tech* (14 June 2016), <http://www.breitbart.com/milo/2016/06/14/milo-orlando-left-got-choose-gay-rights-islam/> [accessed 14 March 2018].

16 There is good reason to doubt whether liberal attitudes towards sexual self-determination, claimed as common to all Europeans are shared by a majority of them. As late as 2001, after Social Democrats in Berlin cheered their new party-chief Klaus Wowereit's public acknowledgement of his homosexuality, 80 % of German men responded in a representative survey "that they would react negatively or very negatively if they themselves had a lesbian or gay child" (Klauda 2008, 125).

characteristic. Customs that protect women from violence (among which veiling is sometimes counted, see Merry 2003, 64) are often similarly overlooked. It is not decisive for the selection and elevation of symbols around which to organize exclusionary discourses that they constitute real or acute problems that required redress, but rather that they are well suited to the dramatization of radical cultural alterity between new migrants and host societies. There is a major public debate on the legal status of the burka in Germany, for example, despite it being a very marginal phenomenon in the German public sphere.

Characteristic of the strategic deployment of these cultural symbols is the resort to tendentious interpretations, which are enabled by their consistent decontextualization. Anachronisms abound. Former French Prime Minister Manuel Valls juxtaposed the naked bosom of Marianne with the much-discussed burka, which meanwhile has been elevated in Europe to the core symbol of disrespect for, and oppression of, women in the world of the illiberal Other and is banned in countless public spaces. He expressed, with pathos: "Marianne's breasts are naked because she is feeding the people; she is not covered up because she is free! That is the republic!" (Böhmer 2016). The blatant sexism of reducing women to their role as nourishers of children aside, the interpretation of nakedness as freedom in Valls' pathetic comparison is striking. One must ask: would such an interpretation be possible were it not for the purpose of demarcation from the Islamic world?

The tendency to culturalize masks the fact that many of these allegedly 'significant' and culturally 'typical' practices are, as a rule, recent developments. The extended use of the headscarf among Muslim migrants in Europe is a development of only the last decades. In the 1980s, the headscarf was common only among older women. Not long ago, European-ness decorated itself in very different feathers. Although citizens in European metropolises today celebrate their openly gay mayors and politicians (and in so doing, celebrate their own liberalism), and express their postmodern commitment to diversity through the introduction of gay and lesbian figures in traffic lights, it is regularly forgotten that this liberal attitude toward sexual self-determination is decidedly new. The decriminalization of homosexuality (let alone its equality with heterosexuality) did not take place, even in some of the core countries of the European West, until very late in the 20th century. For centuries, the West claimed masculinity for itself and identified the Orient as synonymous with sexual deviance and promiscuity, effeminate men and disreputable women. Now that the West has discovered sexual self-determination for itself – and, similar to the criticism of anti-Semitism, not long ago – the East is accused of heteronormativity and the rigid disciplining of female sexuality. This retention of the boundary by means of a complete inversion of values and norms reveals that the celebratory tolerance afforded sexual self-determination has less to do with an eternal liberal Western culture and more with Orientalizing exclusionary practices.

4 Conclusion: Entanglements in Past and Present

This chapter addresses an old issue, namely the Orientalist juxtaposition of the superior West and the inferior East, albeit one which has been reconfigured anew. Despite the sweeping social changes in the 19th and 20th centuries, this binary discursive structure remains robust. The ambivalent use of the concept of culture in the public discourse is left unchanged, depending on who is talking about whose culture. Nevertheless, the discursive means with which traditional boundaries and oppositions are reaffirmed is novel. The close coupling of the discourse on migration with Orientalism, i.e. the figuration of the 'Undesirable' as the quintessential 'Other,' is also recent. This configuration is essential to cultural fundamentalism. The cultural fundamentalist view of the incompatibility of the culture of migrants with that of the host society is premised on the assumption of radical alterity characteristic of Orientalism. The focus on gender and sexuality is the main symptom of the Orientalization of migration discourse. Although gender relations also colored the earlier juxtaposition of the West and East, the centrality of gender and sexuality to the organization of this dichotomy is new.

An anthropological perspective could help uncover the historical contingency of cultural constructions. It would reveal that the world is not a natural mosaic of distinct, mutually isolated cultures distributed across space, more or less (in-) compatible with one another. Rather, cultural alterity is constructed and reconstructed time and again. Cultural differences are not only products of particular historical circumstances and open to continuous transformation. But also, as postcolonial perspectives remind us, this history is a *shared history*, one that both unites and separates Europe from its former colonies (Randeria 1999). Cultural fundamentalism derives its persuasiveness by eclipsing, or even denying, this mutual imbrication of the European Self and the non-European Other. Radical alterity, and the political demand for exclusion predicated on such difference, become untenable from a perspective that foregrounds past and present entanglements (Conrad, Randeria, and Römhild 2013).

Global power relations, capitalist expansion, economic inequality, war, mass mobility, the exchange of ideas, images, goods, and technology, are inherent to the construction of what today is often perceived as 'culture.' Cultural processes/ transformations within the Islamic world, as well as the hegemonic representations of Islam, are decisively shaped by the position of the Islamic world in the global system. Whatever is considered by culturalist discourses to be backward and typical of a culture is a product of a complex dialogue, unfolding under conditions of global politico-economic inequality. Whereas culturalists and cultural fundamentalists in the West argue for the radical alterity of Islam and its incompatibility with the Western values by drawing on stereotypes of Islamic treatment of women, a focus on their entangled histories may reveal the colonial and bourgeois (i.e. Western) origins of this female ideal that today is regarded as 'genuinely' Islamic (Abu-Lughod 1998). Similarly, cultural fundamentalist interpretations of fundamental homophobia in Islam become less convincing once we recognize that a genuinely Western concept of pathological sexual identities was adopted in the Islamic world as a result of Western expansion. The fact that Muslim elites are essentially more strongly homophobic than non-elites is, therefore, not a function of their distance from the West, but rather a consequence of Western influence (Klauda 2008).

Rather than expecting anthropology to provide explanations of cultural particularity, it could be used instead to interrogate difference as a natural, ahistorical given. Cultural differences would then be seen as consequences of specific socio-political processes and not as their causes. Thus, the decisive anthropological question is not what effects cultural differences produce, but rather how particular differences come about, and how they come to be regarded as 'cultural' (Gupta and Ferguson 1997; Bunzl 2004a). With regard to migration, the key questions could be: what are the processes involved in the construction of a person or group as 'migrant'? And how are such constructions made and deployed?

Our chapter has attempted to address these very questions. By pointing to 'undesirability' as the essential defining characteristic of migrants, exclusion was identified as the central structuring principle of the discourse on migration. The Orientalization of migration, i.e. the framing of migrants as the quintessential Other and as a danger to the liberal European Self, is the most recent effort to justify such exclusion. This construction of radical cultural alterities goes hand in hand with the establishment of the border regime known as Fortress Europe. It remains to be seen whether such a move will prove to be more effective. But European democracies will be poorer because of these practices of exclusion.

References

Abu-Lughod, Lila. "Writing Against Culture." *Recapturing Anthropology: Working in the Present.* Ed. Richard Fox. Santa Fe, NM: School of American Research Press, 1991. 137–154.

Abu-Lughod, Lila. "The Marriage of Feminism and Islamism in Egypt: Selective Repudiation as a Dynamic of Postcolonial Cultural Politics." *Remaking Women: Feminism and Modernity in the Middle East.* Ed. Lila Abu-Lughod. Princeton: Princeton University Press, 1998. 243–269.

Benthall, Jonathan, and John Knight. "Ethnic Alleys and Avenues." *Anthropology Today* 9.5 (1993): 1–2.

Boggs, James P. "The Culture Concept as Theory, in Context." *Current Anthropology* 45.2 (2004): 187–199.

Böhmer, Christian. "Ihre Brust ist nackt, weil sie das Volk nährt." *Wiener Zeitung* (30 August 2016).

Brightman, Robert. "Forget Culture: Replacement, Transcendence, Relexification." *Cultural Anthropology* 10.4 (1995): 509–546.

Brown, Wendy. *Regulating Aversion: Tolerance in the Age of Identity and Empire.* Princeton: Princeton University Press, 2006.

Bunzl, Matti. *Symptoms of Modernity: Jews and Queers in Late-Twentieth-Century Vienna.* Berkeley: University of California Press, 2004.

Bunzl, Matti. "Boas, Foucault, and the 'Native Anthropologist': Notes Toward a Neo-Boasian Anthropology." *American Anthropologist* 106.3 (2004a): 435–442.

Bunzl, Matti. "Between anti-Semitism and Islamophobia: Some Thoughts on the New Europe." *American Ethnologist* 32.4 (2005): 499–508.

Butler, Judith. "Sexual Politics, Torture, and Secular Time". *The British Journal of Sociology* 59.1 (2008): 1–23.

Çağlar, Ayşe. *The Prison House of Culture in the Studies of Turks in Germany* (Sozialanthropologische Arbeitspapiere 31). Berlin: Arabisches Buch, 1990.

Conrad, Sebastian, Shalini Randeria, and Regina Römhild, eds. *Jenseits des Eurozentrismus. Postkoloniale Perspektiven in den Geschichts- und Kulturwissenschaften.* 2nd improved ed. Frankfurt a.M./New York: Campus, 2013.

DISS (Duisburger Institut für Sprach- und Sozialforschung). *SchlagZeilen. Rostock: Rassismus in den Medien* (DISS-Skripte Nr. 5). Duisburg: DISS, 1992.

Eisenstadt, Shmuel N. "Multiple Modernities." *Daedalus* 129.1 (2000): 1–29.

Esser, Hartmut. *Aspekte der Wanderungssoziologie. Assimilation und Integration von Wanderern, ethnischen Gruppen und Minderheiten. Eine handlungstheoretische Analyse.* Darmstadt: Luchterhand, 1980.

Fassin, Eric. "National Identities and Transnational Intimacies: Sexual Democracy and the Politics of Immigration in Europe." *Public Culture* 22.3 (2010): 507–529.

Ferguson, James. *Expectations of Modernity: Myths and Meaning of Urban Life on the Zambian Copperbelt.* Berkeley: University of California Press, 1999.

Fernando, Mayanthi L. "Save the Muslim Woman, Save the Republic: Ni Putes Ni Soumises and the Ruse of Neo-Liberal Souvereignty." *Modern and Contemporary France* 21.2 (2013): 147–165.

Fernando, Mayanthi L. *The Republic Unsettled: Muslim French and the Contradictions of Liberalism.* Durham, NC/London: Duke University Press, 2014a.

Fernando, Mayanthi L. "Intimacy Surveilled: Religion, Sex, and Secular Cunning." *Signs: Journal of Women in Culture and Society* 39.3 (2014b): 685–708.

Friedman, Thomas L. *The World is Flat: A Brief History of the Twenty-First Century.* London et al.: Allen Lane, 2005.

Gaonkar, Dilip Parameshwar, ed. *Alternative Modernities.* Durham, NC/London: Duke University Press, 2001.

Geertz, Clifford. "The Impact of the Concept of Culture on the Concept of Man." *The Interpretation of Cultures: Selected Essays by Clifford Geertz*. New York: Basic Books, 1973. 33–54.

Gupta, Akhil, and James Ferguson. "Beyond 'Culture': Space, Identity, and the Politics of Difference." *Cultural Anthropology* 7.1 (1992): 6–23.

Gupta, Akhil, and James Ferguson. "Discipline and Practice: The Field as Site, Method, and Location in Anthropology." *Anthropological Locations: Boundaries and Grounds of a Field Science*. Eds. Akhil Gupta and James Ferguson. Berkeley: University of California Press, 1997. 1–46.

Hoffmann-Nowotny, Hans-Joachim. *Soziologie des Fremdarbeiterproblems. Eine theoretische und empirische Analyse am Beispiel der Schweiz*. Stuttgart: Enke, 1973.

Karagiannis, Evangelos, and Nina Glick Schiller. "Contesting Claims to the Land: Pentecostalism as a Challenge to Migration Theory and Policy." *Sociologus: Journal for Empirical Social Anthropology* 56.2 (2006): 137–171.

Karagiannis, Evangelos, and Shalini Randeria. "Zwischen Begeisterung und Unbehagen. Ein anthropologischer Blick auf den Begriff der Kultur." *Transkulturelle Politische Theorie: Eine Einführung*. Eds. Sybille de la Rosa, Sophia Schubert, and Holger Zapf. Wiesbaden: Springer VS, 2016. 63–83.

Klauda, Georg. *Die Vertreibung aus dem Serail. Europa und die Heteronormalisierung der islamischen Welt*. Hamburg: Männerschwarm Verlag, 2008.

Meillassoux, Claude. "Gegen eine Ethnologie der Arbeitsmigration." *Dritte Welt in Europa*. Eds. Jochen Blaschke and Kurt Greussing. Frankfurt a.M.: Syndikat, 1980. 53–59.

Mepschen, Paul, Jan Willem Duyvendak, and Justus Uitermark. "Progressive Politics of Exclusion: Dutch Populism, Immigration and Sexuality." *Migration and Citizenship* (Newsletter of the American Political Science Association, Organized Section on Migration and Citizenship) 2.1 (2013): 8–12.

Merry, Sally Engle. "Human Rights Law and the Demonization of Culture (And Anthropology along the Way)." *Political and Legal Anthropology Review* 26.1 (2003): 55–76.

Radtke, Frank-Olaf. "Fremde und Allzufremde. Zur Ausbreitung des ethnologischen Blicks in der Einwanderungsgesellschaft." *Das Fremde in der Gesellschaft. Migration, Ethnizität und Staat*. Eds. Hans-Rudolf Wicker, Jean-Luc Alber, Claudio Bolzmann, Rosita Fibbi, Kurt Imhof, and Andreas Wimmer. Zurich: Seismo Verlag, 1996. 333–352.

Randeria, Shalini. "Jenseits von Soziologie und soziokultureller Anthropologie. Zur Ortsbestimmung der nichtwestlichen Welt in einer zukünftigen Sozialtheorie." *Soziale Welt* 50.4 (1999): 373–382.

Said, Edward W. *Orientalism*. New York: Pantheon Books, 1978.

Scheibelhofer, Paul. *Integrating the Patriarch? Negotiating Migrant Masculinity in Times of Crisis of Multiculturalism*. Doctoral thesis submitted to Central European University (Comparative Gender Studies). Budapest, 2013.

Schiffauer, Werner. *Die Gewalt der Ehre. Erklärungen zu einem türkisch-deutschen Sexualkonflikt*. Frankfurt a.M.: Suhrkamp, 1983.

Sökefeld, Martin. "Das Paradigma kultureller Differenz. Zur Forschung und Diskussion über Migranten aus der Türkei in Deutschland." *Jenseits des Paradigmas kultureller Differenz. Neue Perspektiven auf Einwanderer aus der Türkei*. Bielefeld: transcript, 2004. 9–33.

Spiegel, Hubert. "Theaterskandal in Hamburg. Was erlauben Hermanis?" *Frankfurter Allgemeine Zeitung* (9 December 2015).

Spivak, Gayatri Chakravorty. “Can the Subaltern Speak? Speculations on Widow Sacrifice.” *A Critique of Postcolonial Reason: Toward a History of the Vanishing Present.* Cambridge, MA/London: Harvard University Press, 1999.
Stocking, George. *Race, Culture and Evolution: Essays in the History of Anthropology.* Chicago: University of Chicago Press, 1968.
Stolcke, Verena. “Talking Culture: New Boundaries, New Rhetorics of Exclusion in Europe.” *Current Anthropology* 36.1 (1995): 1–13.
Strobel, Margaret. “Gender, Sex, and Empire.” *Islamic and European Expansion: The Forging of a Global Order.* Ed. Michael Adas. Philadelphia: Temple University Press, 1993. 345–375.
Therborn, Göran. “Routes To/Through Modernity.” *Global Modernities.* Eds. Mike Featherstone, Scott Lash, and Roland Robertson. London et al.: Sage, 1995. 124–139.
Therborn, Göran. “Entangled Modernities.” *European Journal of Social Theory* 6.3 (2003): 293–305.
Ticktin, Miriam. “Sexual Violence as the Language of Border Control: Where French Feminist and Anti-Immigrant Rhetoric Meet.” *Signs: Journal of Women in Culture and Society* 33.4 (2008): 863–889.
Uitermark, Justus, Paul Mepschen, and Jan Willem Duyvendak. “Populism, Sexual Politics, and the Exclusion of Muslims in the Netherlands.” *European States and their Muslim Citizens: The Impact of Institutions on Perceptions and Boundaries.* Eds. John R. Bowen, Christophe Bertossi, Jan Willem Duyvendak, and Mona Lena Krook. Cambridge, MA/London: Cambridge University Press, 2014. 235–255.

Heike Greschke

Family Life in the Digital Age of Globalization

Critical Reflections on 'Integration'

1 Introduction

> Most people think that those who come here achieve everything immediately: work, prosperity, and a life of luxury. But that's not how it is. The people who come here out of necessity are in for a rough time, of – let's just say – sheer survival. When you arrive with a job, it's different, but many aren't that lucky. Arriving without papers, without work, without acquaintances, without friends, many don't understand a thing. And for this reason, many have problems and separate. Because there are times when you don't even have enough to call your kids, and even when you can live with a family member, you have to pay him for the room. And the first need that we have here as immigrants is to secure a room for ourselves, because if I don't have money to pay for my room, where do I go then? No one gives you a thing here. Nothing. So the first thing that we usually do, or have to do, is to make sure we have a place to stay, because we can't live on the streets. And the second thing would be money for the family, for the kids. And some people often can't manage that. They can't manage it. And, well, that was the start of my coming here. Angela, Spain, July 2015[1]

This quote is taken from an interview with a migrant mother who had already been working in Spain for ten years by 2015, when Diana, a fellow of my research team spoke with her about her arrival in that country. Before emigrating to Spain, Angela lived in Paraguay with her husband and her two young children. With a tourist visa and tickets bought with money borrowed from neighbors, Angela

1 Angela is part of a family that shared its stories and everyday life with myself and a group of other researchers in 2015. Along with Angela's family, 68 other transnationally organized families from Poland, Hungary, Spain, Ukraine, and several Latin American countries participated in the project "Mediatization of parent-child relationships" (see Greschke, Dreßler, and Hierasimowicz 2017 for a more detailed research report). These families contributed to our understanding of contemporary migration dynamics, and how advancing communication technologies shape social relationships and care practices in partly migrating families. On behalf of the research team, which consisted of Diana Dreßler, Konrad Hierasimowicz, Tímea Bauman, Jagoda Motowidlo, and myself, I would like to thank Angela and all of the mothers, fathers, sons, daughters, aunts, uncles, grandparents, and friends who shared with us their confidence, hospitality, and collaboration. The names and other personal data of all participants have been anonymized to protect their personality rights. All quotations are originally in Spanish, and were translated into English for this contribution.

https://doi.org/9783110600483-014

managed to reach her destination in Europe, but with little money, no accommodation, and no work, let alone a work permit. During the nearly four years it took Angela to legalize her residential status, she was subjected to precarious working conditions, and was sometimes not even compensated for her labor.

Approaching Angela's immigration experiences from a *methodological nationalist* perspective (Wimmer and Glick Schiller 2002) allows us to conceive of integration in terms of individuals' inclusion in the hosting nation-state. In so doing, it becomes clear that Angela was excluded from her legal right to participation and support upon setting foot in Spain ("No one gives you a thing here"). The fact that Angela was granted a residence and work permit only after considerable time may stem from state authorities' ambivalent and perhaps even calculated disposition towards immigration. Whether intentional or not, this way of dealing with undocumented immigrants reflects a de facto Darwinian integration policy: those who endure the irregularity, exploitation, and denial of rights in their host state prove that they are worthy of acquiring legal rights.

Angela's experiences illustrate how demanding immigration is for individuals who "come [...] out of necessity" and who, without legal support, must fight for legal recognition. Moreover, Angela's account evinces several characteristics of contemporary migration processes, some of which can be better studied through an analytical distance from current political debates on national integration policies. Indeed, a two-fold shift in perspective opens up further insights into these migration processes. 'Zooming in' on Angela's experience reveals that migration is not a process that can be reduced to the movement of individuals across national borders with the aim of permanently changing one's place of residence. Before Angela physically moved from Paraguay to Spain, she had already been part of a transnational migration network. Her brother, who had been in Spain for several years, encouraged her to immigrate there, and when she arrived, he supported her as she established herself. And by borrowing money for her trip to Spain from neighbors in Paraguay, she incorporated these neighbors into her transnational migration network. As a result, they acquired certain claims to the success of her migration project. 'Zooming out' of Angela's migration experience and situating her networks within an overarching global social system reveals how both the position of Paraguay in global markets and its postcolonial relationship to Spain have shaped current living conditions and migration processes.

Using Angela's migration story as the primary source material for this chapter, I will clarify what can be gained by employing a two-fold, shifting perspective on migration processes and practices. I begin with a conceptual clarification of the terms 'integration' and 'migration' in order to analyze contemporary migration processes and practices at a distance from political concerns. The remainder of the chapter is organized according to my approach of shifting between 'zooming in' and 'zooming out.' Returning to Angela's subjective experiences as a migrant, I zoom out to contextualize her account within the broader socioeconomic conditions of her home country. Taking Paraguay as a model for an actor-centered global perspective on migration and integration, two questions arise. The first concerns units of analysis in the study of integration; the second concerns practices of distant care (Baldassar and Merla 2013) and their impact on integration. Zooming into Angela's account illuminates the experience of starting and maintaining a family in contemporary Paraguay, as well as the path to becoming a transnational family. When stating that she had failed to send money to her family or even call her children because she did not earn enough during her first time in Spain, Angela refers to two constitutive features of contemporary transnational family life – remittances and communication technologies – which I examine accordingly. Finally, I summarize my findings by analyzing the social structure of transnational families and the overlapping layers of integration in which they are embedded.

2 Migration and Integration: Conceptual Clarifications from an Actor-Centered Global Perspective

Around the turn of the millennium, an essay by Andreas Wimmer and Nina Glick Schiller (2002) attracted considerable attention and set off a self-critical debate around basic theoretical and methodological premises in the social sciences. With the fundamentally critical concept of 'methodological nationalism,' the authors diagnosed a general and a migration studies-specific lack of reflexive distance from nationalistic ideologies and their constitutive significance for the self-conception and construction of modern societies. The nation-state as an organizational political form is not contextualized as a historical phenomenon, but rather presupposed as a universal social organizational form of modernity, and is reified as such in sociological terms and methods. This is reflected in comparative studies' assumption that nation-states are natural unities, and in the fact that transnational processes are systematically tuned out. In other words, these processes are only ever perceived as cut out from each state territory being researched. As a result, an overview of migration phenomena in both political and social-scientific thinking mostly assumes a spatially connoted social order whose interior and exterior are determined through a sovereign nation's territorial borders. From the perspective of an adoptive nation, migrants are the foreigners who come from elsewhere and plead for entry and participation in that nation, whether in the education system, the job market, or the social sphere. The foreign-cultural baggage that is attributed to immigrants offers locals the opportunity to present themselves as a cultural unity in opposition to the foreigners, and/or to obstruct these foreigners, thereby confirming the members of each group. The proximity of research in the social sciences to nation-state notions of normality in society is also reflected in the prioritization of questions regarding social, or more specifically, cultural integration or segregation.

Upon the development of methodological nationalism, migration and integration became problematic terms in the humanities due to an increasing awareness of how similar scientific terms were to political terms. It is therefore necessary to redefine the pertinent terms and to develop new approaches for researching the contemporary practices and social dynamics of migration and integration. In this respect, I suggest starting with a micro-perspective on individual cases, and tracking the interactions between subjects' movements, networks, practices, and values in order to understand how each of these elements creates and shapes units of integration in their own right. This micro-perspective on individual cases must thus be contextualized within a global frame of analysis to enable scholars to account for how different social units of integration, in which individuals are situated, mutually influence one another. I prefer a restriction of the term 'integration,' and will conceptually differentiate 'migration' into three distinct modes of mobility.

2.1 Integration

The concept of integration has long been a controversial topic in both politics and scientific discourse. Debates on methodological nationalism caused a shift in perspective in some factions of migration studies, resulting, in part, in a critical stance towards integration, and in some cases, in an outright rejection of the term (see Faist and Ulbricht 2014). In the context of contemporary refugee immigration to Europe, the paradigmatic status of integration has been strengthened again, including a return to assimilation claims in the debate (Koopmans 2017). But what does integration mean? Many scholars have attempted to define the concept, and in so doing have made a consistent definition progressively more difficult. Some scholars define the term in accordance with its undisputed etymological core. Deriving from the Latin word *integrare*, integration means the (re)establishment of unity (Scheller 2015, 23). Albrecht Koschorke points out this definition's nostalgic connotations, which is evident not only in the concept itself, but is also reflected in its use. According to him, this longing for an earlier state refers to the *identikit* of a sovereign territorial state, which houses a culturally homogeneous civil society: a form of government that appears threatened by migration and other globalization processes. He concludes that it is actually "a ubiquitous and diffuse fear of disintegration that gives the demand for integration its emotional impact" (Koschorke 2014, 220, my translation). The emotional impact resulting from an experience of the loss of an idealized society burdens the analytical value of this term. From a more objective standpoint, the German sociological lexicon defines integration as the unity of a system established by the binding and consensual determination of positions in the system's structure, as well as by roles in the system's division of labor and duties (Epskamp 2007, 301).

There are three advantages to this definition. The first is its renouncement of a concept of society that makes it apt for every social unit, be that the nucleus of a family, a cross-border social network of mutual support, a regional or global market, or the world system of nation-states as potential units of integration. This quality also enables interactions between these layers of integration to be taken into consideration. Second, this definition does not focus on individuals or groups, like migrants, as the object of integration, but rather focuses on the system itself. The third benefit of this definition is that it takes inequality into account as a potential structural feature of social systems. As such, it does not define any normative claims on inclusion in terms of equal participatory rights for each element of the system, i. e., each member of society, but solely in terms of a consensual and binding social order vis-à-vis social positions and labor division. Indeed, this definition distances itself from any political usage of the term, which is the equation of society to nation-state to unit of integration. Furthermore, it enables the empirical exploration of the relevant units of integration in each individual case, and the study of the dynamics of social positioning in a multilayered social field. In sum, this precise definition is freed from any idealizations about society and may thus serve as a working concept in the following analysis.

2.2 Migration

The present study differentiates migration into three interdependent modes of mobility: *corporeal*, *medial*, and *social mobility*. As mentioned previously, this modification is necessary in order to detach sociological from political relevancies, with the ultimate aim of foregrounding the inner logic of migration processes. Moreover, the constitutive import of 'media' for practical and discursive modes of socialization emerges through the differentiation and analysis of relationships between varying forms of mobility.

Corporeal mobility encompasses all possibilities regarding the decision to select one's location for life, work, study, and/or temporary residence, and with it, the necessity and ability to cross national borders. Here, the administrative classification according to the type of migration and the respective legal provisions and procedures is somewhat decisive. Is a border crossing politically, economically, or touristically motivated, or is it classified in the framework of international agreements on educational exchanges? The geographical origin and socioeconomic status of immigrants is decisive in attributing intentions to migrants, and in judging the legitimacy of their reasons for immigration.

Conversely, *social mobility* refers to the concrete alterations of social status that could be associated with a migratory act. Are qualifications in the destination country recognized as nearly equivalent, or are they devalued or upgraded? How are migration as a practice and migrants as social figures represented in public discourse? And in this process, in what ways are countries of origin, phenotype, religion, reason for migration, and other criteria roughly differentiated? Especially in the context of transnational lifestyles, paradoxical effects often reveal themselves with regard to social positioning.

Finally, *medial mobility* signifies the possibilities of access to and the competent use of media. These include cultural semiotic systems, i. e., language, which enable the crossing of linguistic borders, as well as information and communication technologies (ICT) that are required to cross spatio-temporal borders. In the context of migration, access to ICT is relevant in many regards: first, to obtain information and orientation vis-à-vis possible destinations and migration routes; second, to remain informed from afar on current events in the homeland, and integrated in the relevant discourses in the place or places to which one is socially connected; third, to shape social relationships, the coordination of which shifts in the process of migrating and must be readjusted; and fourth, to expand the possibilities for crossing linguistic borders through, for instance, the use of language-learning and translation applications. Before illustrating how transnational families practice these distinct forms of mobility and how these forms influence each other, I propose that changing perspectives, and zooming out on Paraguay's global social position, will contextualize contemporary Paraguayan living conditions and pinpoint how they shaped Angela and her family's migratory practices.

3 Zooming Out: Paraguay – 'Living Boundaries' and Global Soy Fields

A landlocked country in the center of South America, Paraguay borders Bolivia, Argentina, and Brazil. It is smaller than California with roughly 407,000 square kilometers, and its estimated 6.8 million inhabitants make it one of the least populated nations in Latin America. Despite its status as a relatively unknown and powerless state in cultural, economic, and political terms, Paraguay has acquired an important position in the global market as the world's fourth largest exporter of soy (Guereña 2013, 15). In this context, *Oxfam International* Executive Director Jeremy Hobbs has spoken out on Paraguay for being one of the countries with the least equitable land distribution (see also Bareiro 2004). Indeed, 77 % of Paraguayan land ownership is concentrated in the hands of only 2 % of the population, with 40 % of the population still living in rural areas (Hobbs 2012). In recent years, soy production has rapidly expanded, further deepening the inequality of land distribution. This is reflected in the *Gini index*[2], which showed an increase from 0.91 in 1991 to 0.94 in 2008 (Guereña 2013, 9). Because nearly half a million hectares of land have been turned into soy fields annually, thousands of rural families have been evicted as a result of this elevation in soy production. With the demand for soy still climbing – particularly in China and Europe, where soy is used mainly for cattle feed and biofuel – one may forecast a further concentration of land in the hands of few and the expansion of monocultural soy-based agriculture (Guereña 2013, 9). To be sure, Paraguay is a globalized territory largely controlled by transnational agribusiness, rather than a sovereign nation able to care for its inhabitants. On the contrary, the social and environmental consequences of the expansion of soy cultivation as well as cattle breeding for the world market[3] have increasingly manifested themselves in the rise of sick people, displaced communities, murdered peasants, polluted rivers, logged forests, and, particularly pertinent to this study, growing migration rates. In short, the global fields of soy cultivation cause severe social problems that require transnational solutions, such as migration.

In 2003, the global Swiss agribusiness Syngenta Corporation published an advertisement that cynically captured Paraguay's fateful position in the global

2 The *Gini-Index* or *Gini coefficient* is a statistical measure of inequality (with 1 representing the highest rate of inequality and 0 representing the highest rate of equality). Here the *Gini-Index* refers not to the statistical dispersion of income, but to land distribution.

3 Following *Oxfam International* Paraguay now houses twice as many cattle as humans. <https://www.oxfam.org/en/countries/paraguay> [accessed: 14 March 2018].

market. The advertisement consisted of a map on which a large, shaded area encompassing Argentina, Brazil, Bolivia, Paraguay, and Uruguay was called *The United Soy Republic*. The heading of this advertisement, *Soy does not know borders*, hints at practices of illegal land appropriation from so-called *Brasiguayos*. These Brazilian landowners buy Paraguayan territories in the border zone to cultivate soybeans; in so doing, they effectively expand the Brazilian frontier in the direction of Paraguay. Lenient laws and politics also attract foreign investors. Arantxa Guereña estimates that “at least 25.3 % of the country’s agricultural and livestock land is owned by foreigners” (Guereña 2013, 14). These processes of land concentration and expropriation are connected to the colonial history of both states and reflect a continuance of power relations. Framing Paraguay’s current situation in the context of its colonial and postcolonial development reveals that Paraguay never had the chance to develop into a nation-state following Anthony Smith’s definition: “a named community of history and culture, possessing a unified territory, economy, mass education system and common legal rights” (Smith 2004, 183). As I discuss elsewhere in greater detail (see Greschke 2012), the sovereignty of the Paraguayan state has always been under the demographic, cultural, economic, and political pressure of transnational influences. Although formally acknowledged as a nation-state, Paraguay empirically consists of a territory accommodating a mixture of sociocultural organizing forms, including Japanese, Brazilian, and Mennonite colonies, as well as contemporary translocal migration communities. Its political and economic power relations are strongly shaped by historical transnational relationships and loyalties, which resist attempts at national frontier demarcation, and have instead been creating ‘living frontiers’ (Clementi 1987). These living frontiers ultimately make Paraguay a weak state structured on blatant social inequality, fixing the country’s position quite low in the global social structure of an already “terribly unequal” world (Brock and Blake 2016), and at the same time, expelling Paraguay’s population.

4 Zooming In: Threats to Starting and Maintaining a Family in Paraguay

> Due to all the problems that I’ve had with each birth, we have been financially ruining ourselves. In Paraguay, medical care in hospitals is all at one’s own expense. Because I had three consecutive births, we were stuck with a lot of expenses, and, ever since I had the third one, we’ve remained in ruin. It’s been practically a constant struggle to survive, with three children and losing the house, losing the job, and losing all business. Because everything has gone bankrupt, it’s too much with three children. I had to make the decision

> with my heart in my hand, and decided to look for a better future for my children, and I had to have courage, because there was no way at that time. Angela, Spain, July 2015

When asked about her situation in Paraguay and the reasons for her migration, Angela discusses precarious living conditions and the powerlessness of national policy to support its constituents. According to a report on human rights in Paraguay, nearly half of the nation's inhabitants lived below the poverty line in the year before Angela emigrated, and an increasing number sought *salidas individuales* – individual solutions – to escape the crisis. *Salidas individuales* not only alludes to rising rates of migration, but also to a rising number of suicides (Bareiro 2004, 14). In Angela's account, ironically enough, starting a family is the main reason for the family's economic plight, and Angela's separation from her husband and children seems to be the only way to preserve her family. Angela's account reveals the precarity of human life in Paraguay, as well as state authorities' inability to provide for citizens' basic needs and safety. Her experiences demonstrate the logic of transnational migration practices in terms of strategies for coping with poverty.

The World Bank recently estimated that in 2013, 14.8 % of Paraguayan citizens lived and worked abroad, primarily in neighboring Argentina, followed by Spain, Brazil, and the United States. During the course of the economic crisis in Argentina, a particularly large number of people migrated to Spain, a country known to Paraguayans as the *mother of the fatherland*. But even if Spain is much farther from Paraguay then Argentina, a large part of migrants do not leave their family or local community, they do not emigrate from Paraguay, nor do they immigrate to Spain in the classical sense of the terms. As Angela's own account indicates, a considerable number of people 'transmigrate.' Transnational migration is a concept that was introduced by social anthropologist Nina Glick Schiller and her colleagues in 1994 (Basch, Glick Schiller, and Szanton Blanc 1994). This term refers to cross-border mobility practices that create social structures transcending physical boundaries in the familial, economic, political, health, and symbolic spheres. The transnational social spaces (Pries 1998), within which transmigrants organize their lives (Basch, Glick Schiller, and Szanton Blanc 1994), stretch from the migrant's present place of residence to other places (of origin and/or belonging) in other nation-states. Angela's case distinctly indicates that transnational migration often refers to transnational families, which Deborah Bryceson and Ulla Vuorela (2002) define as families that "live some or most of the time separated from each other, yet hold together and create something that can be seen as a feeling of collective welfare and unity, namely 'familyhood,' even across national borders" (Bryceson and Vuorela 2002, 3). The mobility practices of transmigrants bring national systems of gender-, class-, and ethnicity-related inequality into contact (Gregorio 1998). The nation-states in question are, in turn, usually disparately related to each other through the distribution of material and immaterial resources, political and economic power, prestige, as well as the "development" of and influence on global policy and economy (see Kaneff and Pine 2011). In brief, transnational migration often denotes mobility practices motivated by a prosperity gap between the country of origin and the country of destination. In Angela's case, this prosperity gap manifests itself in the prospect of much higher earnings in Spain than would be possible for her in Paraguay. Accordingly, the logic of transnational family practices that unfolds between nation-states tied together by a wealth gap is as follows: one or more family members move to a wealthier country for work and send part of the money back home to the family. Because the earnings in the wealthier country are relatively high compared to the cost of living in the home country, these so-called remittances often allow the family to meet their basic needs while advancing socially.

To the extent that migration rates have risen over the last few years in Paraguay, remittances have constituted a growing share of the gross domestic product. In 2009, more than 10 % of Paraguayan households had family members abroad, and more than 12 % of these households received remittances from them (Gómez and Bologna 2014, 435). From 2002 until 2012, the number of remittances grew steadily, eventually reaching up to $ 634 million in 2012 (World Bank 2016). Remittances are the third largest source of cash income in the country, following only the export of soy and meat. Whereas the money that is earned with soy and meat remains in the hands of a few landowners and transnational agribusiness, remittances are distributed among the larger population, including the most vulnerable, who would live in extreme poverty without them (Gómez and Bologna 2014).

5 Zooming Out: Remittances as Micropolitics of Economic Redistribution

As the World Bank report from 2016 establishes, the sending and receiving of remittances has markedly increased in Paraguay and worldwide since 1990 (see figure 1). A global perspective on these money transfer flows demonstrates a distinct division between sending and receiving countries in terms of global wealth stratification (Kofman 2008). In other words, remittances can be understood as a micropolitics of economic redistribution, which is significant for reasons beyond

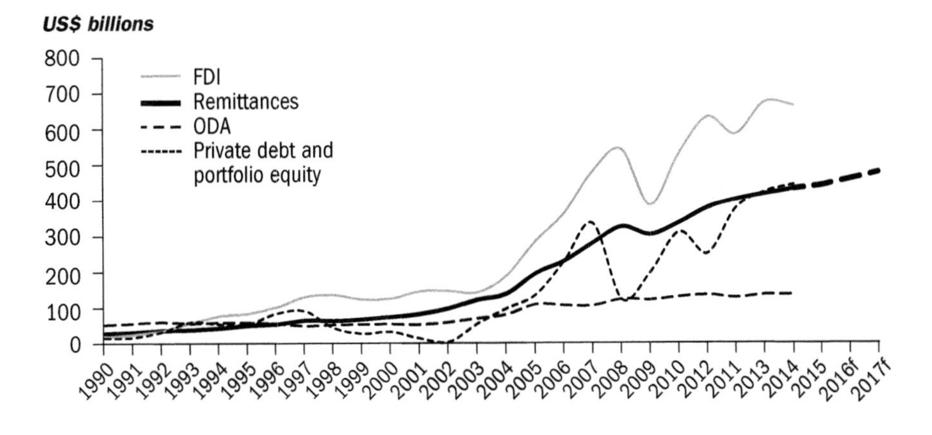

Fig. 1. Remittances to Developing Countries Are Large and More Stable than Other External Financing (World Bank 2016, 17).

the absolute quantity of money transferred. Remittances significantly exceed official development aid and other sources of foreign investment and "constitute reliable sources of foreign exchange earnings" (World Bank 2016, 17). Remittances are therefore classified as a foreign income source that in receiving countries is "less volatile and more stable than all other external flows" (World Bank 2016, 17). Because this micropolitics of redistribution contributes considerably to the mitigation of poverty in receiving countries (Adams and Page 2005; Gómez and Bologna 2014), remittances have attracted the attention of various international organizations in the last few years. In 2015, *The Fund for Agricultural Development* even declared June 16th the *International Day of Family Remittances*, which "is aimed at recognizing the significant financial contribution migrant workers make to the well-being of their families back home and to the sustainable development of their countries of origin."[4] Remittances not only cover the migrant and his or her family's basic needs, but can improve the family's social position in the home country, or at least in their local community, by improving the family's (and especially the children's) access to health care and education, by building houses, and by investing in businesses (World Bank 2016). When migrants spend money on community development, the resulting improvements often impact their families' neighborhoods or social environments (Smith 1998). If, for instance, a family receiving remittances is the first in the village with Internet access, non-transna-

4 <http://www.un.org/en/events/family-remittances-day> [accessed: 14 March 2018].

tional families in the village also benefit from this migration-induced technological development (Greschke, Dreßler, and Hierasimowicz, 2018).

6 Zooming In: Remittances and the Overlapping of Distinct Integration Units

In her account, Angela's understanding of remittances dictates both the priorities she sets and her justification for the use of her earnings. In her view, "money for the family, for the children" is second only to a "place for her to stay." Even so, Angela seems to feel that the order of her priorities needs further clarification, and she spends nearly two full minutes and more than two hundred words attempting to explain why she must first pay for her own room before sending the remainder of her earnings to her family. Why does she make such an enormous effort to rationalize a seemingly obvious matter? To whom does she address her account? One may assume that the interviewer is not the main addressee; rather, the interview offers an opportunity for correcting a misconception of her as a migrant mother who does not care for her children. Her account indicates an overlap of two conflicting integration units: the new one, in which she is involved as an undocumented migrant; and the old one, in which she is involved as a mother who has ostensibly frustrated her family by not sending money immediately upon arriving in Spain. From the family's point of view, the mother must compensate for her physical absence with financial support, since this was the precondition for her migration. To be sure, transnational migration alters the family system's division of labor, defining the mother's role primarily by her work outside of the home, which ensures the family's economic basic security. When the mother's money fails to arrive, this may mean the total absence and exclusion of the mother from the family system.

Earning, saving, and sending as much money as possible for remittances; receiving and spending remittance money on family matters; and accounting for how that money was used are elements of one of the most crucial practices of integration in transnational family systems. This practice restructures the distribution of tasks within the family system, as well as the social positions of the family members within and beyond it. From the migrant mother's point of view, she must take care of herself in order to continue taking care of her family. As Angela says: "The first need that we have here as emigrants is to secure a room for ourselves [...] We can't live on the streets." Without a doubt, living on the streets is uncomfortable and, particularly for women, unsafe. Furthermore, the deviance associated with homelessness attracts the attention of authorities, potentially jeopardizing the undocumented migrant's stay. Since migrants without papers constantly run the risk of being identified and expelled as 'illegal,' they must remain as invisible as possible, which naturally entails not attracting attention as a homeless person, but also ensures minimal self-inclusion in the labor and housing markets. Indeed, undocumented migrants generally have no legal rights to personhood in their host countries. Upon first arriving in Spain, Angela had to totally exclude herself from her family system in favor of the host country's work and housing subsystems in order to remain unnoticed by the authorities. During that time, she experienced social descent within both integration units, as well as strong restrictions in terms of corporeal and mediated cross-border mobility. She was not able to travel, to send money, or even to phone home, since, as she phrases it, "sometimes one does not even have enough to call one's children."

Besides indicating her lack of mobility, Angela's statement suggests the importance of communication technologies for the emergence and maintenance of transnational families, with migrant mothers taking on the additional responsibility of maintaining a communicative relationship with their children. In her comparative study of undocumented and documented migrant women in Spain, Asunción Fresnoza-Flot finds that "[a]ll respondents have made sure that each member of their family possesses a cellular phone and are usually the ones paying for the cell phone bills of their children by including this amount in their monthly remittance" (Fresnoza-Flot 2009, 260). By thus facilitating a communicative relationship and providing media equipment for their children, migrant parents set up the infrastructure for modes of remote parenting which include and go far beyond the aforementioned economic care practices, as will be discussed in the following section.

7 Being Here and There: Communication Technologies and Family Integration

On May 16, 2007, on the occasion of Mother's Day, an article on the feminization of migration in Paraguay was published in the online version of the Paraguayan newspaper *Ultima Hora*.[5] The article caught my attention due to two quotes. The first one was from children whose mothers worked in Spain. When asked what they would do on Mother's Day, one child responded: "We will send her an e-mail.

---

5 *Ultima Hora*, 16 May 2007. <http://www.ultimahora.com>; quotations originally in Spanish, my translation.

Dad is scanning a card with a heart that I made. I say to her there that I love her and miss her and lots of other things. [...] We go as often as possible to the Internet cafe to go online and talk to her and see her (webcam). She got thinner." The second quote was from educators reporting on a project seeking to improve children's rights. The participating children were asked to draw a picture of the ideal community in which they would like to live. The educators said: "They mostly drew telephone booths and explained to us that they would like to have one so that they can speak to their mother who is in Spain." In both instances, the essential role that media-based communication, as well as telephone and Internet use, plays in the organization of transnational family life comes to light. Steven Vertovec (2004) describes an accordingly significant increase in international phone calls, which he attributes to a parallel increase in transnational migrants. While general cost and tariff reductions and migration-related services would have made it easier to maintain transnational relationships in the long term, the advent of the Internet and, even more critically, the smartphone accelerated this process of mutual influence between corporeal and mediated forms of mobility and family care. It is, indeed, this same generation of transnational families that we encountered in our field research on the mediatization of parent-child-relationships in Spain in 2015.

> When I arrived, there was already Orkut. First came Orkut; you would get in and open and it would be full of messages. Then came Messenger and there you could already be seen in the Messenger, after that came Facebook and then Skype [...] and now WhatsApp [...] We use everything and my kids over there have everything too [...] Now it is easier because everything is on the phone [...] So I am here with my body, but my whole mind, my heart, my thoughts, are there as if I was there. And it is a continuous, steady condition, it is never broken.
>
> Alicia, Spain, July 2015

In the above quote, Alicia, another Paraguayan migrant mother who had been living in Spain for nine years by the time of our interview, describes this accelerated phase of 'mediatization' (Krotz 2007) in which she participated when migrating to Spain. Despite being separated by more than 5,500 miles from her four underage children in Paraguay, she does not believe herself to be distant from her children. The high degree of media mobility she achieves through a combination of different platforms – namely, *Facebook*, *Skype*, and *WhatsApp* – and the possibility of always being available for her children allows Alicia and her family to mutually participate in one another's daily lives. However, this very technological connectedness makes her feel a decoupling of her working body and caring mind, with her working body being housed by her employer in Spain and "my whole mind, my heart, my thoughts" by her beloved family in Paraguay. As in Alicia's case, technological advancements have enabled a growing number of transnational parents around the world to maintain their care duties in their children's daily lives. They can be present when their children need advice, someone to talk to, or someone to play with. They are responsible for sending enough money and providing the technological infrastructure of family life, but they are also entitled to monitoring financial expenditures, the children's educational progress, and their well-being. Sending remittances has also been streamlined through smartphone technologies, as telecommunication enterprises now offer money transfer services with mobile remittance applications that are cheaper than traditional ones. In brief, digital technologies can facilitate the integration of the family system by granting the migrant parent a more sophisticated role in his or her children's lives, including but not limited to the duty of providing for their economic security. However, these technologies are no guarantor of family integration, despite what the advertising campaigns of companies such as *Skype*[6] and *Ria*[7] suggest. Since the rise of transnational families promises to open lucrative market segments, interested companies tend to produce idealized imaginings that do not necessarily meet the reality of transnational family life.

According to our findings, next to the aforementioned technologies, human 'media' are immensely important for the success of familial integration under transnational conditions. The younger the children are, the more the family depends upon caretakers in the children's place of residence who not only ensure their care, but model the involvement of the physically absent parent. Technology alone does not enable the parent's presence in the children's daily lives; producing and regulating "connected presence" (Licoppe 2004) is an elaborate and specialized process, which, the younger the children are, depends all the more strongly on a third person. Following her temporal exclusion from the family system, Angela had to endure the bitter experience of no longer being allowed her role as wife and mother. As she reports, her husband extracted her from the lives of her children as a person and as an actor playing the culturally prescribed motherly role. When, after her initial difficulties in Spain, Angela began to send money and to try to restore her communicative relationship with her children, her husband accepted the money and used it to provide for the children. However, contact with the children proved to be difficult, and Angela held her husband responsible, blaming him for having hindered her communication with her family. It was only after Angela traveled to Paraguay without warning that

6 See *Skype*'s "Stay Together" campaign on *YouTube*: <https://www.youtube.com/watch?v=Qx7Mmwnstqw> [accessed: 14 March 2018].

7 See *Ria*'s "El Mariachi" video on *YouTube*: <https://www.youtube.com/watch?v=HUCT3mg6P8Q> [accessed: 14 March 2018].

the situation became clear to her. She learned from her children that their father had led them to believe that their mother had left them, and that he had earned the money on which the family lived. By depicting Angela thus, the father succeeded in excluding the mother from the family system, simultaneously claiming her money as his and using her migration to rehabilitate his own faltering social status. In their internal relationship and with respect to the direct social environment of the family, he could meet the expectations of a conventional patriarchal family arrangement, which encourages the man alone to occupy the role of provider. When Angela visited Paraguay, she showed her children copies of the money transfers that she had carefully retained to prove to them that she had fulfilled her maternal responsibilities the entire time she had been away. After the passage of so many years, she reports, it was nevertheless very difficult to regain her children's trust and to rebuild her relationship with them. Once again, communication technologies – above all, the smartphone – have facilitated Angela's attempts to reestablish a trusting relationship with her children.

> They have grown older [...] they already have their own mobile phones and they have WhatsApp, and although they often do not answer my calls, well, I know that they are there, and if anything happens, they will let me know immediately. It's not like before, when they had to depend on their dad to talk to me. Angela, Spain, July 2015

In the above quotation, Angela indicates the difference that her children's ages and their personal access to communication technologies make in her situation. When the children were younger, their father was the coordinator who could enable, or disable, an intimate relationship between Angela and her children. When the children grew older, Angela had the opportunity to establish a more exclusive relationship with them. However, this was only possible under three conditions: first, she had to be physically present at least once to reconnect with her children and to provide evidence – with her body and copies of money transfers – that she had always complied with her care duties, despite her husband's assertion of the opposite. Second, she had to and still has to provide them with their own smartphones and permanent Internet access to facilitate their communicative mobility and independence from the father. Third, she must continually exert herself to maintain their confidence in her presence as a mother. While she describes the relationship with her son as intense yet conflicted, because "teenager are a bit rebellious," she explains how she has been working towards a stable, close relationship with her daughter:

> Well, I have been conquering my daughter with stories. I talk about when I was pregnant with her, the things I wanted, the things I did. I told her the beginning [...] of how much I wanted to have a child, I even dreamed of it. [...] The conversation I have been generat-

> ing with her has changed her [attitude] towards me [...] So what I do is talk, constantly be connected, tell her what I'm doing, where I am, even up to what I'm eating or what I'm preparing. I leave on the speaker [of the smartphone] or I record [videos] of the things that I'm doing and send them. Then she's sure that I'm not here wasting time at nightclubs.
>
> Angela, Spain, July 2015

Angela describes two modes of reconnecting with her daughter. The first consists of an invitation on an imaginative trip to the conception of the relationship between mother and daughter, Angela's firstborn. Locating this conception even before the mother was pregnant, the quality of their relationship is granted through the mother's great desire to have a child, which manifests itself well into her dreams. This communicative relationship, which she very appropriately denotes as *conversación*, virtually counteracts the daughter's memory of being left by her mother as a small child and, thus, of not being wanted or loved by her. With her *cuentos* (stories), Angela transcends the child's memory while simultaneously effacing the father's role in the conception of the mother-child-relationship. He is not presented in her stories, since her stories embody *her* wish to have children, and *her* pregnancy. But while this *conversación* transforms the relationship between mother and daughter, it alone does not suffice, and their relationship requires constant updating: the mother must perpetually reassert her maternal presence and orientation towards her children's wellbeing. She is always 'on,' so to speak, painstakingly documenting and letting her daughter continually follow her daily routine in order to prove to her that she is not in Spain to have fun, as the father claims, but to provide for the well-being of her children.

8 Conclusion

In this chapter, I have discussed the significance of integration in transnational families. By approaching an individual case through an actor-centered global perspective, I have demonstrated how transnational migration creates parallel processes of inclusion and exclusion in different social systems, which are asymmetrically intertwined within one overarching global social system. Social positioning has become a multilayered process producing social decline and rise, a phenomenon Rhacel Salazar Parreñas (2001) coined as "contradictory class mobility." When analyzing the social structure of transnational families, it is therefore important to consider that a great deal of the world and its subsystems are already highly globalized, through global markets and production processes and transnational companies and organizations. The infrastructure for global communication and mobility is also highly developed, at least in technological terms. On the other hand, the degrees of physical freedom to move have always been distributed disparately and are strongly contingent upon nationality. Since the global social system consists of nation-states, citizenship becomes a kind of social escort for positioning individuals within a global social structure. Bearing this in mind, strategies of family welfare can be interpreted as modes of inclusion in a globalized society, the borders of which do not necessarily correspond to the formal borders of nation-states. Transnational family care practices are provoked by growing social inequalities on a global scale and are reinforced by advancing communication technologies that enable the compression of relationships over vast geographical distances. Ironically, this results in the effect that geographical distances might develop into something more virtual, and virtual or mediated relationships might become something more real. Maintaining a family under transnational conditions, however, takes much more than a smartphone and Internet access.

As I have established in this chapter, being a virtually-present-while-physically-absent parent necessitates a cooperative family system, which, with the help of technology, enables the absent parent to be there with his or her family and to assume responsibilities for care. Presence, to put it in Christian Licoppe's words, is nothing physically given or taken by migration; rather, it is "something to be worked on and is based on skills, as well as dispositions, mechanisms, resources and constraints" (Licoppe 2015, 97). In a severely unequally globalized and mediatized world, the ability to acquire resources, develop skills, and learn to manage dispositions and constraints in order to expand the reach of one's presence turns out to be essential when striving for a decent social position in global society.

References

Adams, Richard H., and John Page. "Do International Migration and Remittances Reduce Poverty in Developing Countries?" *World Development* 33.10 (2005): 1645–1669.

Baldassar, Loretta, and Laura Merla. *Transnational Families, Migration and the Circulation of Care: Understanding the Mobility and Absence in Family Life*. New York: Routledge, 2013.

Bareiro, Line. "Paraguay empobrecido: Análisis de coyuntura política 2004." *Derechos Humanos en Paraguay 2004*. Ed. CODEHUPY. Asunción: Editora Litocolor, 2004. 13–28.

Basch, Linda, Nina Glick Schiller, and Cristina Szanton Blanc, eds. *Nations Unbound: Transnational Projects, Postcolonial Predicaments and Deterritorialized Nation-States*. Amsterdam: Gordon and Breach, 1994.

Brock, Gillian, and Michael Blake. "Global Justice and the Brain Drain." *Ethics & Global Politics* 9.1 (2016) <http://dx.doi.org/10.3402/egp.v9.33498> [accessed: 14 March 2018].

Bryceson, Deborah Fahy, and Ulla Vuorela. *The Transnational Family: New European Frontiers and Global Networks*. New York: Berg, 2002.

Clementi, Hebe. *La frontera en America. Una clave interpretativa de la historia americana 1.* Buenos Aires: Leviatan, 1987.
Epskamp, Heinz. "Integration." *Lexikon zur Soziologie*. Eds. Werner Fuchs-Heinritz et al. Wiesbaden: Springer VS, 2007. 301.
Faist, Thomas, and Christian Ulbricht. "Von Integration zu Teilhabe? Anmerkungen zum Verhältnis von Vergemeinschaftung und Vergesellschaftung." *Sociologia Internationalis* 52.1 (2014): 119–147.
Fresnoza-Flot, Asunción. "Migration Status and Transnational Mothering: The Case of Filipino Migrants in France." *global networks* 9.2 (2009): 252–270.
Gómez, Pablo Sebastián, and Eduardo Bologna. "Pobreza y remesas internacionales Sur-Sur en Paraguay." *Revista Brasileira de Estudos de População* 31.2 (2014): 431–451. <http://www.scielo.br/pdf/rbepop/v31n2/a10v31n2.pdf> [accessed: 14 March 2018].
Gregorio Gil, Carmen. *Migración femenina. Su impacto en las relaciones de género*. Madrid: Narcea Ediciones, 1998.
Greschke, Heike. *Is There a Home in Cyberspace? The Internet in Migrants' Everyday Life and the Emergence of Global Communities*. New York/London: Routledge, 2012.
Greschke, Heike, Diana Dreßler, and Konrad Hierasimowicz. "Die Mediatisierung von Eltern-Kind-Beziehungen im Kontext grenzüberschreitender Migration." *Mediatisierung als Metaprozess*. Eds. Friedrich Krotz, Cathrin Despotović, and Merle-Marie Kruse. Wiesbaden: Springer VS, 2017. 59–80.
Greschke, Heike, Diana Dreßler, and Konrad Hierasimowicz. "Im Leben kannst Du nicht alles haben – Soziale Ungleichheit im digitalen Zeitalter. Elternschaft auf Distanz in teilweise migrierten Familien." *Mediatisierte Gesellschaften. Medienkommunikation und Sozialwelten im Wandel*. Eds. Andreas Kalina, Friedrich Krotz, Matthias Rath, and Caroline Roth-Ebner, 2018 (in print).
Guereña, Arantxa. *The Soy Mirage: The Limits of Corporate Social Responsibility – the Case of the Company Desarollo Agrícola del Paraguay.* Oxfam Research Reports August 2013. <https://d1tn3vj7xz9fdh.cloudfront.net/s3fs-public/file_attachments/rr-soy-mirage-corporate-social-responsibility-paraguay-290813-en_2.pdf> [accessed: 14 March 2018].
Hobbs, Jeremy. "Paraguay's Destructive Soy Boom." *New York Times* 2 July 2012 <http://www.nytimes.com/2012/07/03/opinion/paraguays-destructive-soy-boom.html> [accessed: 14 March 2018].
Kaneff, Deema, and Frances Pine. "Emerging Inequalities in Europe: Poverty and Transnational Migration." *Global Connections and Emerging Inequalities in Europe: Perspectives on Poverty and Transnational Migration*. Eds. Deema Kaneff and Frances Pine. London: Anthem Press, 2011. 1–37.
Kofman, Eleonore. "Stratifikation und aktuelle Migrationsbewegungen." *Transnationalisierung sozialer Ungleichheit*. Eds. Anja Weiß and Peter A. Berger. Wiesbaden: Springer VS, 2008. 107–135.
Koopmans, Ruud. *Assimilation oder Multikulturalismus? Bedingungen gelungener Integration*. Berlin: LIT Verlag, 2017.
Koschorke, Albrecht. "Ordnungen der Vielfalt. Integration." *Das neue Deutschland. Von Migration und Vielfalt*; anlässlich der Ausstellung *Das Neue Deutschland. Von Migration und Vielfalt*. Eds. Özkan Ezli and Gisela Staupe. Konstanz: Konstanz University Press, 2014. 220–223.
Krotz, Friedrich. *Mediatisierung. Fallstudie zum Wandel von Kommunikation*. Wiesbaden: Springer VS, 2007.

Licoppe, Christian. “Connected Presence: The Emergence of a New Repertoire for Managing Social Relationships in a Changing Communication Technoscape.” *Environment and Planning D: Society and Space* 22 (2004): 135–156.

Licoppe, Christian. “Contested Norms of Presence.” *Präsenzen 2.0, Medienkulturen im digitalen Zeitalter*. Eds. Kornelia Hahn and Martin Stempfhuber. Wiesbaden: Springer VS, 2015. 97–112.

Parreñas, Rhacel Salazar. *Servants of Globalization: Women, Migration and Domestic Work.* Stanford, CA: Stanford University Press, 2001.

Pries, Ludger. “‘Transmigranten’ als ein Typ von Arbeitswanderern in pluri-lokalen sozialen Räumen. Das Beispiel der Arbeitswanderungen zwischen Puebla/Mexiko und New York.” *Soziale Welt* 49 (1998): 135–150.

Scheller, Friedrich. *Gelegenheitsstrukturen, Kontakte, Arbeitsmarktintegration. Ethnospezifische Netzwerke und der Erfolg von Migranten am Arbeitsmarkt.* Wiesbaden: Springer VS, 2015.

Smith, Anthony D. *The Antiquity of Nations*. Cambridge: Polity Press, 2004.

Smith, Robert. “Transnational Localities: Community, Technology and the Politics of Membership within the Context of Mexico and US Migration.” *Transnationalism from Below*. Eds. Michael P. Smith and Luis E. Guarnizo. New Brunswick, NJ: Transaction Publishers, 1998. 196–240.

Vertovec, Steven. “Cheap Calls: The Social Glue of Migrant Transnationalism.” *Global Networks* 4.2 (2004): 219–224.

Wimmer, Andreas, and Nina Glick Schiller. “Methodological Nationalism and Beyond: Nation-State Building, Migration and the Social Sciences.” *Global Networks* 2.4 (2002): 301–334.

World Bank Group. *Migration and Remittances Factbook 2016*. 3rd ed. Washington, DC: World Bank, 2016. © World Bank. <https://openknowledge.worldbank.org/handle/10986/23743> License: CC BY 3.0 IGO.

Doris Bachmann-Medick

Migration as Translation

1 Some Preliminary Positionings

This article aims to contribute to the broader critical field of migration or postmigration studies by employing 'translation' as a differentiating analytical lens for a new view on migration. In so doing, it seeks to overcome the confinements of some current macro-theoretical assumptions:

- first, its translational perspective avoids approaching migration as mainly a regime of governance and management that aims to control, normalize, and discipline (see the critique by Castro Varela 2013);
- second, it follows an actor-oriented approach that challenges the dominant 'methodological nationalism' in migration studies (see the critique by Wimmer and Glick Schiller 2002; on its problematization from a translational perspective, see Sakai 2013) by focusing on migration in its multi-sited transnational movements and plurilingual conditions;
- third, it shakes up the dichotomy of the dominant majority society versus the immigrant reflected in many approaches. It achieves this by pointing to overlapping belongings or affiliations and to multiple linguistic entanglements in processes of migration, and by drawing on agency-based approaches that accentuate the making and shifting of boundaries by migrants themselves (see Wimmer 2008);
- and fourth, it questions the all-too-general assumptions of the colonial/postcolonial trajectories of contemporary migration, which could be reflected anew and more precisely as translational displacements (see the critique by Gutiérrez Rodríguez 2006).

In this sense, a translational view suggests paying more attention to micro-theoretical approaches, and proposes focusing on small-scale units – for example, on migrational urban communities and their language use in multilingual conditions (see Simon 2008, 2012). In fields like these, 'translation' could be used as a productive analytical concept that enables transparency. It could reach beyond analyzing the obvious challenges presented by the multilingual conditions of migration scenarios. It could also help to reveal power relations in the linguistic and discursive sphere that shape the process of 'making' migrants. A translational approach thus includes attention to forms of linguistic resistance by drawing on critical self-designations of migrants against their official label-

https://doi.org/9783110600483-015

ings – as refugees, as non-citizens, and as people with no rights (see Doppler and Vorwergk 2014, 51–52).

Thus, 'translation' will be emphasized in this article as a methodological tool for breaking scenarios of intercultural encounter into smaller units, thereby taking notice of actors, mediators, and translators, and of their practices, steps, and emotional involvements – always considering starting points of cultural intervention and participation. From this stance translation can not only be seen as a research category that is "good to think with" (Edmund Leach) but also as a social condition and mode of existence of migration itself. An approach like this helps to acknowledge and analyze above all the disturbing dimensions of migration processes.

2 Translational Conditions of Migration

Migration in recent times has often been viewed as a global flow of people, as a process of transnational mobility, and therefore as an important element of an emerging cosmopolitan society that celebrates the crossing of borders. This view – as present reality has shown – is in many ways too simple and too smooth (for a critique, see Glick Schiller and Salazar 2014). Migration could and should be understood rather as both a field of border politics and a multiplicity of often conflicting translation activities. This comes to the fore in an approach that refers to the specific relevance of borders in migration contexts, in their function of differentiating between inside and outside, and of demarcating in this way who is to be reckoned as a stranger and who is not. In contrast to this notion of borders as "functions of warfare," not least against migrants, Etienne Balibar has emphasized the potential of borders as "functions of translation" (Balibar 2010, 317), stressing the importance of discourse vs. power for the construction of political space.

This article suggests replacing this still-abstract bird's-eye perspective with a stronger actor-oriented view that examines translation as a practice for managing the necessary but often precarious shifts between different contexts. In taking this approach it follows the claims of Néstor García Canclini, who summarized his research experience of examining migratory processes from Mexico to the US with these words: "migration implies a radical way of experiencing uncertainty and the passage from one way of naming and speaking to another" (Canclini 2011, 24). Such "displacements of meanings" (Canclini 2011, 24) are existential movements of cultural translation – tense, conflicted, but also potentially productive. Translation must be seen here as a specific mode of cross-cultural disruption and enduring uncertainty. Certainly, there can be global flows of people here, and (successful or less successful) linguistic and other forms of integration or assimilation there. But a greater challenge for the cultural analysis of migration processes – in their interactive, personal, and affective aspects, as well as their abstract and structural ones – seems to be presented by the constantly occurring ruptures, frictions, and insecurities. The immigrant experience in particular, including such irritations and the "pain of immigration" (Hron 2009, x; Sayad 2004), has often been dismissed in discourses on migration. A translational lens that is associated with those 'nonlinearities' can make these dimensions accessible.

The disrupting dimensions of migration constitute a central and defining force. I am referring here on the one hand to the active power of migrating individuals: the power to deal with cultural displacements, discontinuities, interventions, and shifting social contexts; to be confronted with misunderstandings and obstructions; and even to exercise agency in triggering social transformations. We have come in this way to see migrants as translators themselves, as mediators and active agents of translation. But on the other hand, when I speak of disruptions, I am also referring to the alienation and degradation of migrants, to their definition by others. I am referring to migrants who are – as Salman Rushdie (1991, 17) put it – "lost in translation" as "translated men": exposed to circumstances in which they lack any translational competence, in which they instead become translated by others or are forced to undergo the painful process of self-translation, of redefining themselves as persons. In many cases such complex conditions of translation and translatedness find expression in oral utterances, unreflected social and linguistic categorizations, or hidden, unofficial articulations (see Gentzler 2013, 343–344). This dimension of "hidden translations" and of being translated should by no means be underestimated; it can be seen at work even when migrants actively engage themselves as translators and cultural brokers.

Could inspiration from translation studies hope to open up new horizons here, especially as this discipline has come to consider translation itself as a mode of "cultural migration" (see Polezzi 2012a, 102–104)? The emerging approaches of looking at migration from the perspective of translation studies have certainly uncovered new areas for research (see Vorderobermeier and Wolf 2008; Polezzi 2006, 2012; Wolf 2012). It seems, however, that too often these approaches reproduce many jargonish key terms – coming mostly from postcolonial theory, and including the usual categorical suspects: hybridity, in-between state, cultural displacement, and negotiation – that are then applied rather uncritically to the field of migration.[1] What we should strive for is a more empirically grounded and less metaphorical understanding of translation activities in migration contexts. It is in this direction that I try later in this essay to offer some tentative but potentially operative models of translation.

Yet: in which respect do we speak of 'translation' here? The understanding of translation has itself migrated during the last decades from the linguistic and textual sphere to that of cultural transmissions and transfers, as well as to the field of social practices.[2] Indeed, translation originally developed from a migratory practice itself, one that aimed at transferring meaning from one context to another. It has expanded into a category of social practice and thus has become a "cultural technique," a "modus operandi of our times" (Young 2011, 59), and even "an international civil rights issue" (Gentzler 2013, 342). Translation has helped to manage shifts between and across different social contexts or lifeworlds; it has also been methodologically refined as an analytical category, and as such can be used as a new tool for the analysis of migration processes. But why use translation in this way? In short, because it allows us to dissect, examine, and understand instances of migration as individual or collective steps of translation. The door now seems open for the discovery of pragmatic modes of translation, on a local or micro level, that were previously invisible or overlooked in the analysis of migration practices, interactions, and self-narrations. But can we really dare to conceptualize migration as translation?

The need for new descriptive and analytical categories when we face migration on a global scale seems uncontested. In the words of Nikos Papastergiadis: "In an age of global migration we also need new social theories of flow and resistance and cultural theories of difference and translation" (Papastergiadis 2000, 20). 'Translation' as a means of dealing with differences can be applied here to the various forms of migration – whether they are processes of individual or collective migration into exile, re-migration after World War II, or recent global mass migrations. But dealing with differences via translation often means dealing with asymmetries, as the unequal conditions of global knowledge circulation reveal. In fact, in the global sphere we are facing a collision or convergence of different, asymmetrical, and hierarchical knowledge orders, which tend to marginalize "foreign" migrant knowledge and expertise. As Simone Lässig and Swen Steinberg claimed in a recent special issue on "Migration and Knowledge" of the journal

1 See, to some extent, the debate on "Translation and Migration" in *Translation Studies* 5.3 (2012): 345–368; 6.1 (2013): 103–117; 6.3 (2013): 339–351.

2 On the extended understanding of translation in the wake of a translational turn in the humanities and social sciences, see Bachmann-Medick 2016, 175–209; 2016a; 2014.

Geschichte und Gesellschaft, it is exactly this asymmetrical hegemonic condition that demands a new recognition of "migrants as subjects of the history of knowledge and (of) the change, translation, and genesis of knowledge 'in movement' from one place to another" (Lässig and Steinberg 2017, 323–324).

Translation can be an appropriate analytical tool for exploring in closer detail the occluded or "hidden" contributions of migrants to this broad field of knowledge formation, as, for instance, in their ability to challenge established boundaries in a society or to unsettle its conventional classifications. Above all, translation can be a tool for identifying elements of a possible reciprocal enrichment between the languages and cultural articulations of migrants on the one hand, and established languages and knowledge orders on the other. A complex field of translation is opening up here, in which migrants act as definers, but at the same time also belong to the defined, in continually being translated and having to struggle with hegemonic discourses and exclusion. This translational field of migration thus seems constantly to be on edge, facing obstructed communication and troubled integration. A failure of translation might even occur deliberately here, when it is connected to a refusal of translation with regard to migrants – as translation scholar Edwin Gentzler foresaw a few years ago:

> The problem is that many groups, and here I think of the very strong English-only factions in the United States or of many German and Austrian anti-Muslim voices, resist new forms of communication and linguistic and cultural differences. This leads to national policies of non-translation, of sequestering immigrants, programs of deportation, and, for those who remain, discrimination and mistrust. (Gentzler 2013, 344)

But even in hostile situations of "non-translation," immigrants still embody a permanent translational challenge for changing the host country's language and being changed by it, since they can operate through productive 'detours' and exercise a "large range of translation activity that is carried out behind the scenes" (Gentzler 2013, 344).

Translation activities "behind the scenes," resistances, and non-translations refer back to a field of informal communication that is actively in place before any actual translation endeavor gets underway, where the course is set for such a process and where common reference points are located or new reference points are produced. I refer here to the often-overlooked pivot points in each translation process, a closer knowledge of which allows us access to the important dimension of "pre-translations" or "hidden translations." By drawing on such antecedent dimensions in trajectories of migration, we can no longer rely on a linear, bipolar framework that assumes unidirectional movement between a country of origin and a host country. We should instead search for multi-polar transnational and translational reference points that open up third spaces of back-and-forth translations between multiple spheres of belonging. Perhaps in this understanding, the tool of 'translation' could be made fruitful for the analysis of "super-diversity conditions," as the migration anthropologist Steven Vertovec calls it (Vertovec 2007), or of "transmigration," in the sense of nonlinear movements between specific localities and not between entire nations (Alvarez 2014).

Conceptualizing migration as a multi-polar process of translation can only gain analytical force when one leaves behind the mere metaphorical understanding of translation. But the established linear model of 'translation as representation' – that depends on its reference and fidelity to an original – also seems inappropriate for the analysis of complex multi-polar migration processes. What is needed, then, is an attempt to develop other 'models of translation' that are more suitable for analyzing the multi-directional scenarios of migration, whether they exist as exile, diaspora, labor migration, flight, forced migration, or other forms of displacement. But how do they address the living conditions of migrants between disruption, displacement, and re-placement in the wake of the long colonial trajectories of migration (Bandia 2014)? And could they even unfold as "modi operandi," as practices or operative models of translation that open up possible paths for action in migrational circumstances themselves?

3 Models of Translation

3.1 The Model of Negotiating "In-between" Positions

> This liminality of migrant experience is no less a transitional phenomenon than a translational one; there is no resolution to it because the two conditions are ambivalently enjoined in the 'survival' of migrant life. [...] The migrant culture of the 'in-between,' the minority position, dramatizes the activity of culture's untranslatability; and in doing so, it moves the question of culture's appropriation [...] towards an encounter with the ambivalent process of splitting and hybridity that marks the identification with culture's difference. (Bhabha 1994, 224)

Homi Bhabha, the author of this somewhat cryptic statement, describes the migrant's in-between position as a liminal, dangerous, ambivalent, and critical one. It is at the same time a "translational" position, with the potential to overcome appropriation and assimilation and to resist a culture's staging of seemingly untranslatable cultural difference. From this perspective, migrants dwell in a split space, a "third space." They appear to be a threat, as they hold the potential to translate and subvert dominant cultural discourses, even if they serve here only as a surface onto which collective fears and uncertainties are projected.[3]

At first glance, the characterization of migration as an in-between state that interrogates the assumption of assimilation or nationalization sounds plausible; it proposes a new liminal state of hybridity in which the migrant lives with a split personality and multiple belongings. The study of culture, in this way, seems to celebrate in-betweenness in its potential to break open the homogeneity of a national culture. It fosters a difference-oriented concept of culture to be derived from the coexistence of contradictory norms. But one might ask whether migration really corresponds to a condition of "in-betweenness," as Bhabha claims – a condition that, after all, still indicates a binary system of polarities.

In her manifesto entitled "Against Between," Leslie Adelson critically questions the worn-out claim that migrants oscillate in a vague intermediate state between two worlds (Adelson 2001, 245). This positioning, Adelson states – using the example of Turkish-German migrant literature – still takes as a given a territorialized concept of home and belonging, and with it, the idea of separate lifeworlds. In her view, migrants should instead be ascribed the capacity to create "Sites of Reorientation" (*Orte des Umdenkens*, 247) in which reimaginations of traditional national orders and belongings take place. But what does this mean in a concrete sense? We certainly cannot find an answer by adhering to the rhetorics of belonging, hybridity, and translation as a mode of difference, or to other formulas of global encounter. Instead, we need a more precise, non-metaphorical, and methodologically elaborated approach that considers translation as a social practice – as a specific form of communicating that could evoke a changed concept of "integration."

3.2 The Model of Integration or Assimilation

From Joachim Renn's extended study on this subject (Renn 2006), we have learned that a society in its multiple voices cannot be grasped through its holistic 'representation,' but only through a dynamic of "translational relations." Translation can become effective here not by aiming at any equivalence of meaning (as it would be the claim of 'representation'), but as a vehicle for producing a specific "unity" of society in its coexisting plurality: This means that "the unity of society [...] can only be *realized* in permanent translation in a variety of contexts" (Renn

3 On the historical dimension of this assumed threat or subversive potential of challenging norms, see, in the context of early modern migration, Schunka 2014, 33ff.

2002, 209). Integration is thus conceived as a matter of constant translation activities between non-generalizable, divergent perspectives, interests, positions, and forms of life. As such, integration is seen as the outcome of pragmatic exchanges and negotiations, unfolding as a "countermeasure against disintegrative tendencies in concrete actions and situations" (Renn 2002, 211). Renn uses the category of translation to show how the fragmented units of a society do not necessarily remain stuck in parallel worlds, but from the beginning are infused by "translational relations." In this sense, the notion of translation can be made productive for a sociological "revision of the concept of integration" (Renn 2006, 8).

Renn's methodology finds expression in the investigation of a concrete historical example of migration, available in Peter Conolly-Smith's book *Translating America* (2004), that also uses translation as a suitable tool for looking at integration or even assimilation. In this detailed study, Conolly-Smith explores how German immigrants to America at the turn of the 20th century cultivated an assimilation-oriented practice of self-translation and self-integration. These immigrants borrowed then-innovative American models of visual representation (such as cartoons or comic strips) for their own German-American newspapers (*Hearst's German Journal*, the *Staats-Zeitung*, and others) that functioned as mediating agents of cultural translation. How, specifically, did this work? As Conolly-Smith explains, "The *process* that led […] to a state of cultural incorporation was […] a process of translation that, for immigrants, necessarily had to *precede* incorporation" (18). This finding evokes a general insight: we can only grasp the complexities of translation processes in the course of migration by searching for traces of those 'pre-translations' that precede any accomplishment of integration or incorporation. Should we, then, try to reconstruct decisive pivot points through which a particular migration process receives its direction?

Conolly-Smith's historical example gives an explicit answer, revealing a truly translational 'method' in the increasing efforts of German immigrants to contribute to American culture. The 'method' lies in the practice of a "gradual translation" (19), which can be examined by considering its successive steps and agents: by adopting the new visual codes of the emerging American mass culture and entertainment industry, and at the same time by transforming or subverting them, the German immigrants gained a new tool for translating German culture. It allowed an overwriting of their 'original' idea of German *Hochkultur* and constituted a more modern practice of self-positioning in popular culture. The daily German-American newspapers served as crucial mediators, as agents of translation, by promoting and achieving a "subsequent incorporation of their readers" (50) and of the German immigrant community on the whole into American public life.

This mode of a "translational assimilation" (Cronin 2006, 52), however, seems to have a clear deficiency: it is one-sided and based on the condition of surrender to the dominant norms of the host society. In recent years, more attention has been paid to reciprocal processes of translation and the capacity to create a shared 'third culture' beyond any claims of national hegemony. This appeal for mutual translations is demanding for a society in its translational relationships between different social groups, between migrants and other parts of the population – certainly much more demanding than a one-sided integration offer from the dominant society as represented, for instance, by the highly controversial *Einbürgerungstest* (test for German citizenship) for immigrants in contemporary Germany. As Boris Buden and Stefan Nowotny have pointed out, this test exposes migrants to political control through cultural inclusion or exclusion, by their getting "culturally translated into 'being German'" (Buden and Nowotny 2009, 197), and by being translated into German citizenship. There is no doubt about the location of the hegemonic reference point in this act of translation: it is the unquestioned canon of asserted German values.

Thus, in such cases, it can be disputed whether immigration should be considered a mode of assimilation at all, as a mode of becoming similar to the destination society by reducing distinctions and differences (see Bartram et al. 2014, 16 on 'assimilation' vs. 'integration'). Isn't it instead a permanent process of transformation and, in this sense, a massive challenge for the nation-state? The process of transformation applies first to the migrant him- or herself: as a continuous demand for self-translation.

3.3 The Model of Self-Translation and Appropriation

If migration is not to be understood as celebrating some vague sense of in-between, then perhaps it constitutes a state in which one is "lost in translation." In the area of "migrant writing" (Gallagher 2014, 130ff.), of life writing in "borrowed tongues" (Karpinski 2012), a telling example is certainly Jewish journalist Eva Hoffman's autobiographical report detailing the complicated accommodation process in her Canadian and US-American exile after leaving Poland in 1959: *Lost in Translation: A Life in a New Language* (1990). What comes to the fore here is not only the linguistic confrontation with a new language, but even more a "sense of absolute division, between the past and the present, between my old self and the new self, etc." (Phoenix and Slavova 2011, 340). It marks a biographical rupture that should also become an immense challenge for migration researchers who practice micro-sociology by biographical interviews (Lutz 2011). In Hoffman's case we can observe the migrant's effort to gain access to a new level of agency by practicing a cultural code-shifting and self-translation in the course of the migration process: "I have to translate myself" (211), Hoffman notes in describing the disturbing experiences she had to deal with after leaving her Polish home country and settling in the New World.

Self-translation means in this case not only translating one's own works into another language, but translating oneself from one context into another – certainly not in a linear way, but as a multilayered "process of triangulation" (Frittella 2017, 372), trying to find a position as a person between two different frames of reference.[4] Translation is further considered as a conscious effort to reach beyond words or beyond verbal equivalence and grasp the hidden pre-translations present in a society – by shifting gradually, though always incompletely, to meet the unfamiliar "common agreements of a society," trying to adapt to their unspoken cultural connotations and "subtle signals" through which social meanings are signified (Hoffman 1990, 211, 172). Practicing self-translation in this way can indicate an active personal engagement with an enduring process of social integration; the effort of self-translation will not necessarily lead to integration or even assimilation, but also to displacement and alienation. Translation can even convey "an acute sense of dislocation," an active "refusal to assimilate" (197). In this respect, Hoffman seems to assume the notion of a pluralized and fragmented society, by asking: "In a splintered society, what does one assimilate to? Perhaps the very splintering itself" (197). But her individual, emotional response exceeds any sociological approach by pointing to the unsettling personal effects of being drawn into a liminal form of existence. She describes her own experience with linguistic uncertainty that was accompanied by psychotherapy in the sense of what she calls "translation therapy" (271). Witnessing how this specific mode of "translation therapy" is also considered a generation-specific attitude to exile existence facilitates an important insight for further analyses of migration situations. It is always important to ask if there are generation-specific forms of translation and self-translation – for instance, the contemporary phenomenon of translating migration experiences with regard to the digital sphere, to social networks, and to a communicative exchange through cellular phones.

It is furthermore important to reflect upon the sequence of translation activities in migration processes beyond the rather narrow understanding of adapting to the conditions of the host society. The 'regime' of translation here is much more complex. As Cecilia Alvstad has shown in the letters of Swedish transatlantic migrants to Latin America in the early 20th century, it already defines earlier

4 On self-translation, see Cordingley 2013; on translation as "triangulation," see Hoffman 1990, 276.

stages of the extended transit experience: "the translational process started long before the migrant arrived in the new continent" (Alvstad 2013, 103). And again, with regard to Eva Hoffman's memoir one should take into account – as Alvstad claims by referring to the letters of these migrants – that translation in the course of migration requires two audiences: the addressees of the home country, and those of the host country (107). This insight is certainly eye opening, above all with respect to the ambivalent translational 'management' of the split affiliations that signify the life stories of contemporary refugees.

3.4 The Model of Positioning between Multilingualism and Labeling

Of course, we cannot neglect the fact that those and other migration situations are – apart from the socio-cultural dimension – always characterized by complex multilingual conditions. We should not underestimate the impact of "translational relations" (Renn 2006) in the field of language use, as well as in the multilingual challenges for a homolingual nation-state. Such translational relations and circumstances can be made visible by following the practice of translation as a "bottom-up localization" (Cronin 2006, 30), as a tool for dealing with migrational spaces of multilingualism in small-scale communities and communication units: "With the intensification of migration, diasporal communities and hybridity, translation operates increasingly across small spaces, 'at home'" (Simon 2009, 209). A transcultural approach is thus not confined to the big movements and context shifts between and across cultures, but can instead focus even on shared, "'internal' spaces," for instance on specific urban migrational spaces that reveal "the complexities of translation across the shared spaces of today's cities" (Simon 2009, 209). The practice of translation in such shared migrational spaces can be conceived as a small-scale "model of intense, though conflictual, cultural movements across language lines" (Simon 2008, 8): a translational laboratory of linguistic movements in processes of 'internal migration' that can also be found on a larger scale, that is, in the national sphere.

With regard to a national context, Boris Buden (2012) discussed artist Alexander Vaindorf's stimulating video installation, *Detour* (2006), that can be seen as a great example for translating migration into images, media, and artistic forms.[5] The film deals with the migration of Russian women to work as housekeepers in

5 On the activating "role of translation in its political transfer of migration into the arts," see Rizzo 2017, 53.

Italy. Because of their initially insufficient language skills, these women started in the south of Italy; after improving their Italian they moved northward in a kind of internal migration *within* their host society, and in so doing were able to achieve a better standard of living. This migration scenario demonstrates the translation of migrants into citizens – by gradually accommodating with a homogeneous national language. Another way of turning migrants into citizens is represented in the previously mentioned German *Einbürgerungstest* (test for German citizenship). Here we see that migrants are not only translators themselves, but often desperately need translators and interpreters to come to their aid, for instance, in asylum hearings, in legal trials, in health care institutions, and in the multitude of naturalization procedures (*Einbürgerungsverfahren*) (see Mokre 2015; Inghilleri 2005). It is an example, however, that does not apply to other forms of migration, such as those cosmopolitan migrants with apriori privileged positions who draw their translational competence from their acclaimed intellectual or professional authority – performing a widely known mode of academic nomadism.

The example of a hierarchical south-to-north migration within Italy again shows that social accommodation depends on the institutional, political, and economic 'pre-translations' at work therein: on the invisible intersection of language skills, social hierarchies, and institutional translations as an "instrument of political exclusion and control" (Buden 2012, 366). Whether migrants are labeled as refugees, as economic migrants, as temporary labor migrants, as climate change migrants, or even as potential terrorists makes a difference. And through this bureaucratic process of designation we experience how "stereotyped identities are translated into bureaucratically assumed needs" (Zetter 2007, 39): translated into the institutional needs of specific programs, for housing and food aid, but also for control. These institutional needs are mostly incongruent with the self-perceptions and responses of the migrants themselves, who are in this case dealt with as objects of a translation from individual identities into stereotyped identities or even from "foreign others into political categories" (Giordano 2014, 10). They are objects also of a translation between the new bureaucratic language, in which they are categorized, and their own social language, which draws on different past norms and experiences. Their translation movement can, in this way, be understood not as a reference to an ethno-national "origin," but rather as a social "movement between and across" (Bassnett 2011, 74) that is experienced as a spatial movement between lifeworlds and different conditions of living. Shifting between the migrants' self-perceptions and the categorizations imposed on them externally, translation often becomes an important instrument of converting migrants into citizens, as Cristiana Giordano has shown in her ethnography of migrants in Italy:

> Through conceptual acts of translation, a complex interplay of therapeutic, bureaucratic, and religious apparatuses transform foreign others into political categories – the 'migrant,' 'refugee,' and 'victim' – that the state can recognize and use to legitimize their difference. (Giordano 2014, 10)

The transformation, however, is ambivalent, even as "this translation gives foreigners access to services and rights" (10). But beyond this accessability, it continually demands a differentiating question: "How is the voice of the migrant rendered, translated, heard, erased, and produced in different institutional settings" (Giordano 2008, 597)?

3.5 The Model of Reference to a "Third Idiom"

Today, we are concerned by racist anti-refugee demonstrations and the struggles against the allotment of homes for asylum seekers occurring in Germany and other European countries. Can a reference to translation really prove its worth in situations like these, where stateless and rights-less refugees are attacked and their claims for social and legal recognition are at stake? Isn't their liminal state – without passports, without rights, without work, and without recognition – a signal that all possible attempts to translate are bound to fail? Even so, we can point to critical public statements such as "Refugees Welcome!" signs. And we could even go further and try to explore hidden dimensions of the possible translation process to be activated, whether in the critical reexamination of polity categories or the relation to third, common reference points in order to gain recognition. The advocacy of pro-asylum groups and the migrants/refugees themselves often refer to human rights as a shared normative point of reference (see Mokre 2015, 56). I would call this practice translation through the use of a detour. This strategy can also be applied to translate one's own immediate social or political claims into a "third idiom."

For a better understanding, we can turn to the thought-provoking findings that sociologist Martin Fuchs has drawn from research on the Indian Dalits, a marginalized group of Untouchables living mainly in the slums of Indian megacities. The Dalits, Fuchs maintains, need translation existentially, as an everyday tool for survival: to make themselves recognized in their marginal position, the Dalits translate their discriminating experiences explicitly into a universal "frame of reference," into global, religious or civil society idioms. In this case, the religious language of Buddhism is utilized for its promise of social recognition: "The participants undertake what can be called a 'translation' of their claims and concerns into a new or 'third idiom,' which ideally is not owned by any one side [...]" (Fuchs 2009, 30–31). At the same time, the shift to shared idioms activates an effective social context where translation serves "to open the self towards the other, thus extending and developing target and source languages" (Fuchs 2009, 24). Making use of reference systems like this one is more than a mere reference back to an original. Instead, translation as a social practice is, in this case, a forward-facing model of social addressing, a means of pushing for action by an "attempt to reach out to others" (Fuchs 2009, 30). With this claim, Fuchs describes translation as an intentional, goal-oriented social practice for self-empowerment. This exact practice can also be found in migration scenarios. The same model of translation applies especially to those situations where migrants do not have sufficient agency to achieve self-translation, but attempt to translate nonetheless, by making detours.

A detour like this can use the reference to universal norms, such as human rights, as a strategy for gaining recognition and winning new regional or particular rights. Sociologist Walter Nicholls has shown a similar modus operandi in the context of "undocumented immigrants" in the US who also use translation as detour. They rely on "support organizations" and alliance networks that "possess the knowledge and culture needed to translate immigrant claims into powerful mobilizing frames that resonate with the norms of the national political field" (Nicholls 2013, 93). Here too, the translation of equal rights claims into "mobilizing frames" is used as a strategic detour to "gain recognition for undocumented immigrants" as legitimate subjects.

With regard to migrants and asylum seekers in general, it could be argued that these social groups might adopt translation in the form of a pragmatic detour. Their struggle for recognition can also engage with a third idiom – in this case, by making an explicit link between human rights and the right to claim asylum. Here too, the model of address can, in fact, become a model of action.

3.6 The Model of Transformation and Creating New Contexts

A quite different example is the explicit future-oriented mode of translation that can be accessed through the case of Erich Auerbach, the German Jewish philologist and humanist, who arrived as an exile in Istanbul in 1936. He was one of numerous exiled academics from many disciplines who fled Germany to teach at Turkish universities or practice at hospitals. Kader Konuk describes how Auerbach "found himself at home" in Turkey (Konuk 2010, 17), a country that today forces numerous academics from its own universities into exile for political reasons (see Konuk's contribution to this volume). It was in Istanbul that Auerbach wrote *Mimesis,* his pathbreaking book of literary criticism, by which he initiated the translation of Western humanism into the reform process of Turkish Westernization (Konuk 2010, 2008). Though still speaking and teaching in German, while constantly relying on a Turkish translator, he thus translated an established Western cultural habitus – thereby helping to trigger the transformation of a whole nation. Could we thus say, with Andreas Langenohl's cultural theory of translation, that Auerbach followed a practice of translation by referring to and opening up a future context that did not yet exist (Langenohl 2014, 24)?

In Auerbach's case, translation means transformation – not from a vague in-between state, not through a "detachment" from the German origin(al), as exile has been conventionally seen, but through the adoption of a state of "multiple attachments" in exile, as Kader Konuk concludes (Konuk 2010, 13)[6]. This situation of multiple affiliations and associations provided exactly the right conditions for a forward-facing, active effort of cultural translation. Auerbach's example of migrant translation shows how specific migrant situations can generate cultural works (like *Mimesis*) that have the potential to set a transformative force for an entire society in motion. This future-oriented view of migrants' ability to initiate cultural translation and to create new cultural contexts differs from a view that always falls back to originals or fixed, already-established systems of reference. Andreas Langenohl's notion of the context-creating force of translation applies here: translation effects "a configuration of the target context, in the sense of a perspectivization, anticipation, or envisioning. Together, text, speech act, and act of address assemble hopes, as it were, of what they might become in the target context" (Langenohl 2014, 24). This reference to future contexts accentuates the agency of translators, like Auerbach, who generated new idioms and cultural values in the target society (insofar as he contributed to the development of a Turkish humanism). Such far-reaching translation endeavors indicate that migration need not necessarily amount to *dis*location, but can also become a fruitful *re*location.

In the specific case of exile, translation thus can become effective as a cultural technique for supporting transnational knowledge transmission or even the transformation process of a society.[7] This can be seen to operate in a broader translational field that is open to the reception of more than just a transfer of

6 On a translational approach to the analysis of exile in general, see Kliems 2007.

7 A further example for a context-creating force of transnational translation by immigrants is the case of German Jewish refugees to Palestine in the 1930s, who there translated concepts, practices, and curricula of social work education. Through this kind of "knowledge translation" (Gal and Köngeter 2016, 264) they helped to establish a professional social welfare system in Palestine: "Forced to flee Germany, these social workers were recruited [...] to form the backbone of the social welfare system and the social work training institution in the country" (267).

ideas. Such an expanded horizon is marked in a letter to Walter Benjamin that Auerbach wrote during his Istanbul exile in January 1937:

> We teach all the European philologies here [...]. We try to influence the instructional life and the library and to Europeanize the administrative management of scholarship all the way from the instructional grid down to the card catalog ("vom Stundenplan bis zum Zettelkasten"). That is naturally absurd, but the Turks want it, even if they occasionally try to get in the way. (3 January 1937, Auerbach et al. 2007, 750)

Translation – in this rather optimistic statement, with its hegemonic overtones – appears to be an all-encompassing cultural practice, transmitting not only European scholarly ideas and results, but also the practices, instruments, and equipment of European-style research. Nowadays this could be interpreted with reference to the insights of Bruno Latour and Actor-Network-Theory, which reflects translation as a collaborative network including the instruments of research themselves as well as the instrumental facilities in acts of translation.

It is certainly remarkable that intellectual migrants, émigrés, and refugee scholars establish their translational impact not only with the transfer of ideas, but with the transfer of material and economic developments as well. Gottfried Bermann Fischer, for instance, made an impact on the re-education and democratization process in postwar Germany by way of a specific medium: the import of the American pocket paperback. This led to the development of the *Taschenbuch* in Germany, which can be seen as the starting point for a transformation of the German publishing industry. Here, translation worked in the sense of what we would today call 'innovation management.' An analogous process of a materially based 'knowledge transfer' can be observed in the case of Jewish refugee scholars to the United States. As émigrés and active cultural translators in the 20th century, they contributed to a transnational modernism in America. Walter Gropius, for instance, translated the "Bauhaus ideals to the world of American business" (Logemann 2017, 425), paving the way for a marketization of these ideals through American design companies.

Examples like these could help to support an understanding of translation in migration contexts as not only a single practice, but rather, in the sense of Pierre Bourdieu's field theory, as a broader translational "field." This too could perhaps be expanded further toward a practical understanding of translation as a specific kind of network, represented by Actor-Network-Theory. Here again, we see more than a 'cultural translation' at work, but rather a 'creative adaptation' pushing toward future developments.[8] Nonetheless, even migrants with a high degree of potential intellectual, academic, and social agency might continue to be caught in a hazardous state of liminality. Even as they promote new developments or transformations in their respective fields, their positions as translators could often be precarious, and even quite risky.

4 Some Concluding Remarks

The different models of translation and migration that I have considered here vary, without doubt, when other determining factors such as class, gender, religion, age, ethnicity, family, disability are taken into account. It always depends on whether we are dealing with migrants as bourgeois members of the intellectual elite, as in the case of Erich Auerbach, or whether we face, as in contemporary society, the mass migration of people with a lack of agency, such as boat people, refugees, diasporic, and displaced groups. By approaching these different forms of migration with the tool of translation, perhaps we can find new angles from which to reevaluate existing case studies of migration that so far have been presented in the disciplines of history, sociology, or cultural anthropology. What we need, however, are further concrete and detailed translational analyses via empirical studies, analyses that include a consideration of gender as a defining dimension of difference. Ultimately, translation is only *one* methodological tool to use to unpack the complexities of migrational worlds that are not merely in flow and in motion, but also infused by multiple ruptures, shifts, and criss-crossings of translation activities. The recent translational turn could certainly elucidate this field, though it should in no way be exaggerated or isolated, but instead utilized as a research perspective to be combined with economic, political, social, and psychological analyses. Only when embedded in a wider multidisciplinary approach can a translational perspective gain its full analytical force for the exploration of decisive pivot points in migration processes: of their context shifts, negotiation procedures, inclusions and exclusions, but also of the articulations and multiple linguistic affiliations and conflicts of the migrants themselves.

Extending beyond the analytical perspective, an encompassing field of translation as a social practice is opening up. Here, we come to recognize that previ-

8 Jan Logemann also stresses the methodological productivity of seeing migration processes from the perspective of (cultural) translation: "By embracing such notions of cross-cultural translation and creative adaptation, the well-worn field of emigration history can offer fresh, actor-centered perspectives on the history of transnational knowledge transfer" (2017, 417).

ously unexposed explicit or hidden elements of cultural disruption may yet be discovered, even in cases of seemingly successful practices of translation, migration, re-migration, and reconciliation. This focus on active attempts of 'translation' and their simultaneous irritations can shed new light on the ambivalences between the role of migrants as agents of translation and their day-to-day struggles as translated individuals and groups.

References

Alvarez, Sonia E. et al., eds. *Translocalities/Translocalidades: Feminist Politics of Translation in the Latin/a Américas*. Durham, NC/London: Duke University Press, 2014.

Alvstad, Cecilia. "The Transatlantic Voyage as a Translational Process: What Migrant Letters Can Tell Us." *Tales of Transit: Narrative Migrant Spaces in Atlantic Perspective, 1850–1950*. Eds. Michael Boyden, Hans Krabbendam, and Liselotte Vandenbussche. Amsterdam: Amsterdam University Press, 2013. 103–119.

Adelson, Leslie A. "Against Between: A Manifesto." *Unpacking Europe: Towards a Critical Reading*. Eds. Salah Hassan and Iftikhar Dadi. Rotterdam: Museum Boijmans Van Beuningen, 2001. 244–255.

Auerbach, Erich, Martin Elsky, Martin Vialon, and Robert Stein. "Scholarship in Times of Extremes: Letters of Erich Auerbach (1933–46), on the Fiftieth Anniversary of His Death." *PMLA* 122.3 (2007): 742–762.

Bachmann-Medick, Doris, ed. *The Trans/National Study of Culture: A Translational Perspective*. Berlin/Boston: De Gruyter, 2014.

Bachmann-Medick, Doris. *Cultural Turns: New Orientations in the Study of Culture*. Transl. by Adam Blauhut. Berlin/Boston: De Gruyter, 2016.

Bachmann-Medick, Doris. "The Transnational Study of Culture: A Plea for Translation." *The Humanities between Global Integration and Cultural Diversity*. Eds. Birgit Mersmann and Hans G. Kippenberg. Berlin/Boston: De Gruyter, 2016a. 29–49.

Balibar, Etienne. "At the Borders of Citizenship: A Democracy in Translation?" *European Journal of Social Theory* 13.3 (2010): 315–322.

Bandia, Paul F. "Translocation: Translation, Migration, and the Relocation of Cultures." *A Companion to Translation Studies*. Eds. Sandra Bermann and Catherine Porter. West Sussex: Wiley, 2014. 273–284.

Bartram, David, Maritsa V. Poros, and Pierre Monforte. *Key Concepts in Migration*. Los Angeles/London/New Delhi/Singapore/Washington, DC: Sage, 2014.

Bassnett, Susan. "From Cultural Turn to Translational Turn: A Transnational Journey." *Literature, Geography, Translation: Studies in World Writing*. Eds. Cecilia Alvstad, Stefan Helgesson, and David Watson. Newcastle: Cambridge Scholars Publishing, 2011. 67–80.

Bhabha, Homi K. *The Location of Culture*. London/New York: Routledge, 1994.

Buden, Boris. "Response." *Translation Studies* 5.3 (2012): 364–368 (Forum: Translation and Migration).

Buden, Boris, and Stefan Nowotny. "Cultural Translation: An Introduction to the Problem." *Translation Studies* 2.2 (2009): 196–208 (Forum: Cultural Translation).

Canclini, Néstor García. "Migrants: Workers of Metaphors." *Art and Visibility in Migratory Culture: Conflict, Resistance, and Agency* (Thamyris/Intersecting: Place, Sex and Race No. 23). Eds. Mieke Bal and Miguel À. Hernández-Navarro. Amsterdam/New York: Rodopi, 2011. 23–36.
Castro Varela, María do Mar. *Ist Integration nötig? Eine Streitschrift.* Freiburg: Lambertus, 2013.
Conolly-Smith, Peter. *Translating America: An Immigrant Press Visualizes American Popular Culture, 1895–1918.* Washington, DC: Smithsonian Books, 2004.
Cordingley, Anthony, ed. *Self-Translation: Brokering Originality in Hybrid Culture.* London/New Delhi/New York/Sydney: Bloomsbury, 2013.
Cronin, Michael. *Translation and Identity.* Abingdon/New York: Routledge, 2006 (Chapter: Translation and Migration 43–74).
Doppler, Lisa, and Friederike Vorwergk. "Refugees und Non-Citizen im Streik. Sprache als Ort des Widerstandes in Flüchtlingsprotesten." *Migration, Asyl und (Post-) Migrantische Lebenswelten in Deutschland. Bestandsaufnahme und Perspektiven migrationspolitischer Praktiken.* Eds. Miriam Aced et al. Berlin: LIT Verlag, 2014. 47–66.
Feld, Natalia. "Von der Migrationsliteratur zu translationswissenschaftlichen Entwürfen." *Texturen – Identitäten – Theorien. Ergebnisse des Arbeitstreffens des Jungen Forums Slavistische Literaturwissenschaft in Trier 2010.* Eds. Nina Frieß, Inna Ganschow, Irina Gradinari, and Marion Rutz. Potsdam: Universitätsverlag, 2011. 443–458.
Foglia, Cecilia. "Tracking the Socio-graphical Trajectory of Marco Micone: A Sociology of Migration by Way of Translation." *Translators Have Their Say: Translation and the Power of Agency.* Ed. Abdel Wahab Khalifa. Berlin/Münster/Vienna/Zürich/London: LIT Verlag, 2014. 20–41.
Frittella, Francesca Maria. "Cultural and Linguistic Translation of the Self: A Case Study of Multicultural Identity Based on Eva Hoffman's *Lost in Translation.*" *Open Cultural Studies* 1 (2017) (special issue *Migration and Translation*): 369–379.
Fuchs, Martin. "Reaching Out, Or, Nobody Exists in One Context Only: Society as Translation." *Translation Studies* 2.1 (2009) (special issue *The Translational Turn*): 21–40.
Gal, John, and Stefan Köngeter. "Exploring the Transnational Translation of Ideas: German Social Work Education in Palestine in the 1930s and 1940s." *Transnational Social Review* 6.3 (2016): 262–279.
Gallagher, Mary. "Lost and Gained in Migration: The Writing of Migrancy." *From Literature to Cultural Literacy.* Eds. Naomi Segal and Daniela Koleva. Basingstoke/New York: Palgrave Macmillan, 2014. 122–137 (Part II: Migration and Translation).
Gentzler, Edwin. "Response." *Translation Studies* 6.3 (2013): 342–347 (Forum: Translation and Migration).
Glick Schiller, Nina, and Noel B. Salazar. "Introduction: Regimes of Mobility Across the Globe." *Regimes of Mobility: Imaginaries and Relationalities of Power.* Eds. Noel B. Salazar and Nina Glick Schiller. Abingdon/New York: Routledge, 2014. 1–18.
Giordano, Cristiana. "Practices of Translation and the Making of Migrant Subjectivities in Contemporary Italy." *American Ethnologist* 35.4 (2008): 588–606.
Giordano, Cristiana. *Migrants in Translation: Caring and the Logics of Difference in Contemporary Italy.* Oakland: University of California Press, 2014.
Gutiérrez Rodríguez, Encarnación. "Translating Positionality: On Post-Colonial Conjunctures and Transversal Understanding." *Translate* 6 (2006). <http://eipcp.net/transversal/0606/gutierrez-rodriguez/en/base_edit> [accessed: 11 March 2018].
Hoffman, Eva. *Lost in Translation: A Life in a New Language.* New York: Dutton, 1990.

Hron, Madelaine. *Translating Pain: Immigrant Suffering in Literature and Culture*. Toronto/Buffalo/London: University of Toronto Press, 2009.

Inghilleri, Moira. "Mediating Zones of Uncertainty: Interpreter Agency, the Interpreting Habitus and Political Asylum Adjudication." *The Translator* 11.1 (2005): 69–85.

Inghilleri, Moira. *Translation and Migration*. Abingdon/New York: Routledge, 2017.

Karpinski, Eva C. *Borrowed Tongues: Life Writing, Migration, and Translation*. Waterloo, Ontario: Wilfrid Laurier University Press, 2012.

Kliems, Alfrun. "Transkulturalität des Exils und Translation im Exil. Versuch einer Zusammenbindung." *Exilforschung. Ein internationales Jahrbuch* 25 (2007): *Übersetzung als transkultureller Prozess*. 30–49.

Konuk, Kader. "Erich Auerbach and the Humanist Reform to the Turkish Education System." *Comparative Literature Studies* 45.1 (2008): 74–89.

Konuk, Kader. *East West Mimesis: Auerbach in Turkey*. Stanford: Stanford University Press, 2010.

Lässig, Simone, and Swen Steinberg. "Knowledge on the Move: New Approaches toward a History of Migrant Knowledge." *Geschichte und Gesellschaft* 43.3 (2017) (special issue *Knowledge and Migration*): 313–346.

Langenohl, Andreas. "Verknüpfung, Kontextkonfiguration, Aspiration. Skizze einer Kulturtheorie des Übersetzens." *Zeitschrift für Interkulturelle Germanistik* 5.2 (2014): 17–27.

Logemann, Jan. "Consumer Modernity as Cultural Translation: European Émigrés and Knowledge Transfers in Mid-Twentieth-Century Design and Marketing." *Geschichte und Gesellschaft* 43.3 (2017) (special issue *Knowledge and Migration*): 413–437.

Lutz, Helma. "Lost in Translation? The Role of Language in Migrant's Biographies: What Can Micro-Sociologists Learn from Eva Hoffman?" *European Journal of Women's Studies* 18.4 (2011) (special issue *Living in Translation: Voicing and Inscribing Women's Lives and Practices*): 347–360.

Mokre, Monika. *Solidarität als Übersetzung. Überlegungen zum Refugee Protest Camp Vienna*. Ed. Andrea Hummer. Vienna et al.: transversal, 2015.

Nicholls, Walter J. "Making Undocumented Immigrants into Legitimate Political Subjects: Theoretical Observations from the United States and France." *Theory, Culture and Society* 30.3 (2013): 82–107.

Papastergiadis, Nikos. *The Turbulence of Migration: Globalization, Deterritorialization, and Hybridity*. Cambridge/Malden, MA: Polity, 2000.

Phoenix, Ann, and Kornelia Slavova. "'Living in Translation': A Conversation with Eva Hoffman." *European Journal of Women's Studies* 18.4 (2011): 339–345.

Polezzi, Loredana. "Translation, Travel, Migration." *The Translator* 12.2 (2006) (special issue *Translation, Travel, Migration*, ed. Loredana Polezzi): 169–188.

Polezzi, Loredana. "Translation and Migration." *Translation Studies* 5.3 (2012): 354–356 (Forum: Translation and Migration).

Polezzi, Loredana. "Migration and Translation." *Handbook of Translation Studies* 3. Eds. Yves Gambier and Luc van Doorslaer. Amsterdam: John Benjamins, 2012a. 102–107.

Renn, Joachim. "Die Übersetzung der modernen Gesellschaft. Das Problem der Einheit der Gesellschaft und die Pragmatik des Übersetzens." *Übersetzung als Medium des Kulturverstehens und sozialer Integration*. Eds. Joachim Renn, Jürgen Straub, and Shingo Shimada. Frankfurt a.M./New York: Campus, 2002. 183–214.

Renn, Joachim. *Übersetzungsverhältnisse. Perspektiven einer pragmatistischen Gesellschaftstheorie*. Weilerswist: Velbrück, 2006.

Rizzo, Allessandra. "Translation as Artistic Communication in the Aesthetics of Migration: From Nonfiction to the Visual Arts." *Ars Aeterna: Literary Studies and Humanity* 9.2 (2017): 53–70 (De Gruyter Open).
Rushdie, Salman. *Imaginary Homelands: Essays and Criticism 1981–1991.* London: Granta, 1991.
Sakai, Naoki. "Transnationality in Translation." *Translation: A Transdisciplinary Journal* (Spring 2013): 15–31.
Sayad, Abdelmalek. *The Suffering of the Immigrant.* Cambridge/Malden, MA: Polity Press, 2004.
Schunka, Alexander. "Normsetzung und Normverletzung in Einwanderungsgesellschaften der Frühen Neuzeit." *Normsetzung und Normverletzung. Alltägliche Lebenswelten im Königreich Ungarn vom 18. bis zur Mitte des 19. Jahrhunderts.* Ed. Karl-Peter Krauss. Stuttgart: Steiner, 2014. 29–55.
Simon, Sherry. "Cities in Translation: Some Proposals on Method." *Doletiana: Revista de Traducció, Literatura i Arts* 2 (2008): 1–12.
Simon, Sherry. "Response." *Translation Studies* 2.2 (2009): 208–213 (Forum: Cultural Translation).
Simon, Sherry. *Cities in Translation: Intersections of Language and Memory.* London/New York: Routledge, 2012.
Steiner, Tina. *Translated People, Translated Texts: Language and Migration in Contemporary African Literatur* (2009). Abingdon/New York: Routledge, 2014.
Translation Studies 5.3 (2012): 103–117, 6.1 (2013): 345–368, 6.3 (2013): 339–351 (Forum: Translation and Migration).
Vertovec, Steven. "Super-Diversity and Its Implications." *Ethnic and Racial Studies* 30.6 (2007): 1024–1054.
Vorderobermeier, Gisella, and Michaela Wolf, eds. *„Meine Sprache grenzt mich ab.…" Transkulturalität und kulturelle Übersetzung im Kontext von Migration.* Vienna: LIT Verlag, 2008.
Wimmer, Andreas. "Elementary Strategies of Ethnic Boundary Making." *Ethnic and Racial Studies* 31.6 (2008): 1025–1055.
Wimmer, Andreas, and Nina Glick Schiller. "Methodological Nationalism and Beyond: Nation-state Building, Migration and the Social Sciences." *Global Networks* 2.4 (2002): 301–334.
Wolf, Michaela. "Cultural Translation as a Model of Migration?" *Translatio/n: Narration, Media and the Staging of Differences.* Eds. Federico Italiano and Michael Rössner. Bielefeld: transcript, 2012. 69–88.
Young, Robert J. C. "Some Questions about Translation and the Production of Knowledge." *Translation: A Transdisciplinary Journal* (2011) (inaugural issue): 59–61.
Zetter, Roger. "More Labels, Fewer Refugees: Remaking the Refugee Label in an Era of Globalization." *Journal of Refugee Studies* 20.2 (2007): 172–192.

Notes on Contributors

Doris Bachmann-Medick is Senior Research Fellow at the International Graduate Centre for the Study of Culture (GCSC) of the Justus Liebig University Giessen. She held numerous appointments as a Visiting Professor, recently at the universities of Graz, Göttingen, UC Irvine, Cincinnati, and Georgetown University. Her main fields of research are cultural theory, *Kulturwissenschaften*, literary anthropology, and translation studies. Her recent book publications include *Cultural Turns. Neuorientierungen in den Kulturwissenschaften* (6th ed. Rowohlt, 2018 [2006]), revised English edition *Cultural Turns: New Orientations in the Study of Culture* (De Gruyter, 2016), and her edited volumes "The Translational Turn" (special issue of the journal *Translation Studies*, 2009) and *The Trans/National Study of Culture: A Translational Perspective* (De Gruyter, 2014). She serves on the Editorial Board of *Translation Studies* (since 2008).

Christine Bischoff is Research Fellow at the Institute for Cultural Anthropology at the University of Hamburg. Her research and teaching encompass the following areas: theories and methods of visual anthropology, media theory and media communication, migration research, post-colonial studies, ethnographic research on religiosity and spirituality (especially conversion research), and qualitative methods in cultural studies. Her recent book publications include *Blickregime der Migration. Images und Imaginationen des Fremden in Schweizer Printmedien* (Waxmann, 2016), *Images of Illegalized Immigration: Towards a Critical Iconology of Politics* (transcript, 2010), and *Methoden der Kulturanthropologie* (UTB, 2014). She is currently a scholar at the *Isa Lohmann-Siems Foundation* and preparing a conference and publication on *Confessions: Forms and Formulas*. Her habilitation project is about *Religious Mobility: Conversion as a Social and Cultural Resonating Space.*

Friederike Eigler is Professor of German at Georgetown University and has widely published on 20th- and 21st-century literature and culture with special foci on memory, space/place, and gender. She was Editor of *The German Quarterly* from 2004 to 2006 and she is the author of *Gedächtnis und Geschichte in Generationenromanen seit der Wende* (Schmidt, 2005). Recent publications include the volume *Heimat: At the Intersection of Memory and Space* (De Gruyter, 2012), which she co-edited with Jens Kugele, a special issue of *German Politics and Society* on "German-Polish Border Regions in Literature and Film" (2013) co-edited with Astrid Weigert, and a monograph titled *Heimat, Space, Narrative: Toward a Transnational Approach to Flight and Expulsion* (Camden House, 2014). One of Eigler's current research projects looks at the responses of literature and theater to the influx of refugees into Europe.

Heidrun Friese is an anthropologist and Professor of Intercultural Communication at Chemnitz Technical University. She held numerous appointments as Researcher and Visiting Professor at, for instance, the École des Hautes Études en Sciences Sociales, Paris, the Scuola Superiore di Studi Universitari e di Perfezionamento Sant'Anna di Pisa, the European University Institute, Florence, and the institute HyperWerk at Hochschule für Gestaltung und Kunst, Basel. Her research interests focus on social and political theory, (cultural) identities, borders and transnational practices, hospitality and undocumented mobility (Mediterranean), and digital anthropology. Publications include *Flüchtlinge: Opfer – Bedrohung – Helden. Zur politischen Imagination des Fremden* (transcript, 2017), *Grenzen der Gastfreundschaft. Die Bootsflüchtlinge*

https://doi.org/9783110600483-016

von Lampedusa und die europäische Frage (transcript, 2014). She serves on the Editorial Board for the *European Journal of Social Theory* and *Time and Society*.

Heike Greschke is Professor of Sociology with a focus on Comparative Cultural Studies and Qualitative Research at the Technische Universität Dresden, where she also serves as Chairperson of the Executive Committee of the Centre for Integration Research. Prior to that, she was Junior Professor of Sociology with a focus on Media Sociology at Justus Liebig University Giessen. From 2015–2017 she lead the research project "The Mediatization of Parent-Child-Relationships in Transnational Migration" funded by the DFG-Priority Program 1505 "Mediatized Worlds." Since 2018 she is Principal Investigator of the Collaborative Research Center (Sonderforschungsbereich) 1285 "Invectivity: Constellations and Dynamics of Disparagement," leading the subproject "Invektive Kodierungen von Interkulturalität. Ethnografische Situationsanalysen in interkulturellen Trainings- und Integrationskursen." Her main interests in research and teaching are related to migration, media, family, globalization and culture, and methodological issues of cultural studies. Her recent publications in English include the monograph *Is There a Home in Cyberspace? The Internet in Migrants' Everyday Life and the Emergence of Global Communities* (Routledge, 2012, reprinted as paperback in 2014) and the edited volume *Grounding Global Climate Change: Contributions from the Social and Cultural Sciences* (Springer, 2015, with Julia Tischler).

Encarnación Gutiérrez Rodríguez is Professor of Sociology at Justus Liebig University Giessen. Previous to her appointment in Giessen she was a Senior Lecturer in Transcultural Studies in the Spanish, Portuguese, and Latin American Studies Department at the University of Manchester, UK, and an Assistant Professor (*wissenschaftliche Assistentin*) in Sociology in the Institute of Sociology at the University of Hamburg. Her teaching and research engages with questions of global inequalities and their local articulation particularly in Germany, Spain, and the UK. She is interested in (post)Marxist and decolonial perspectives on feminist and queer epistemology and their application to the field of migration, labor, and culture. This is particularly reflected in her monograph *Migration, Domestic Work and Affect* (2010) and the co-edited collection *Decolonizing European Sociology* (2010, with Manuela Boatcă and Sérgio Costa). Some of her recent publications are the co-edited volume *Creolizing Europe: Legacies and Transformations* (Liverpool University Press, 2015, with Shirley Anne Tate), "Sensing Dispossession: Women and Gender Studies Between Institutional Racism and Migration Control Policies in the Neoliberal University" (*Women's Studies International Forum*, 2016), "Affektive Materialität. Transversale Trauer und darstellbare Gerechtigkeit," *Verantwortung und Un/Verfügbarkeit*, ed. Corinna Bath et al. (Westfälisches Dampfboot, 2017).

Sabine Hess is Professor at the Institute for Cultural Anthropology/European Ethnology, University of Göttingen. Her main areas of research and teaching are migration and border regime studies, anthropology of globalization and transnationalism, anthropology of policy and Europeanization, gender studies, and anthropological methodologies. She was the scientific curator of several interdisciplinary research and exhibition projects on the history of immigration to Germany and director of several research projects on the European border regime funded by national and European research foundations. She is co-founder of the interdisciplinary European-wide "Network for Critical Migration and Border Regime Studies" (kritnet) and member of the German-wide "Rat für Migration" (Council for Migration) as well as a Board Member of the Göttingen Center for Gender Studies. Her recent publications in the field of border and migration in English are: “Turkey’s Changing Migration Regime and its Global and Regional Dynamics” (co-edited issue of *Movements: Journal for Critical Border and Migration Regime Research*, 2017), “Tracing the Effects of the EU-Turkey Deal: The Momentum of the Multi-Layered Turkish Border Regime” (with Gerda Heck, *Movements: Turkey’s Changing Migration Regime within Global and Regional Dynamics*, 2017), and “Under Control? Or: Border (as) Conflict!” (with Bernd Kasparek, *Social Inclusion*, 2017).

Wulf Kansteiner is Professor of History at Aarhus University. He is a cultural historian, historical theorist, and memory studies scholar. His research focuses on collective memories of Nazism and the Holocaust in film, television, and digital culture, the narrative structures of historical writing, and the methods and theories of memory studies. He is the author of *In Pursuit of German Memory: History, Television, and Politics after Auschwitz* (Ohio University Press, 2006), co-editor of *The Politics of Memory in Postwar Europe* (Duke University Press, 2006), “Historical Representation and Historical Truth” (Theme Issue of *History & Theory* 47, 2009), *Den Holocaust erzählen. Historiographie zwischen wissenschaftlicher Empirie und narrativer Kreativität* (Wallstein, 2013), and *Probing the Ethics of Holocaust Culture* (Harvard University Press, 2016). He is also co-founder and co-editor of the Sage-Journal *Memory Studies* (since 2008).

Evangelos Karagiannis is Senior Assistant at the Department of Social and Cultural Anthropology of the University of Vienna. He has held positions at the University of Zurich, the Max Planck Institute for Social Anthropology, and the University of Osnabrück and has participated in numerous research projects. His research interests encompass issues of nationalism and minorities, modernity and secularism, globalization, and migration. He has published a monograph on the Pomak minority in post-socialist Bulgaria and is currently working on a project on the tension between globalism and nationalism in Greek Pentecostalism.

Kader Konuk is Professor and Chair of Turkish Studies at the University of Duisburg-Essen in Germany. She is the Director of the newly founded Academy in Exile, located in Berlin and Essen. Between 2001 and 2013 she was Assistant and Associate Professor of Comparative Literature and German Studies at the University of Michigan. Trained as a comparatist in German, Turkish, and English literature, Konuk focuses on the disciplinary nexus between literary criticism, cultural studies, and intellectual history. Her research is situated at the intersections between religious and ethnic communities, beginning with the Ottoman modernization reforms and continuing on to Turkish-German relations in the 21st century. Her work examines cultural practices that evolve in the context of East-West relations (travel, migration, and exile). In her monograph *East West Mimesis: Auerbach in Turkey* (Stanford UP, 2010), she investigates the relationship between German-Jewish exile and the modernization of the humanities in Turkey. In her current book project, Konuk questions the common equation of secularism with Western modernity and is interested in the connection between religious critique, the freedom of speech, and literary discourses in the history of republican Turkey.

Jens Kugele is Head of Research Coordination at the International Graduate Centre for the Study of Culture (GCSC) of the Justus Liebig University Giessen. He is one of the research center’s Principal Investigators and has been a member of the center’s Executive Board since 2014. Previously, he held appointments as Visiting Researcher and Assistant Professor at the LMU Munich, the Berkley Center for Religion, Peace and World Affairs at Georgetown University, and the Institute for German Cultural Studies at Cornell University. His research interests include intersections of literary and cultural history with the history of religion; German-Jewish literature and culture; constructions of belonging, memory, and space. Recent publications include the volume *Heimat: At the Intersection of Memory and Space* (De Gruyter, 2012), which he co-edited with Friederike Eigler, and a special issue of the *Journal of Religion in Europe* on "Relocating Religion(s)" in museal spaces (Brill, 2011), which he co-edited with Katharina Wilkens. One of Kugele's current research projects looks at "sacred space" as a concept for the interdisciplinary study of culture. He is co-founder and co-editor of the peer-reviewed interdisciplinary open access journal *On_Culture*.

Paul Mecheril is Professor for Migration and Education at the Department for Pedagogy of the Carl von Ossietzky University of Oldenburg. He is Head of the Center for Migration, Education and Cultural Studies. Prior to that, he was Professor for Intercultural Education and Social Change at the University of Innsbruck. His main research areas include migration and pedagogy, racism, and cultural studies. He is the (co-)author of nine and (co-)editor of 26 books. He recently edited *Handbuch Migrationspädagogik* (Beltz, 2016), co-edited *Dämonisierung der Anderen. Rassismuskritik der Gegenwart* (transcript, 2016, with María do Mar Castro Varela), *and Resistance. Subjects, Representations, Contexts* (transcript, 2017, with Martin Butler and Lea Brenningmeyer).

Charlton Payne is DAAD P.R.I.M.E. Fellow in the Literature Department at the University of Erfurt in Germany and a Visiting Researcher in the German Department at UC Berkeley. He has published essays on epic narration and cosmopolitanism in the works of Goethe and Wieland, Kleist and the problem of asylum, the representation of passports and statelessness in 20th-century literature, archiving displacement in postwar Germany, and refugee legibility and the traces of the (in)human in contemporary literature. His book-length publications include *The Epic Imaginary: Political Power and Its Legitimations in Eighteenth-Century German Literature* (De Gruyter, 2012) as well as the edited volumes *Kant and the Concept of Community* (Rochester, 2011), *Niemandsbuchten und Schutzbefohlene. Flucht-Räume und Flüchtlingsfiguren in der deutschsprachigen Gegenwartsliteratur* (V&R, 2017), and, with Jesper Gulddal, a special issue on "Passports" for the journal *symplokē* (2017). He is completing a book on the relationship between fiction and documental identity in the 20th and 21st centuries.

Shalini Randeria is Rector of the Institute for Human Sciences (IWM) in Vienna, Professor of Social Anthropology and Sociology at the Graduate Institute of International and Development Studies (IHEID) in Geneva, as well as the Director of the Hirschman Centre on Democracy. Before joining the IHEID she was Professor of Social Anthropology at the University of Zurich, as well as Professor and Founding Chair of the Department of Sociology and Social Anthropology of the Central European University Budapest. She has published widely on the anthropology of globalization, law, the state, and social movements. Her empirical research on India addresses issues of post-coloniality and multiple modernities. Recent edited volumes include: *Migration and Borders of Citizenship* (with Ravi Palat, special issue of *Refugee Watch: A South Asian Journal on Forced Migration*, 2017), *Politics of the Urban Poor* (with Veena Das, special issue of *Current Anthropology*, 2015), *Anthropology, Now and Next: Diversity, Connections, Confrontations, Reflexivity. Essays in Honour of Ulf Hannerz* (with Thomas Hylland Eriksen and Christina Garsten, Berghahn Publishers, 2014), and *Jenseits des Eurozentrismus. Postkoloniale Perspektiven in den Geschichts- und Kulturwissenschaften* (with Sebastian Conrad und Regina Römhild, 2nd edition, Campus Verlag, 2013).

Werner Schiffauer is Professor Emeritus of Comparative Social and Cultural Anthropology at the European University Viadrina Frankfurt (Oder). His main fields of interest are anthropology of migration, the organization of societal heterogeneity, Islam in Europe, and anthropology of the state. He has published several books, among them *Parallelgesellschaften. Wieviel Wertekonsens braucht unsere Gesellschaft? Ethnographische Überlegungen* (transcript, 2008), *Nach dem Islamismus. Die Islamische Gemeinde Milli Görüş. Eine Ethnographie* (Suhrkamp, 2010), and *Schule, Moschee, Elternhaus. Eine ethnologische Intervention* (Suhrkamp, 2015).

Index